In Situ Analysis of Cellular Functional Molecules

Detection Science Series

Titles in the Series:

1: Sensor Technology in Neuroscience
2: Detection Challenges in Clinical Diagnostics
3: Advanced Synthetic Materials in Detection Science
4: Principles and Practice of Analytical Techniques in Geosciences
5: Microfluidics in Detection Science: Lab-on-a-chip Technologies
6: Electrochemical Strategies in Detection Science
7: Peroxynitrite Detection in Biological Media: Challenges and Advances
8: Biological Fluid-Surface Interactions in Detection and Medical Devices
9: Advanced Environmental Analysis: Applications of Nanomaterials, Volume 1
10: Advanced Environmental Analysis: Applications of Nanomaterials, Volume 2
11: Quenched-phosphorescence Detection of Molecular Oxygen: Applications in Life Sciences
12: Carbon-based Nanomaterials in Analytical Chemistry
13: Forensic Analytical Methods
14: Immunosensors
15: Analytical Electrogenerated Chemiluminescence: From Fundamentals to Bioassays
16: *In Situ* Analysis of Cellular Functional Molecules

How to obtain future titles on publication:
A standing order plan is available for this series. A standing order will bring delivery of each new volume immediately on publication.

For further information please contact:
Book Sales Department, Royal Society of Chemistry, Thomas Graham House, Science Park, Milton Road, Cambridge, CB4 0WF, UK
Telephone: +44 (0)1223 420066, Fax: +44 (0)1223 420247
Email: booksales@rsc.org
Visit our website at www.rsc.org/books

In Situ *Analysis of Cellular Functional Molecules*

By

Huangxian Ju
Nanjing University, China
Email: hxju@nju.edu.cn

Bo Tang
Shandong Normal University, China
Email: tangb@sdnu.edu.cn

and

Lin Ding
Nanjing University, China
Email: dinglin@nju.edu.cn

ROYAL SOCIETY OF CHEMISTRY

Detection Science Series No. 16

Print ISBN: 978-1-78801-722-0
PDF ISBN: 978-1-83916-045-5
EPUB ISBN: 978-1-83916-135-3
Print ISSN: 2052-3068
Electronic ISSN: 2052-3076

A catalogue record for this book is available from the British Library

The Royal Society of Chemistry is a charity, registered in England and Wales, Number 207890, and a company incorporated in England by Royal Charter (Registered No. RC000524), registered office: Burlington House, Piccadilly, London W1J 0BA, UK, Telephone: +44 (0) 20 7437 8656.

For further information see our website at www.rsc.org

Printed in the United Kingdom by CPI Group (UK) Ltd, Croydon, CR0 4YY, UK

Preface

Cells are the basic units of life. Developing sensing platforms for probing the spatiotemporal characteristics of functional molecules in/on living cells is one of the basic goals in understanding the intricate biological processes and identifying early disease signatures for clinical diagnosis, therapy and prognosis. *In situ* sensitive analysis and imaging of cellular functional molecules can provide spatially or temporally resolved information on these essential cellular molecules in a non-invasive manner, and this approach has emerged as a cornerstone solution in the quantitative and/or functional analysis of different types of cellular molecules by the combination of nanotechnology and biotechnology with chemistry, biology, physics, engineering and medicine. With the achievements of nanotechnology, a wide variety of nanoscale materials have been used for the design of nanoprobes and biosensing nanodevices with lower cytotoxicity, high sensitivity and long-term stability, and a large number of research papers related to this topic have been published. Hence this cross-cutting research field has become one of the hottest topics in analytical chemistry, biosensing, nanoscience and biomedicine.

To offer researchers, especially beginners, a survey of the novel principles and detection strategies for the *in situ* analysis of cellular functional molecules, this book introduces the tailor-made design of detection probes and schemes from a top-down perspective according to the unique characteristics of cellular functional molecules, based on our research experience. The latest methodological developments, including the enhancement of detection sensitivity and specificity, precision localization, the implementation of dynamic tracking and the acquisition of quantitative functional information for various important functional molecules are discussed in detail. These functional molecules include cell surface and intracellular glycans, intracellular mRNAs, microRNAs, proteins (telomerase, caspases,

Detection Science Series No. 16
In Situ Analysis of Cellular Functional Molecules
By Huangxian Ju, Bo Tang and Lin Ding

Published by the Royal Society of Chemistry, www.rsc.org

etc.) and reactive oxygen, nitrogen and sulfur species. Methods for the visualization of glycosylation, imaging of tumour microenvironments and visualization of dynamic intermolecular interactions within living cells are also covered. The focus is on how to achieve an assay in an *in situ*, sensitive and specific way.

The material is presented in 10 chapters, covering all the authors' study topics of *in situ* analysis methodology. Some experts who obtained their PhD degrees in Ju's and Tang's groups or who are working in these groups participated in the writing of some chapters, namely Dr Haifeng Dong (Chapter 3), Dr Ruocan Qian (Chapter 4), Dr Jiangwei Tian (Chapter 5) and Dr Yunlong Chen (Chapter 6) in Ju's group, and Dr Ping Li (Chapter 7), Dr Xu Wang and Fanpeng Kong (Chapter 8), Dr Wen Gao and Zhenhua Liu (Chapter 9) and Dr Xilei Xie and Xiaonan Gao (Chapter 10) in Tang's group. We are very grateful to all these colleagues for their contributions.

This book is one of a sequence of books published by the first author (Huangxian Ju) or with the co-authors, including *Electroanalytical Chemistry and Biosensing Technologies* (Science Press, 2006, in Chinese), *Bioanalytical Chemistry* (Science Press, 2007, in Chinese), *Electrochemical Sensors, Biosensors and Their Biomedical Applications* (Elsevier Academic Press, 2007, in English, and Chemical Industry Press, 2009, in Chinese), *NanoBiosensing – Principles, Development and Application* (Springer, 2011, in English, and Science Press, 2012, in Chinese), *Nucleic Acid Detection: Methods for Analysis of DNA and MicroRNA* (Intellectual Property Press, 2015, in Chinese) and *Immunosensing for Detection of Protein Biomarkers* (Elsevier, 2017, in English).

This book provides a comprehensive overview for a broad audience in academia and professionals working on different aspects of cellular analysis and cell biology. It is hoped to bridge the common gap between the research literature and new research ideas to develop the methodology for the *in situ* sensitive analysis of cellular functional molecules. We believe that it will be of great interest to readers and researchers in the fields of analytical chemistry, cell biology, clinical laboratory studies, clinical medicine and oncology. We are fortunate to have had the opportunity to undertake this important project. We warmly acknowledge the gracious support of our families. Finally, we thank the publishers for doing a remarkable job on this book.

Huangxian Ju, Nanjing, China
Bo Tang, Jinan, China
Lin Ding, Nanjing, China

Contents

Chapter 1 Design of *In Situ* Cytosensing Strategies **1**

1.1 General Principles of *In Situ* Cytosensing 1
1.2 Enhancement of Detection Specificity 2
1.3 Design of Off On Signal Switch 4
1.4 Design of Signal Amplification Strategy 7
1.5 Design of Multi-channel Analysis 13
1.6 Perspective 17
References 18

Chapter 2 *In Situ* Detection of Cell-surface Glycans **21**

2.1 Introduction to Cell-surface Glycans 21
2.1.1 Structures and Classification of Glycans 21
2.1.2 Functions of Glycans 22
2.2 Labelling and Recognition of Glycans 22
2.2.1 Lectin Recognition 22
2.2.2 Chemoselective Labelling 24
2.2.3 Metabolic Oligosaccharide Engineering 25
2.2.4 Chemoenzymatic Glycan Labelling 27
2.3 Global Analysis of Cell-surface Glycans 27
2.4 Competition Systems for Quantification of Cell-surface Glycans 33
2.5 Protein-specific Imaging of Cell-surface glycans 35
2.6 Perspective 43
References 44

Detection Science Series No. 16
In Situ Analysis of Cellular Functional Molecules
By Huangxian Ju, Bo Tang and Lin Ding

Published by the Royal Society of Chemistry, www.rsc.org

Chapter 3 *In Situ* Detection of Intracellular Messenger RNA and MicroRNA 47

3.1 Introduction 47
3.2 Biofunction of mRNA and miRNA 48
3.2.1 Biofunction of mRNA 48
3.2.2 Biofunction of miRNA 49
3.3 Viral Gene Vectors 50
3.3.1 Lenti-/Retrovirus-mediated Vectors for Gene Delivery 51
3.3.2 Adenovirus-mediated Vectors for Gene Delivery 51
3.3.3 Adeno-associated Virus Vectors for Gene Delivery 51
3.4 Organic Gene Vectors 53
3.4.1 Cationic Polymers 53
3.4.2 Cationic Lipids 54
3.5 Inorganic Gene Vectors 55
3.5.1 Gold-based Nanomaterials as Gene Vectors 55
3.5.2 Two-dimensional GO and TMD Nanosheets as Gene Vectors 56
3.6 Intracellular mRNA Detection 57
3.7 Intracellular miRNA Detection 59
3.7.1 "One-to-One" Signal Detection 59
3.7.2 Sensitive Detection Based on Signal Amplification 59
3.8 Perspective and Challenges 61
Acknowledgement 61
References 61

Chapter 4 *In Situ* Analysis of Intracellular Telomerase Activity 67

4.1 Clinical Significance of Telomerase 67
4.2 Traditional Analytical Methods for Telomerase 69
4.3 *In Situ* Imaging of Intracellular Telomerase Activity Based on Functional Nanoprobes 70
4.4 *In Situ* Imaging of Intracellular Telomerase Activity Through FRET 72
4.4.1 FRET Based on Organic Quenchers 73
4.4.2 FRET Based on AuNPs 75
4.5 Raman Spectroscopy-based Intracellular Telomerase Activity Sensing 76

4.6 *In Situ* Monitoring of Intracellular Telomerase Activity Using AIEgen-based Bioprobes 78
4.7 Perspective 79
Acknowledgement 79
References 80

Chapter 5 *In Situ* Analysis of the Intracellular Caspase Family 82

5.1 Introduction to the Intracellular Caspase Family 82
5.2 Imaging of Caspase Activity in Living Cells 83
5.2.1 FRET-based Sensors for Imaging of Intracellular Caspase Activity 83
5.2.2 Fluorescence Lifetime Imaging Microscopy for Imaging of Intracellular Caspase Activity 87
5.2.3 AIE Probe for Imaging of Intracellular Caspase Activity 90
5.2.4 Two-photon (TP) Imaging of Intracellular Caspase Activity 92
5.2.5 Nanomaterial-based Probe for Imaging of Intracellular Caspase Activity 94
5.2.6 Bioorthogonal Cyclization-mediated *In Situ* Self-assembly of Small-molecule Probes for Imaging of Caspase Activity 97
5.2.7 Bioluminescence Imaging of Intracellular Caspase Activity 99
5.3 Simultaneous Imaging of Caspases and Other Proteins 101
5.4 Real-time Monitoring of Caspase Cascade Activation 104
5.5 Perspective 108
Acknowledgement 108
References 108

Chapter 6 Visualization of Intracellular Glycosylation 112

6.1 Introduction to Intracellular Glycosylation 112
6.1.1 *O*-GlcNAcylation 112
6.1.2 Sialylation 113
6.2 Visualization of Intracellular *O*-GlcNAcylation 114
6.2.1 Protein-specific *O*-GlcNAcylation 114
6.2.2 General *O*-GlcNAcylation 119

6.3 Visualization of Intracellular Sialylation-related Enzymes 125
6.3.1 Intracellular STs 125
6.3.2 Intracellular Neus 133
6.4 Perspective 139
Acknowledgement 139
References 139

Chapter 7 Real-time Monitoring of Intracellular Reactive Oxygen Species 144

7.1 Fluorescent Probes for Superoxide Anion 144
7.1.1 Phenol–Quinone Interconversion 144
7.1.2 Privileged Hydrogen Abstraction Reaction of Benzothiazoline 147
7.1.3 Chemiluminescence Method 148
7.1.4 Nucleophilic Substitution Reaction 149
7.2 Fluorescent Probes for Hydrogen Peroxide 149
7.3 Fluorescent Probes for Hydroxyl Radical 154
7.4 Fluorescent Probes for Lipid Peroxy Radical 157
7.5 Fluorescent Probes for Other Reactive Oxygen Species 160
7.5.1 Fluorescent Probes for HClO 160
7.5.2 Fluorescent Probes for HOBr 163
7.5.3 Fluorescent Probes for $^1O_2$ 165
7.5.4 Fluorescent Probes for O_3 165
Acknowledgement 166
References 166

Chapter 8 Imaging of Intracellular Reactive Nitrogen Species and Reactive Sulfur Species 170

8.1 Introduction 170
8.2 Imaging of Intracellular Reactive Nitrogen Species 171
8.2.1 Probes for Nitroxyl (HNO) 171
8.2.2 Probes for Nitric Oxide (NO) 175
8.2.3 Probes for Peroxynitrite ($ONOO^-$) 182
8.3 Fluorescent Probes for GSH, Cys and Hcy 187
8.3.1 Fluorescent Probes for GSH 187
8.3.2 Fluorescent Probes for Cys 189
8.3.3 Fluorescent Probes for Hcy 190
8.3.4 Fluorescent Probes for Two or All of GSH, Cys and Hcy 191
8.4 Fluorescent Probes for H_2S 192
8.4.1 Nucleophilic Addition of H_2S 193
8.4.2 H_2S-induced Thiolysis 195

8.4.3 Azide Group Reduction 196
8.4.4 Other Assay Principles 197
8.5 Fluorescent Probes for H_2S_n 197
8.6 Fluorescent Probes for Cys-SSH 199
8.7 Fluorescent Probes for SO_2 199
Acknowledgement 202
References 202

Chapter 9 Imaging of the Tumour Microenvironment 211

9.1 Introduction 211
9.2 Fluorescent Probes for Intracellular pH and Hypoxia Imaging 212
9.3 pH and Hypoxia Activatable Tumour Imaging 215
9.4 Tumour pH and Hypoxia Targeting Nanotechnology 220
9.5 pH and Hypoxia Imaging for Tumour Therapeutic Monitoring 222
9.6 Perspective 227
Acknowledgement 228
References 228

Chapter 10 Visualization of Dynamic Intermolecular Interactions in Living Cells 232

10.1 Introduction 232
10.2 Dynamic Interactions among Active Small Molecules 233
10.2.1 Dual Detection of ROS 233
10.2.2 Dual Detection of ROS and RSS 235
10.2.3 Dual Detection of ROS and pH 237
10.2.4 Dual Detection of ROS and MicroRNA 237
10.2.5 Dual Detection of RNS and RSS 238
10.3 Dynamic Interactions among Biomacromolecules 238
10.3.1 Proteins 239
10.3.2 DNAs 240
10.3.3 RNAs 242
10.4 Dynamic Intermolecular Interactions Between Active Small Molecules and Biomacromolecules 242
10.4.1 Dual Detection of Phosphatase and H_2S 242
10.4.2 Dual Detection of miRNA and mRNA 244
10.5 Perspectives 244
Acknowledgement 244
References 245

Subject Index 248

CHAPTER 1

Design of In Situ *Cytosensing Strategies*

1.1 General Principles of *In Situ* Cytosensing

Owing to the continuing attention devoted to human health and life, recent decades have witnessed tremendous developments in analytical strategies for the *in situ* probing of chemical and biological events at the molecular level in living cells. These tools provide spatiotemporal information on cellular molecules, biochemical reactions and signalling and communication processes,[1–3] thus contributing to the better understanding of the sophisticated mechanisms of life and to improvements in or even transformation of disease diagnosis and treatment concepts and techniques.[4,5] To analyse cellular functional molecules *in situ*, one needs to design a detection system composed of a recognition module, a signal transduction module and a signal output module. The recognition module achieves the differentiation of the targets by specific binding and the signal transduction module converts the recognition event into a signal, transformation or reaction to indicate the existence of the target. The latter can be integrated with the capability to amplify the signal or implement multiplexed analysis. The last module performs the final collection and reporting of the signals that correspond to the target. Owing to the low expression levels of the targets and the complex biological environment, four key aspects should be taken into consideration when designing the system: recognition/binding specificity, signal generation pathway, signal amplification and multichannel analysis capability. These issues play crucial roles in the analytical performance and the comprehensive consideration of these aspects is challenging. This chapter presents a review of the most promising trends regarding these major concerns.

Detection Science Series No. 16
In Situ Analysis of Cellular Functional Molecules
By Huangxian Ju, Bo Tang and Lin Ding

Published by the Royal Society of Chemistry, www.rsc.org

1.2 Enhancement of Detection Specificity

Specific detection depends on the recognition of a biosensing interface towards the target. Common recognition motifs include antibodies, lectins, peptides, nucleic acids (including aptamers), biomimetic polymers, artificial receptors and small molecules. To improve the detection sensitivity, it is beneficial to use recognition motifs with high specificity and avidity towards the target, which are often intrinsic properties of the individual recognition pairs. In spite of this, one can still tailor the detection schemes to enhance the binding intensity and specificity. For example, to improve the avidity of lectins towards their corresponding glycan epitopes, the glycans can be assembled in a multivalent fashion to accelerate the binding events. Based on this strategy, Ding *et al.* fabricated a mannan-modified gold electrode to yield a carbohydrate monolayer with multivalent binding capability with the lectin concanavalin A (Con A), thus permitting efficient competition between the mannan monolayer and cell-surface mannose.[6]

To increase the specificity for the identification of a given type of cells, Rudchenko *et al.* developed automata to perform cascade strand-displacement reactions directed by antibody recognition towards cell-surface markers (Figure 1.1).[7] The automata checked the existence of each marker successively according to an "if yes then proceed" logic. Only when all the markers of the given set were present on the cell surface could the automata complete the computation and output a final fluorescence signal, thus achieving the identification of a specific population of lymphocytes within human blood cells.

Concerning the targeting delivery, most current cell targeting strategies rely on only one receptor on the cell surface and therefore usually suffer from the issue of high non-specific interactions. The incorporation of two factors in a targeting system to function in a serial manner may improve the site-specific delivery of the probe. On the basis of this concept, Ren *et al.* used an auto-cleavable hairpin structure acting as "smart keys" to modify the siRNA-loaded oligonucleotide nanovehicle (siRNA-ONV) and bound two kinds of aptamers, sgc8c and sgc4f, on the cell surface to act as "double locks" (Figure 1.2).[8] These "locks" could be opened sequentially by reacting with the "key" in a serial manner. The "dual lock-and-key" mode controlled the cell "locked-open" status and thus achieved cell-subtype-specific recognition and precise probe delivery.

In addition to dual specificity control of the same type, different types of specificity control can also be integrated into one system. For intracellular sensing, considering the complex intracellular environment, it is beneficial to turn off the response capability of the recognition motif of the detection probe until it reaches the desired locations, otherwise the recognition and the subsequent signal generation may take place during the cellular delivery and uptake process, leading to impaired detection accuracy and low signal-to-noise ratio. In this context, Zhao *et al.* designed a nanodevice containing two dependent components: a UV light-activatable DNA aptamer probe and

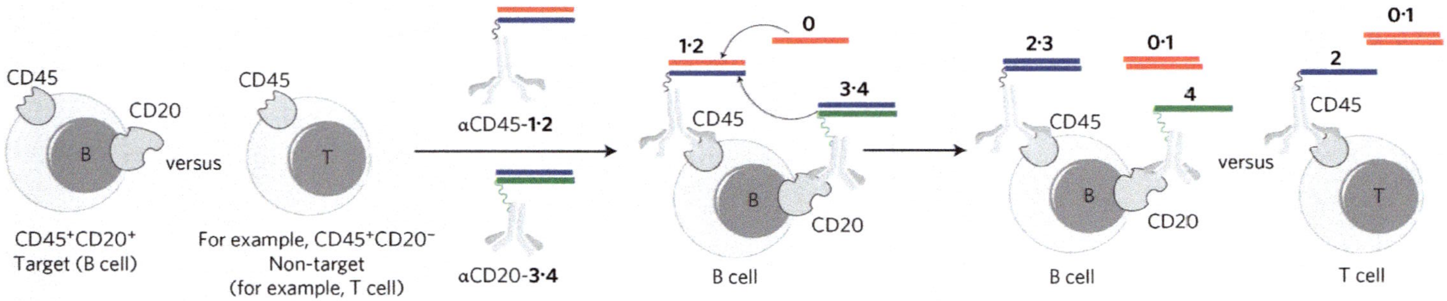

Figure 1.1 Scheme of automata operating on a B cell with a $C45^{+}CD20^{+}$ phenotype (target) and on an example of a non-targeted cell with a $CD45^{+}CD20^{-}$ phenotype.[7]
Reproduced from ref. 7 with permission from Springer Nature, Copyright 2013.

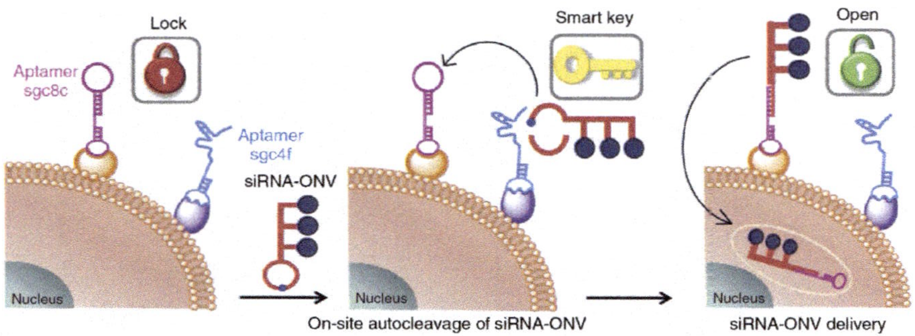

Figure 1.2 Scheme of the working principle of the siRNA-ONV nanotube.[8]
Reproduced from ref. 8, https://doi.org/10.1038/ncomms13580, under the terms of the CC BY 4.0 licence, http://creativecommons.org/licenses/by/4.0/.

lanthanide-doped upconversion nanoparticles (UCNPs).[9] The Cy3-modified aptamer strand was initially locked by a quencher-bearing complementary DNA containing a photocleavable (PC) group, which was denoted a "PC inhibitor", hence the aptamer could not bind to ATP in its locked state and displayed a low fluorescence signal background. When the nanodevice reached the desired region, upon irradiation with near-infrared (NIR) light the PC inhibitor would be cleaved at the PC site by the UV light converted by UCNPs and the capability of the aptamer to switch its structure to bind ATP would be restored. Thus, in the presence of ATP, the dissociation of the cleaved PC inhibitor led to a substantial increase in the fluorescence signal, achieving ATP sensing with high specificity.

1.3 Design of Off–On Signal Switch

When the recognition motif for the target is fixed, a signal transduction pathway must be designed to indicate the occurrence of the recognition/binding events. To achieve high-sensitivity detection and to lower the background signals, a key point is the design of a signal off–on switch that can be specifically triggered by the target or target-associated events to reflect the existence of the target in the analysis scheme. The tailoring of signal generation methods depends on the action modes of the recognition processes, for example, cleavage by an enzyme or hybridization with a DNA strand.

With regard to cellular enzymes with cleaving capability, it is convenient to design signal switches by use of responsive substrates. One can label the substrate with a fluorophore at one end and use the other end to link to a nanomaterial with desired optical properties, thus constituting a certain type of energy transfer pathway, such as fluorescence resonance energy transfer (FRET). The reason for the introduction of nanomaterials lies in the synergistic employment of their unique characteristics not only from an optical properties perspective but also with regard to the accelerating capability for cellular delivery.[10,11] However, FRET-based detection platforms often suffer from a high background signal due to acceptor bleed-through. A smart solution for this issue is the design of a time-resolved fluorescence resonance energy transfer (TR-FRET) system using persistent luminescence nanoparticles (PLNPs), in which the luminescence of the fluorophore is collected only after a delay time, thus suppressing the acceptor bleed-through. Zhang *et al.* demonstrated this concept by developing a TR-FRET platform for the *in situ* lifetime quantification of intracellular caspase-3 (Figure 1.3).[12]

Another common design for enzymes with cleaving capability is to load a peptide-based fluorophore-labelled substrate on nanomaterials with fluorescence quenching ability, such as graphene oxide (GO),[13] gold nanorods (AuNRs),[14,15] manganese dioxide[16] or a metal–organic framework (MOF).[17] The cleavage by the corresponding enzymes leads to a signal-on switch by releasing the fluorophore into solution. For example, Zhang *et al.* fabricated

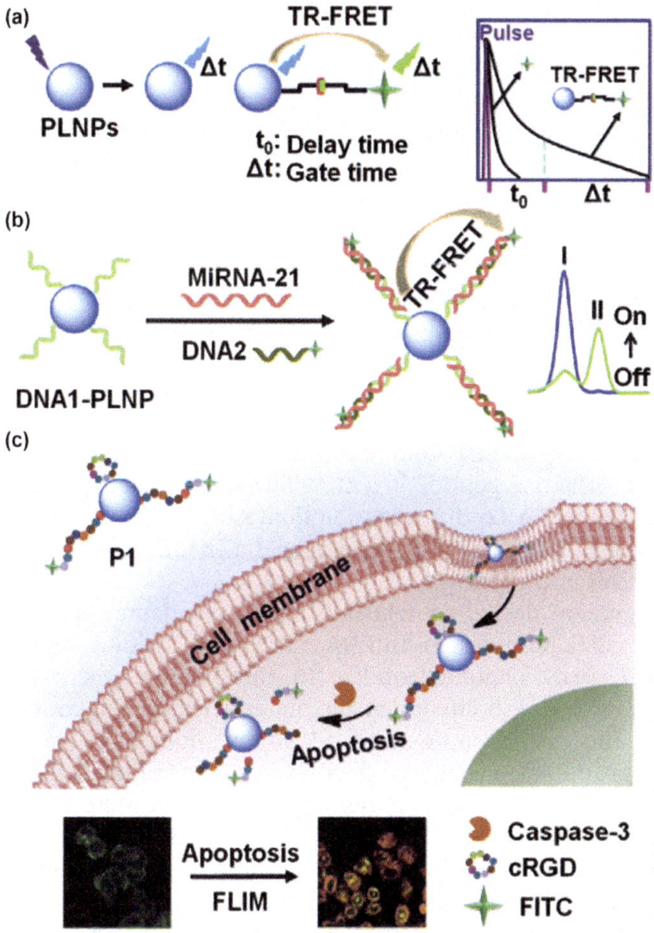

Figure 1.3 Schematic illustration of (a) the PLNP-based TR-FRET principle, (b) TR-FRET-based detection of miRNA-21 using caspase-specific DNA1-functionalized PLNPs and (c) lifetime imaging of intracellular caspase-3 activity during cell apoptosis. The blue and green lines in (b) represent the time-resolved fluorescence curves before and after response to the target, respectively.[12]
Reproduced from ref. 12 with permission from Elsevier, Copyright 2015.

a caspase-3-responsive multifunctional nanoprobe by assembling a porphyrin, a folate targeting motif and a dye-labelled peptide substrate in an MOF cage. Upon executing enhanced photodynamic therapy (PDT), owing to the integration of porphyrin and MOF, this probe allows *in situ* cell apoptosis monitoring by displaying caspase-3 activation-induced fluorescence recovery, thus underlining its great promise in precision cancer therapy.[17]

Cathepsin B (CaB) is another analyte that has attracted much attention owing to its crucial roles in cancer progression and diagnosis, and displays a specific peptide-cleaving capability.[18] Taking CaB as the target,

Tian *et al.* developed a multi-functional theranostic nanoplatform by the non-covalent assembly of 1,2-distearoyl-*sn-glycero*-3-phosphoethanolamine-*N*-{folate[poly(ethylene glycol)]-2000} (DSPE-PEG2000-FA) and chlorin e6 (Ce6)-labelled CaB-specific peptide substrate (Ce6-GRRGKGGFFFF, Ce6-Pep) on the surface of the GO sheet.[13] Ce6, a fluorescent dye and a highly efficient photosensitizer,[19] can be inhibited when in close proximity to the GO sheet. Upon folate receptor-mediated endocytosis into cells, the peptide on the GO sheet could be specifically cleaved by CaB to release the Ce6 into solution with recovery of fluorescence. This signal could be used not only for CaB imaging but also for monitoring the therapeutic efficacy under 660 nm irradiation, because the diffusion of the lighted Ce6 into cytoplasm suggested lysosomal destruction, which is a key step in the lysosomal cell-death pathway. To decrease further the background signals coming from incomplete quenching of the fluorophore by the single use of nanomaterial, Shen *et al.* encapsulated gold nanoparticles (AuNPs) functionalized by Cy3-labelled substrate peptide in a ZIF-8 MOF shell to achieve the synergetic use of AuNPs and MOF for superior fluorescence quenching. Here, only upon release from MOF by a lysosomal acidic environment can the peptide encounter CaB and be cleaved.[20]

These signal-on design principles can also be adapted for cellular functional molecules (telomerase, miRNA, mRNA, *etc.*) with a specific binding capability towards oligonucleotides. In this context, the target-induced conformation or length change of the oligonucleotide substrate by hybridization (in the case of miRNA or mRNA) or elongation (in the case of telomerase) can lead to a signal off–on switch. For example, although it has been widely recognized that telomerase plays an essential role in ensuring the normal telomere length,[21] the *in situ* monitoring of telomerase activity has long been regarded as a challenge. To resolve this issue, Qian *et al.* used telomerase primer to wrap fluorescein-loaded mesoporous silica nanoparticles (MSNs), which were also covalently modified with a black hole fluorescence quencher on their inner walls, for *in situ* "off–on" imaging of intracellular telomerase activity.[22] The primer could be specifically extended by telomerase and formed a rigid hairpin-like DNA structure, which could detach from the MSN surface and result in the release of originally loaded fluorescein and thus recovery of the fluorescence. In a later study by the same group,[23] this signal switch was upgraded to reduce the background signal coming from the non-specific dissociation of the primer from the MSN surface in cytoplasm by designing an AuNP-based vesicle kit. This kit contained a telomerase primer and a Cy5-tagged molecular beacon (MB)-functionalized AuNP. The hybridization of the primer extension product with the MB led to the opening of the MB and thus a signal-on switch. Although this strategy utilized a stable Au–S bond to immobilize MB on the nanoparticles, thus reducing background signals, the complex composition of the kit requires further simplification. Driven by this requirement, the group further designed a nicked MB-functionalized AuNP probe.[24] The telomerase-triggered elongation of the primer, which originally hybridized to

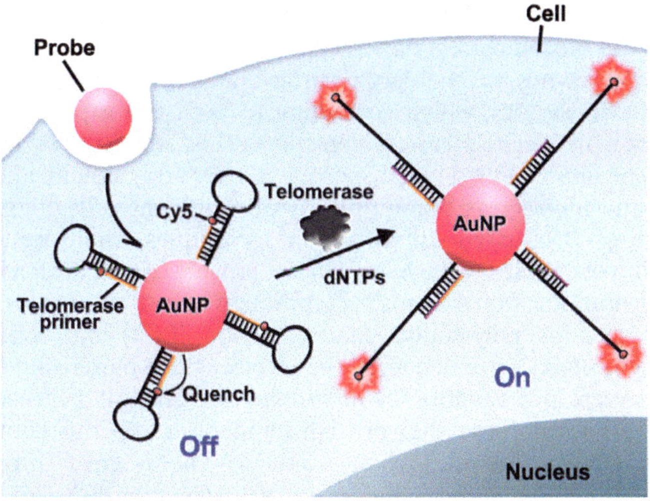

Figure 1.4 Schematic illustration of the nicked MB-functionalized AuNP probe for *in situ* analysis of intracellular telomerase activity.[24]
Reproduced from ref. 24 with permission from American Chemical Society, Copyright 2014.

the stem of the MB, led to inner chain substitution hybridization, opening of the MB and thus a signal-on switch (Figure 1.4).

With respect to miRNA as the target, Liu *et al.* conjugated a dye-labelled MB as the recognition substrate and signal switch, along with folic acid as the targeting motif, on AuNR. The hybridization of target miRNA with the nanoprobe led to the unfolding of MB, resulting in the off–on imaging of miRNA-21 in HeLa cells.[15] Zhang *et al.* also employed their versatile TR-FRET platform for off–on detection of miRNA-21. The signal switch was generated by the sandwich hybridization-triggered FRET from PLNPs to fluorescein isothiocyanate (FITC).[12] Jia *et al.* used a Ce6-labelled ATP aptamer to functionalize MoS_2 nanoplates and developed a nanoprobe for the fluorescence imaging of intracellular ATP.[25]

1.4 Design of Signal Amplification Strategy

Biomolecule-based signal amplification is a major type of amplification scheme that is widely used for *in situ* cellular analysis. Typical representatives include enzymatic catalysis and DNA nanotechnology-based strategies. For cell-surface glycan detection, Cheng *et al.* used horseradish peroxidase (HRP)-labelled lectin for the specific recognition of glycans on living BGC-823 cells which were captured on an electrode modified with RGDS peptide-functionalized single-walled carbon nanotubes (SWNTs).[26] Based on the enzymatic catalytic reaction of HRP towards the oxidation of *o*-phenylenediamine (*o*-PD) by H_2O_2, a highly sensitive cytosensor for the quantitation of cell numbers and cell-surface glycans was developed. Moreover, the biocompatible nanocomposite offered another level of signal amplification owing to the large surface area and excellent electrical conductivity of SWNTs, which contribute to both cell capture and signal amplification in electrochemical detection. This strategy was further developed to be implemented in a disposable sensor array system, allowing the detection and monitoring of multiple glycans on living cells (Figure 1.5).[27]

Concerning DNA-based amplification techniques, the specificity and predictability of nucleic acid base pairing provide a rich design space for developing amplification systems.[28,29] Although several nucleic acid cascade reactions, such as polymerase chain reaction (PCR) and rolling circle amplification (RCA), have benefited the ultrasensitive biosensing of various targets, they are not suitable for *in situ* cellular analysis because of their dependence on exotic enzymes or high temperature. In this context, non-enzymatic catalytic amplification strategies hold great promise for improvement of the performance of *in situ* detection.[30,31] For example, programmable hybridization chain reaction (HCR) and catalysed hairpin assembly (CHA) have been used in intracellular microRNA,[32,33] mRNA[34,35] and telomerase imaging[36] for the enhancement of sensitivity.

Wu *et al.* designed a CHA-based signal amplification for mRNA imaging in living cells.[34] The DNA circuits contained two components: two hairpin-shaped metastable DNA substrates (H1 and H2), whose conformational transformations could be catalytically triggered upon target binding, and a reporter DNA duplex with the two strands respectively labelled with

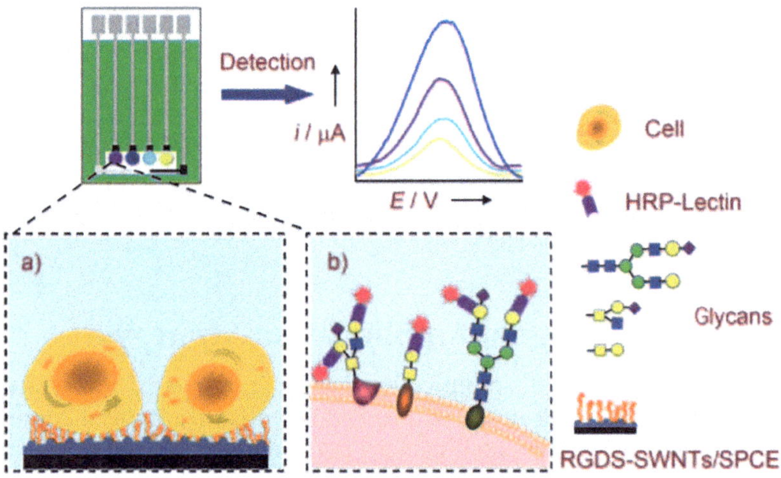

Figure 1.5 Schematic illustration of the electrochemical cytosensor array for cell-surface glycan analysis. (a) Cells captured on RGDS-SWNTs/SPCE (screen-printed carbon electrode); (b) HRP-lectin binding on cell-surface glycans.[27]
Reproduced from ref. 27 with permission from John Wiley and Sons, © 2009 WILEY-VCH Verlag GmbH & Co. KGaA, Weinheim.

fluorophore and quencher. In the presence of target mRNA, a stable hybridized duplex between H1 and H2 could be generated with the target mRNA being released to catalyse the next cycle for H1–H2 duplex formation. This H1–H2 duplex could further dehybridize the reporter, leading to the restoration of the initially quenched fluorescent signal. This design permitted the generation of multiple signal outputs from one mRNA target, thus achieving the sensitive imaging of low-expression mRNA targets.

Inspired by the DNA nature of telomerase primer, Yan *et al.* combined the biological function of telomerase with isothermal nucleic acid-based signal amplification for the *in situ* detection of telomerase activity with enhanced sensitivity.[36] This concept was implemented through the design of a liposome nanoprobe, which consisted of a telomerase-targeting responder–transmitter DNA complex (HPT) module and a CHA signal amplification module. Upon transfection into living cells, the extension of HPT led to the release of the transmitter DNA, which could act as the initiator and catalyst of CHA amplification, resulting in the lighting up of the reporter complex. Because of the recycling use of the transmitter, multiple enhancement of signal outputs from one extension event was achieved. This intracellular CHA-based signal amplification was also integrated on the surface of UCNPs by Huo *et al.* to develop a nanoamplicon comparator (NAC) for miRNA imaging. The catalytic circuit responded to target miRNA and allowed quantitative UCNP-to-organic-fluorophore luminescent resonance energy transfer (LRET) imaging against a native UCNP emission reference channel.[33]

The CHA not only can be used for signal amplification, but also provides a facile solution to reflect the protein-specific glycosylation stoichiometry by anchoring the catalytic strand on the target protein and H1 on the glycans, which allows the cyclic use of protein strands.[37] Similarly, exonuclease[38]- or restriction endonuclease[39]-aided recycling "hybridization and cleavage" processes between the strands respectively labelled on protein and glycans offer an alternative solution to achieve the cyclic use of a DNA probe labelled on protein.

In addition to CHA, HCR, a target-triggered alternating hybridization reaction between two hairpin probes (H1 and H2), is also widely used for signal amplification in cellular analysis. Wu *et al.* developed an electrostatic DNA nanoassembly format that realized intracellular HCR for ultrasensitive mRNA imaging in living cells (Figure 1.6).[35] This design relied on the assembly of a core AuNP, an interlayer of cysteine-terminated cationic peptides and an outer layer of fluorophore-labelled DNA probe H1. The core AuNP permitted efficient quenching of fluorophores labelled at H1 due to surface energy transfer. In the presence of target mRNA and also free-diffused H1 and H2, a chain reaction occurred by alternating hybridization between H1 (using AuNP-carried H1 as the initiator) and H2, generating a rigid double-stranded DNA structure, which could dissociate from the AuNP and activate a FRET signal indicating target mRNA expression. The multiple assembly of the H1 and H2 pair led to efficient signal amplification with a picomolar limit of detection.

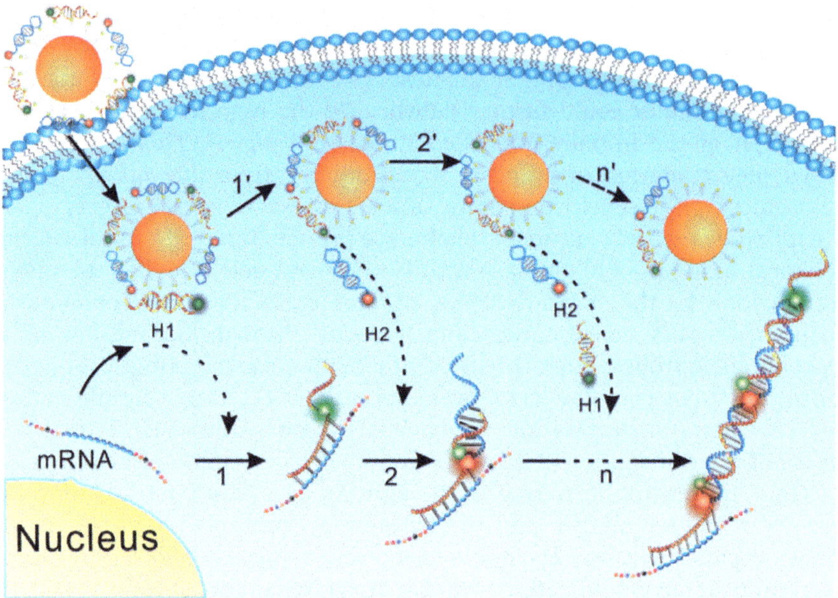

Figure 1.6 Illustration of intracellular HCR for mRNA detection.[35] Reproduced from ref. 35 with permission from American Chemical Society, Copyright 2015.

Despite their powerful capability to enhance sensitivity, the kinetics of these cascade reactions depend on the diffusion of the DNA probes, which prolongs the reaction time and thus compromises the reaction efficiency. We believe that the solution to this problem lies in the confinement of DNA probes in a compact space. As an illustrative model of this concept, Ren *et al.* designed a DNA "nano string light" (DNSL) responsive to target mRNA based on an accelerated DNA cascade reaction (DCR) along a DNA nanowire.[40] This method achieved sensitive mRNA imaging in living cells with a highly enhanced reaction rate and efficiency. The DNSL was constructed by interval hybridization of DNA hairpin probe pairs (self-quenched H1 and H2) to a DNA nanowire with reduplicated sequence segments generated by RCA (Figure 1.7). The hybridization of the intracellular target survivin mRNA with one H1 in DNSL triggered the cascade hybridization of H1 and H2 along the DNA nanowire, which could instantly light up the whole DNSL with a highly amplified signal gain. Notably, the designed DCR could complete the reaction in a time period that is only 15% of that for traditional HCR and the amplification ratio is 20.3 times of that for HCR, demonstrating the superior efficiency of the DCR for signal amplification.

Nanoscale materials (1–200 nm) with unique optical, magnetic, electrical and photothermal properties provide ideal platforms for enhancing electronic or optical signals during *in situ* cellular analysis. Owing to their large surface-to-volume ratio, a wide variety of signal molecules can be

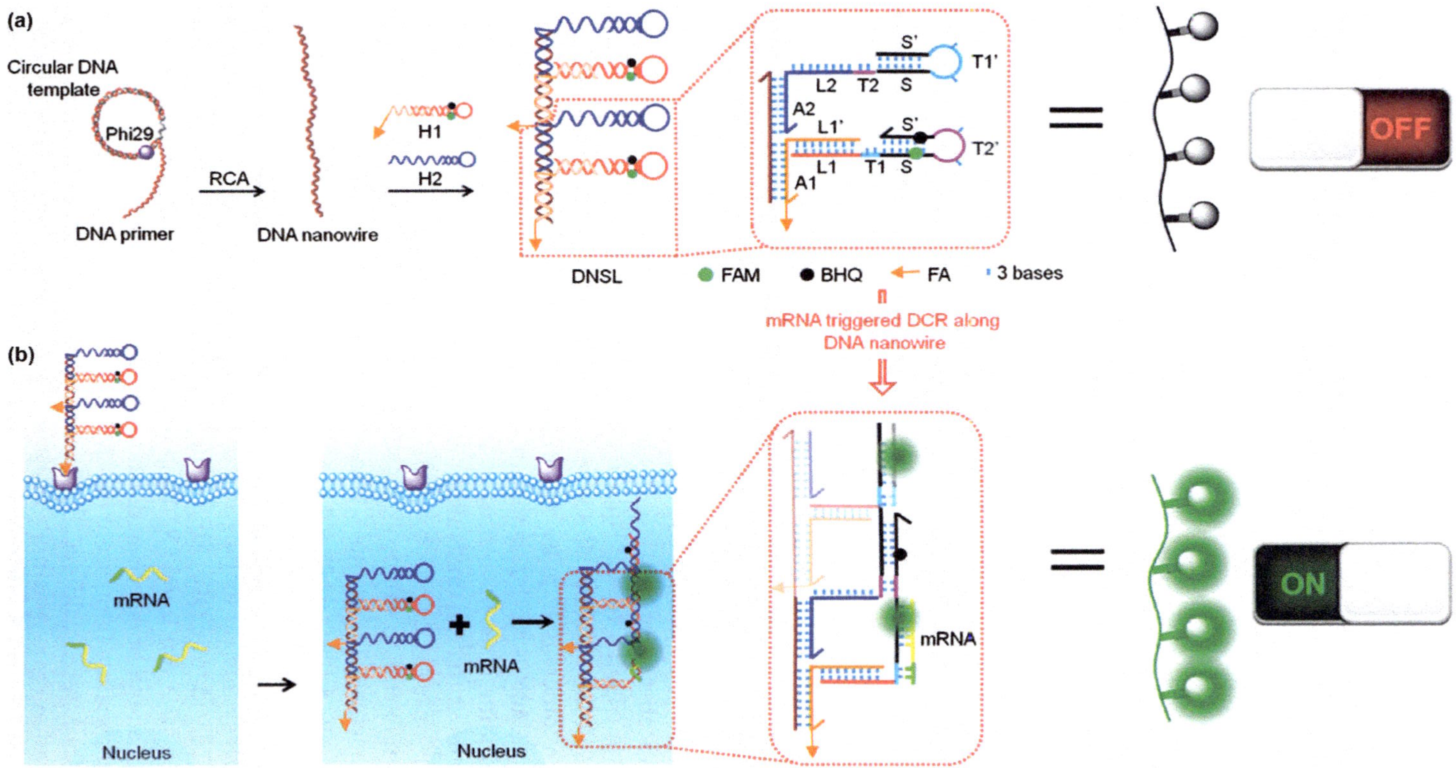

Figure 1.7 Schematic illustration of (a) DNSL synthesis based on interval hybridization of H1 and H2 to a DNA nanowire and (b) targeted delivery of DNSL and imaging of target mRNA in living cells based on accelerated DCR along a DNA nanowire.[40] Reproduced from ref. 40 with permission from American Chemical Society, Copyright 2017.

loaded on/in nanomaterials, such as enzymes, DNA, redox reagents, fluorophores and other types of nanomaterials, thus contributing to the enhancement of the response coming from a single recognition event and improvement of analysis performance. To improve the detection sensitivity for cellular glycans, Ding *et al.* fabricated HRP and lectin co-functionalized AuNPs with a molar ratio of HRP to lectin of 8 : 1, using lectin for recognizing glycans and HRP for amplifying electrochemical signals.[41] The anchoring of the nanoprobes onto electrode-captured cells led to a detection sensitivity that was 34 times higher than that using simple HRP-labelled lectin, verifying the advantages of the designed nanoprobe.

Another type of nanomaterial-based signal amplification is electrochemical stripping analysis of metal element-containing nanoprobes. In this context, the nanoprobe needs to be modified with a recognition motif and the recognition event possesses a positive or negative correlation with the target. The stripping voltammetric measurement of a recognized nanoprobe after dissolution actually brings massive signal species to the electrode surface to implement signal enhancement. As an example, Ding *et al.* developed a one molecule–two surface competition format for reflection of the extent of cell-surface mannose expression by the amount of lectin-functionalized quantum dots (QDs) captured by a mannan-modified electrode.[6] An anodic stripping signal of Cd^{2+} from the dissolved product of QDs was then employed for sensitive quantification. In another example, for the detection of cell-surface sialic acid (Sia), Qian *et al.* used 3-aminophenylboronic acid (APBA) and avidin co-functionalized carbon nanohorn to label the aldehyde groups generated by sodium periodate oxidation of cell-surface Sia *via* a biotin hydrazide linker. The APBA moieties further contributed to the capture of multiple mannan-conjugated AuNPs, which could be released, dissolved and subjected to a pre-electrooxidation and then differential pulse voltammetric detection. Here, the multiple binding of AuNPs to the nanohorn probe amplified the detectable signal, thus providing a sensitive strategy for the detection of Sia and monitoring the dynamic variation on living cells.[42]

To maximize the detection sensitivity, researchers often integrate multiple signal amplification protocols together and design cascade signal amplification schemes. For example, Han *et al.* designed a double signal amplification strategy for cell-surface glycan detection: (1) developing a sandwich binding format based on the recognition of APBA-functionalized QDs towards cell-surface Sia groups and polysialic acid-stabilized AuNPs (PSA-AuNPs); and (2) making use of the sensitizing effect of cadmium cations, dissolved from QDs, on the fluorescence signal of the non-fluorescent metal-sensitive dye Rhod-5N. Thus highly sensitive fluorescent analysis of cells with a limit of detection down to eight human gastric carcinoma (BGC) cells and dynamic fluorescent monitoring of Sia expression variation on the cell surface were achieved.[43] This nanomaterial assembly-based signal amplification strategy can also be used for other imaging modalities. In a surface-enhanced Raman scattering (SERS) imaging strategy for cell-surface

Sia,[44] a phenylboronic acid group-functionalized gold nanoflower was used as a bridge probe for both recognition of target Sia and assembly of poly(*N*-acetylneuraminic acid)-modified gold nanoparticles, leading to plasmonic coupling of two kinds of gold nanoprobes in a single-core–multi-satellite nanostructure. The efficient interparticle plasmonic coupling achieved could serve as a significant contributor to the enhancement of Raman signals.

1.5 Design of Multi-channel Analysis

The simultaneous detection of multiple targets can provide complementary information on multiple targets, thus contributing to a comprehensive understanding of the biological processes and revealing the relationships between different targets. The need to design multiplexed *in situ* analysis strategies is also due to the molecular characteristics shared by the large number of biological samples, hence using the expression pattern as a discriminant criterion is beneficial for precision diagnosis.

For multiplex analysis of cellular functional molecules, electrochemical sensor arrays, with the capability of location-based resolution, offer sensitive and convenient tools by integrating micromachining and nanotechnology. An excellent example of this approach is the disposable electrochemical cytosensor array developed by Cheng *et al.* for the simultaneous analysis of four types of glycans on intact cell surfaces.[27] This cytosensor could be used for monitoring of the dynamic variation of the glycan expression pattern on cancer cell surfaces during drug inducement or erythroid differentiation of human leukaemic K562 cells, suggesting its feasibility for revealing carcinoma cell-surface glycome.

The same group subsequently fabricated a microplate-based cell array to develop a chemiluminescent (CL) imaging strategy for the simultaneous detection of sialyl and galactosyl groups on cell surfaces.[45] These two types of glycans could be substituted with an aldehyde group by chemo- and enzymatic-selective oxidation, respectively, which could be lit up through aniline-catalysed hydrazone ligation with biotin hydrazide, avidin-mediated assembly of HRP and biotin co-functionalized AuNPs and CL imaging. Owing to its high sensitivity, the method achieved the distinguishing of cancer cells from normal cells and monitoring of the dynamic Sia expression on living cells.

Although location-based resolution strategies can provide multiplexed expression extents of cellular functional molecules, different signal channels collect information from cell populations at different locations. Hence these methods fail to reflect the relative quantity of information from the same cell. In this regard, label-based resolution, including spectral resolution and sequence resolution, represents a promising solution.

Spectral resolution-based multi-channel analysis can be designed for systems comprising target species of the same type or different types. For example, for members of the caspase family, which participate at different apoptosis stages/pathways using different caspase enzymes,[46] real-time revealing of the caspase evolution during cell apoptosis is essential for elucidating the cell apoptosis mechanisms. To achieve this, Zhang *et al.* designed a composite nanoprobe by conjugating two types of dye-labelled peptides, specific to upstream caspase-9 and downstream caspase-3, respectively, as the signal switch and folic acid as a targeting motif on an AuNR surface (Figure 1.8).[14] The fluorescence of the two dyes was initially quenched by AuNR. Upon nanoprobe endocytosis and NIR irradiation, cell apoptosis was triggered by the photothermal effect and the two types of peptides could be successively cleaved by the activated caspases from upstream caspase-9 to downstream caspase-3, thus leading to the recovery of the fluorescence of the two dyes. The spectral-resolved turn-on signals permitted the quantification of both caspase-9 and caspase-3 activities in cancer cells and monitoring of their evolution in living mice and also provided an efficient handle to assess therapeutic efficiency.

As an illustration model for the multi-channel analysis of different types of targets, Liu *et al.* developed a black phosphorus–manganese dioxide nanoplatform for monitoring oxygen self-supply, enhancing PDT and feeding back the therapeutic effect (Figure 1.9).[16] The hybridized nanoplatform was

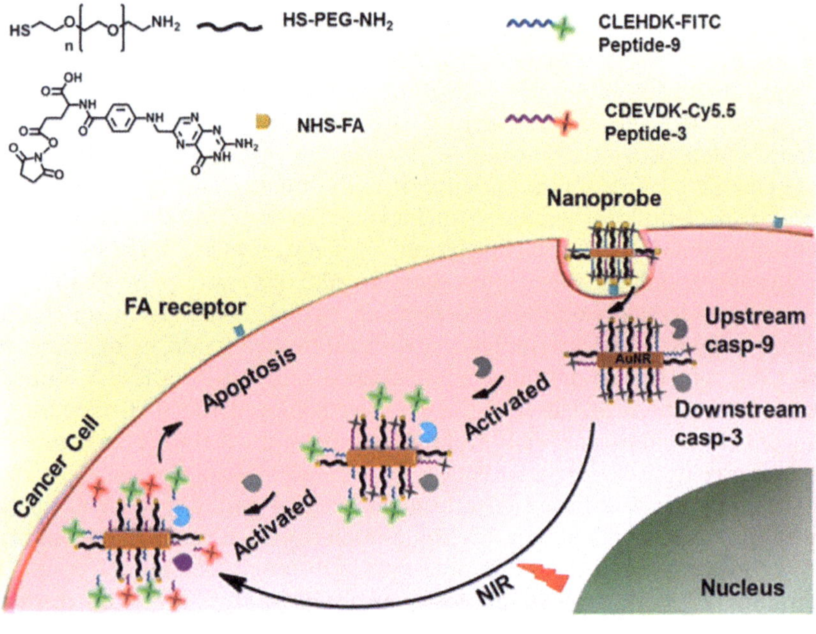

Figure 1.8 Schematic illustration of the integrated AuNR platform for *in situ* monitoring of the evolution of the caspase family activated *via* real-time NIR photothermal therapy.[14]
Reproduced from ref. 14 with permission from the Royal Society of Chemistry.

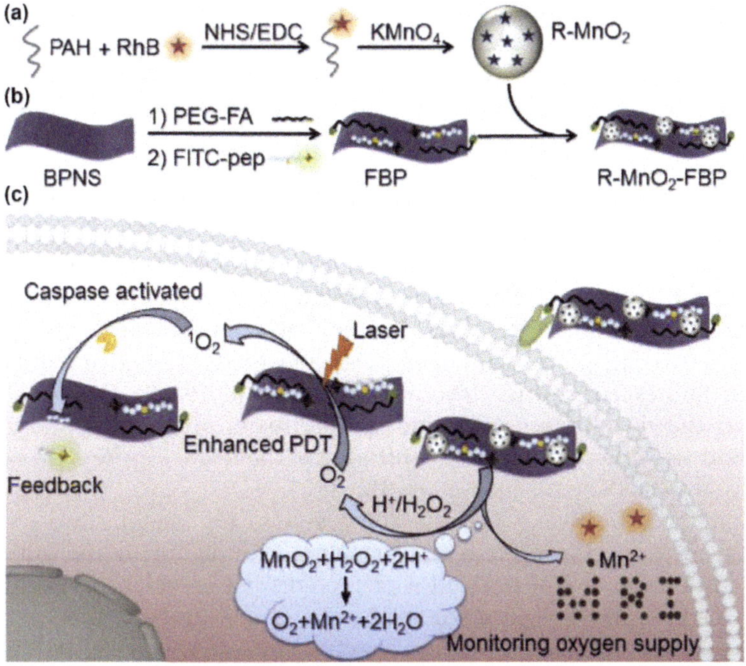

Figure 1.9 Schematic illustration of the synthesis of (a) R-MnO_2 and (b) R-MnO_2 functionalized black phosphorus and (c) the principle for monitoring oxygen self-supply, enhanced PDT and feedback of the therapeutic effect.[16]
Reproduced from ref. 16 with permission from Elsevier, Copyright 2019.

prepared by assembly of rhodamine B (RhB)-encapsulated manganese dioxide (R-MnO_2) as O_2 supplier and indicator and FITC-labelled peptide-functionalized black phosphorus as the theranostic agent. After being delivered into cancer cells, the nanocomposite dissociated in the acidic and H_2O_2-rich environment and generated O_2 to overcome hypoxia-associated PDT resistance. The amount of O_2 released showed a proportional relationship to the liberation of Mn^{2+} and RhB, hence the fluorescence of RhB could be used as an indicator of O_2 self-supply. Upon PDT, the cleavage of peptide by the activated caspase-3 in apoptotic cells occurred, leading to the recovery of the fluorescence of FITC and the achievement of real-time feedback of the therapeutic effect. In addition to PDT, the system featured a dual-monitoring capability, which provided cues for selecting the laser treatment timing and optimizing the therapeutic dose.

Although using different dyes to label different targets is a common approach to achieving the imaging of multiple targets in living cells, it becomes intricate when it comes to visualization of multiple glycans on a given protein at the same time. One needs to label simultaneously both the protein and the target glycans with fluorescence donors and acceptors to design FRET-based imaging systems to report the proximity of the protein and its carried glycans. It should be noted that the number of dyes doubles as the number of glycan target channels increases. However, owing to the difficulty of obtaining appropriate pairs of donors and acceptors without interference and the serious acceptor bleed-through resulting from the discrepancy between protein copy number and glycan abundance,[47] it is not feasible to report the protein-specific glycan expression by multiplexed FRET.

To solve the problem, inspired by the unique polychromatic light-emitting property of UCNPs, Wu *et al.* used UCNPs as a protein-confined single donor to construct a duplexed luminescence resonance energy transfer system (D-LRET) to activate simultaneously multiple acceptors labelled on glycans *via* single near-infrared excitation.[48] Upon 980 nm excitation, the UCNPs exhibited two energy emission bands, each of which matched and lit up one kind of fluorescent dye in close proximity (within 10 nm) through LRET, allowing the visualization of distinct glycoforms of mucin 1 on various cell types and also quantitation and monitoring of the relative expression ratio of termini glycans (Sia and fucose) on mucin 1.

Despite the effectiveness of the above system for imaging two types of glycans on a specific protein, this strategy cannot mirror the authentic hierarchical structure of DNA, RNA or proteins, all of which have extrinsic, single- or multi-level appendage structures emanating from the respective core structures, because relying on spectral resolution will inevitably limit the number of available signalling channels. To meet the challenge, we proposed to encode an arbitrarily expandable number of structural motifs in these biomacromolecules with distinct identification codes. In this regard, the DNA barcoding technique that makes use of the sequence diversity of DNA for identification of different targets provides a useful handle. In addition to the abundant barcodes achievable, it also offers another valuable advantage, namely smooth connection of these codes with various DNA amplification/detection techniques.

As a proof-of-concept demonstration, Li *et al.* developed a hierarchical coding (HieCo) imaging strategy, with DNA coding and decoding of protein and monosaccharides executed in fidelity to the hierarchical order of the target glycoprotein, for mucin 1-specific imaging of terminal Sia and fucose on tumour cells (Figure 1.10).[37] The protein and monosaccharide codes were initially masked and the decoding process began with the timing code and proceeded in a sequential, iterative decoding–unmasking fashion. Specifically, the protein code could decode different monosaccharide codes in a catalytic cycle manner, thus allowing the relative quantification of the extents of expression of different monosaccharides. Distinct monosaccharide patterns for four types of cells were observed and the dynamic monitoring of mucin 1-specific monosaccharide pattern changes associated with both the alteration of cellular physiological states and the occurrence of epithelial-to-mesenchymal transition was achieved. This approach featured expandable monosaccharide identification channels and was regarded as "an important step in the development of a system analogous to the use of green fluorescent proteins for protein tagging" by Suzuki.[49]

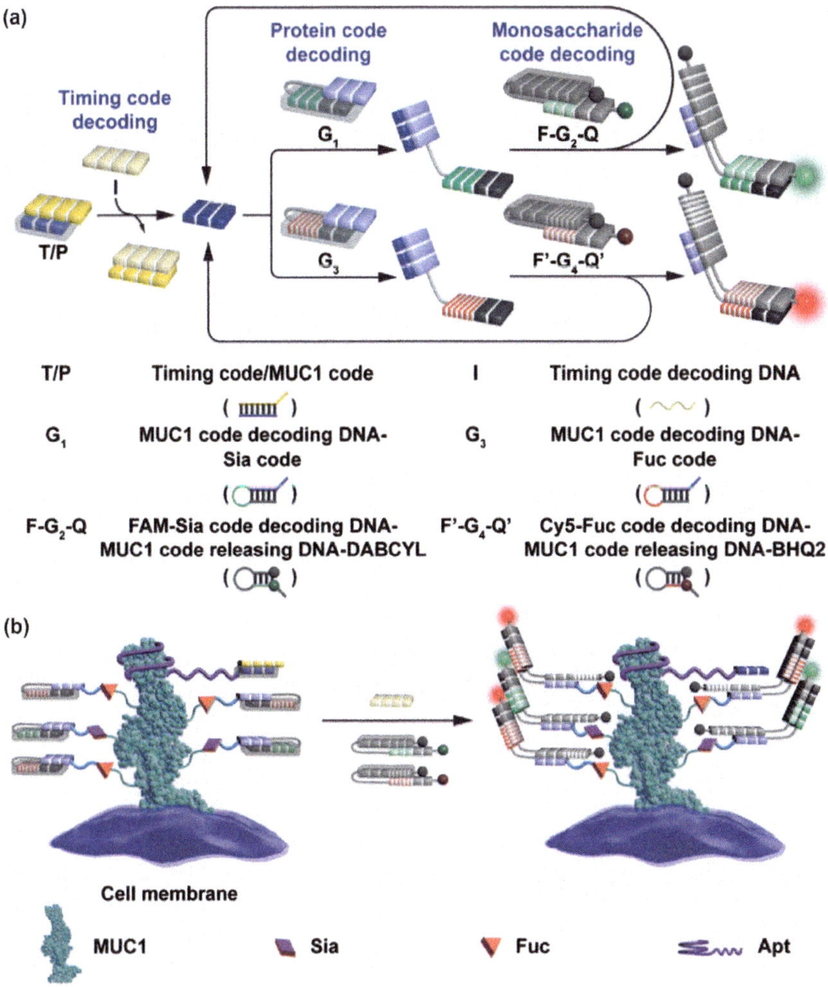

Figure 1.10 The hierarchical coding (HieCo) strategy (a) for live cell imaging of protein-specific glycoform (b).[37]
Reproduced from ref. 37 with permission from John Wiley and Sons, © 2018 Wiley-VCH Verlag GmbH & Co. KGaA, Weinheim.

1.6 Perspective

This chapter summarizes the design concepts for the *in situ* analysis of cellular functional molecules and provides insightful solutions for addressing the challenges in this field. In spite of these advances, further development of *in situ* cytosensing methods is still urgently needed, with numerous opportunities available. For instance, future work will be focused on developing novel nanoprobes by precision synthesis and functionalization. This may endow probes with homogeneous surface properties in addition to site-specific and quantity-predetermined assembly of recognition/signal motifs, and thus will contribute to the decrease of the background signal and enhancement of detection sensitivity. The combined introduction of enzyme-, light- or pH-responsive functional units along with the recognition module into the detection system provides another dimension to control the timing and region of the recognition events, which will facilitate precision detection. To fit the wide dynamic range of some functional molecules, which is due to the biological heterogeneity, it is beneficial to develop multiplexed analysis methods in a multimodal way to combine the inherent strengths of each modality to maximize sensitivity, specificity, spatial resolution, temporal resolution, signal penetration, *etc.*, and obtain extensive complementary information on the targets.

References

1. D. W. Domaille, E. L. Que and C. J. Chang, *Nat. Chem. Biol.*, 2008, **4**, 168–175.
2. H. Kobayashi, M. Ogawa, R. Alford, P. L. Choyke and Y. Urano, *Chem. Rev.*, 2010, **110**, 2620–2640.
3. D. T. Quang and J. S. Kim, *Chem. Rev.*, 2010, **110**, 6280–6301.
4. A. Demchenko, *J. Fluoresc.*, 2010, **20**, 1099–1128.
5. D. E. Lee, H. Koo, I. C. Sun, J. H. Ryu, K. Kim and I. C. Kwon, *Chem. Soc. Rev.*, 2012, **41**, 2656–2672.
6. L. Ding, W. Cheng, X. J. Wang, S. J. Ding and H. X. Ju, *J. Am. Chem. Soc.*, 2008, **130**, 7224–7225.
7. M. Rudchenko, S. Taylor, P. Pallavi, A. Dechkovskaia, S. Khan, V. P. Butler, S. Rudchenko and M. N. Stojanovic, *Nat. Nanotechnol.*, 2013, **8**, 580–586.
8. K. Ren, Y. Liu, J. Wu, Y. Zhang, J. Zhu, M. Yang and H. X. Ju, *Nat. Commun.*, 2016, **7**, 13580.
9. J. Zhao, J. H. Gao, W. T. Xue, Z. H. Di, H. Xing, Y. Lu and L. L. Li, *J. Am. Chem. Soc.*, 2018, **140**, 578–581.
10. M. P. Stewart, A. Sharei, X. Y. Ding, G. Sahay, R. Langer and K. F. Jensen, *Nature*, 2016, **538**, 183–192.
11. Q. H. Sun, Z. X. Zhou, N. S. Qiu and Y. Q. Shen, *Adv. Mater.*, 2017, **29**, 1606628.
12. L. Zhang, J. P. Lei, J. T. Liu, F. J. Ma and H. X. Ju, *Biomaterials*, 2015, **67**, 323–334.
13. J. W. Tian, L. Ding, Q. B. Wang, Y. P. Hu, L. Jia, J. S. Yu and H. X. Ju, *Anal. Chem.*, 2015, **87**, 3841–3848.
14. L. Zhang, J. P. Lei, J. T. Liu, F. J. Ma and H. X. Ju, *Chem. Sci.*, 2015, **6**, 3365–3372.
15. J. T. Liu, L. Zhang, J. P. Lei and H. X. Ju, *ACS Appl. Mater. Interfaces*, 2015, **7**, 19016–19023.
16. J. T. Liu, P. Du, T. R. Liu, B. J. C. Wong, W. P. Wang, H. X. Ju and J. P. Lei, *Biomaterials*, 2019, **192**, 179–188.

17. L. Zhang, J. P. Lei, F. J. Ma, P. H. Ling, J. T. Liu and H. X. Ju, *Chem. Commun.*, 2015, **51**, 10831–10834.
18. M. M. Mohamed and B. F. Sloane, *Nat. Rev. Cancer*, 2006, **6**, 764–775.
19. P. Huang, J. Lin, X. S. Wang, Z. Wang, C. L. Zhang, M. He, K. Wang, F. Chen, Z. M. Li, G. X. Shen, D. X. Cui and X. Y. Chen, *Adv. Mater.*, 2012, **24**, 5104–5110.
20. H. Shen, J. T. Liu, J. P. Lei and H. X. Ju, *Chem. Commun.*, 2018, **54**, 9155–9158.
21. J. Maciejowski and T. de Lange, *Nat. Rev. Mol. Cell Biol.*, 2017, **18**, 175–186.
22. R. C. Qian, L. Ding and H. X. Ju, *J. Am. Chem. Soc.*, 2013, **135**, 13282–13285.
23. R. C. Qian, L. Ding, L. W. Yan, M. F. Lin and H. X. Ju, *Anal. Chem.*, 2014, **86**, 8642–8648.
24. R. C. Qian, L. Ding, L. W. Yan, M. F. Lin and H. X. Ju, *J. Am. Chem. Soc.*, 2014, **136**, 8205–8208.
25. L. Jia, L. Ding, J. W. Tian, L. Bao, Y. P. Hu, H. X. Ju and J. S. Yu, *Nanoscale*, 2015, **7**, 15953–15961.
26. W. Cheng, L. Ding, J. P. Lei, S. J. Ding and H. X. Ju, *Anal. Chem.*, 2008, **80**, 3867–3872.
27. W. Cheng, L. Ding, S. J. Ding, Y. B. Yin and H. X. Ju, *Angew. Chem., Int. Ed.*, 2009, **48**, 6465–6468.
28. Y. J. Chen, B. Groves, R. A. Muscat and G. Seelig, *Nat. Nanotechnol.*, 2015, **10**, 748–760.
29. A. A. Tregubov, P. I. Nikitin and M. P. Nikitin, *Chem. Rev.*, 2018, **118**, 10294–10348.
30. H. M. T. Choi, V. A. Beck and N. A. Pierce, *ACS Nano*, 2014, **8**, 4284–4294.
31. B. Koos, G. Cane, K. Grannas, L. Lof, L. Arngården, J. Heldin, C. Clausson, A. Klaesson, M. K. Hirvonen, F. M. S. de Oliveira, V. O. Talibov, N. T. Pham, M. Auer, H. Danielson, J. Haybaeck, M. Kamali-Moghaddam and O. Soderberg, *Nat. Commun.*, 2015, **6**, 7294.
32. Z. Cheglakov, T. M. Cronin, C. He and Y. Weizmann, *J. Am. Chem. Soc.*, 2015, **137**, 6116–6119.
33. M. Huo, S. Q. Li, P. W. Zhang, Y. M. Feng, Y. R. Liu, N. Wu, H. X. Ju and L. Ding, *Anal. Chem.*, 2019, **91**, 3374–3381.
34. C. C. Wu, S. Cansiz, L. Q. Zhang, I. T. Teng, L. P. Qiu, J. Li, Y. Liu, C. S. Zhou, R. Hu and T. Zhang, *J. Am. Chem. Soc.*, 2015, **137**, 4900–4903.
35. Z. Wu, G. Q. Liu, X. L. Yang and J. H. Jiang, *J. Am. Chem. Soc.*, 2015, **137**, 6829–6836.
36. L. W. Yan, J. J. Hui, Y. R. Liu, Y. H. Guo, L. Liu, L. Ding and H. X. Ju, *Biosens. Bioelectron.*, 2016, **86**, 1017–1023.
37. S. Q. Li, Y. R. Liu, L. Liu, Y. M. Feng, L. Ding and H. X. Ju, *Angew. Chem., Int. Ed.*, 2018, **57**, 12007–12011.
38. Y. L. Chen, L. Ding, W. Y. Song, M. Yang and H. X. Ju, *Anal. Chem.*, 2016, **88**, 2923–2928.

39. Y. R. Liu, L. Liu, S. Q. Li, G. Y. Wang, H. X. Ju and L. Ding, *Anal. Chem.*, 2019, **91**, 6027–6034.
40. K. W. Ren, Y. F. Xu, Y. Liu, M. Yang and H. X. Ju, *ACS Nano*, 2018, **12**, 263–271.
41. L. Ding, Q. J. Ji, R. C. Qian, W. Cheng and H. X. Ju, *Anal. Chem.*, 2010, **82**, 1292–1298.
42. R. C. Qian, L. Ding, L. Bao, S. J. He and H. X. Ju, *Chem. Commun.*, 2012, **48**, 3848–3850.
43. E. Han, L. Ding and H. X. Ju, *Anal. Chem.*, 2011, **83**, 7006–7012.
44. W. Y. Song, L. Ding, Y. L. Chen and H. X. Ju, *Chem. Commun.*, 2016, **52**, 10640–10643.
45. E. Han, L. Ding, R. C. Qian, L. Bao and H. X. Ju, *Anal. Chem.*, 2012, **84**, 1452–1458.
46. G. M. Cohen, *Biochem. J.*, 1997, **326**, 1–16.
47. B. Belardi, A. de la Zerda, D. R. Spiciarich, S. L. Maund, D. M. Peehl and C. R. Bertozzi, *Angew. Chem., Int. Ed.*, 2013, **52**, 14045–14049.
48. N. Wu, L. Bao, L. Ding and H. X. Ju, *Angew. Chem., Int. Ed.*, 2016, **55**, 5220–5224.
49. T. Suzuki, *Nature*, 2018, **561**, 38–40.

CHAPTER 2

In Situ *Detection of Cell-surface Glycans*

2.1 Introduction to Cell-surface Glycans

In eukaryotic cells, there are abundant glycoconjugates,[1] including glycoproteins, proteoglycans and glycolipids. The carbohydrate moieties of these conjugates are commonly called glycans. Glycosylation refers to an enzymatic process that produces glycosidic linkages between saccharides and other saccharides, proteins or lipids. Glycans modify the surface of all mammalian cells, which are high-density information carriers.[2,3] The glycan structures of glycoconjugates show considerable microscopic heterogeneity, which may be related to incomplete glycosylation and the lack of absolute specificity of glycosidases and glycosyltransferases. These microscopic heterogeneities significantly affect the states of the cells and the physiological functions of the organisms and, in particular, are closely related to the generation and development of many diseases.

2.1.1 Structures and Classification of Glycans

Glycans are usually composed of different monosaccharide blocks and these monosaccharides are linked to each other *via* glycosidic bonds formed by the dehydration reaction of a hemiacetal with a hydroxyl group, and this reaction needs to be performed under catalysis by glycosyltransferases. Mammalian cell-surface glycans mainly comprise 10 types of monosaccharides: glucose (Glc), mannose (Man), galactose (Gal), *N*-acetylglucosamine (GlcNAc), *N*-acetylgalactosamine (GalNAc), fucose (Fuc), glucuronic acid (GlcA), sialic acid (Sia), xylose (Xyl) and iduronic acid (IdA).[4] According to the connection modes of glycans with biomolecules, they can be divided into *O*-linked glycans, *N*-linked glycans and sphingolipid glycans, *etc.*[5,6]

Detection Science Series No. 16
In Situ Analysis of Cellular Functional Molecules
By Huangxian Ju, Bo Tang and Lin Ding

Published by the Royal Society of Chemistry, www.rsc.org

Glycoproteins can carry one or more glycans covalently linked to the polypeptide backbone. *N*-Linked glycans are typically covalently linked to the asparagine residues in protein polypeptide chains by GlcNAc. *O*-Linked glycans are typically covalently linked to serine or threonine residues of proteins by GalNAc (Figure 2.1).[7,8]

2.1.2 Functions of Glycans

Glycans play extremely important roles in the processes of cell identification, communication and signalling.[9,10] The glycosylation of cell-surface biomolecules not only regulates the structures and functions of the corresponding glycoconjugates, but also mediates signal transduction, cell differentiation, apoptosis, viral infection, *etc.*, and is closely related to the occurrence and development of many diseases. For example, the glycans attached to the proteins affect the folding and conformational stability of the polypeptides.[11–14] Cell-surface glycans can serve as ligands for proteins to regulate cell adhesion and migration.[15] Moreover, interactions between carbohydrate-carrying biomolecules play important roles in various pathological processes, such as tumour metastasis, viral and bacterial infections and various inflammatory responses.[16–19] One of the research hotspots is the monitoring of the sialylation level, as Sia is associated with cancer progression and holds the potential to be used as a diagnostic marker and therapeutic target.[20–23]

In order to reveal the functions of glycosylation, it is of vital importance to develop *in situ* analytical strategies to track the structures, localization, distribution and dynamic changes of glycans on living cells and establish the relationship between glycosylation and protein activity and function. Owing to the complexity of the structures, the diversity of secondary modifications, the intricacy of biosynthesis and the multi-coordination characteristics of recognition, glycan analysis research lags far behind detection studies of nucleic acids and proteins.

2.2 Labelling and Recognition of Glycans

In recent years, the rapid development of lectin recognition, chemoselective labelling, metabolic oligosaccharide engineering (MOE) and chemoenzymatic glycan labelling (CeGL) methods has provided flexible tools for the labelling and identification of glycans for glycobiology research.

2.2.1 Lectin Recognition

Lectins, a class of non-immunogenic proteins,[24] can specifically, non-covalently and reversibly bind to saccharides and have been widely used in glycan-related research for nearly four decades. Their conjugation with various functional molecules and nanomaterials has allowed the *in situ* optical and electrochemical analysis of cellular glycans.[25–27] For example,

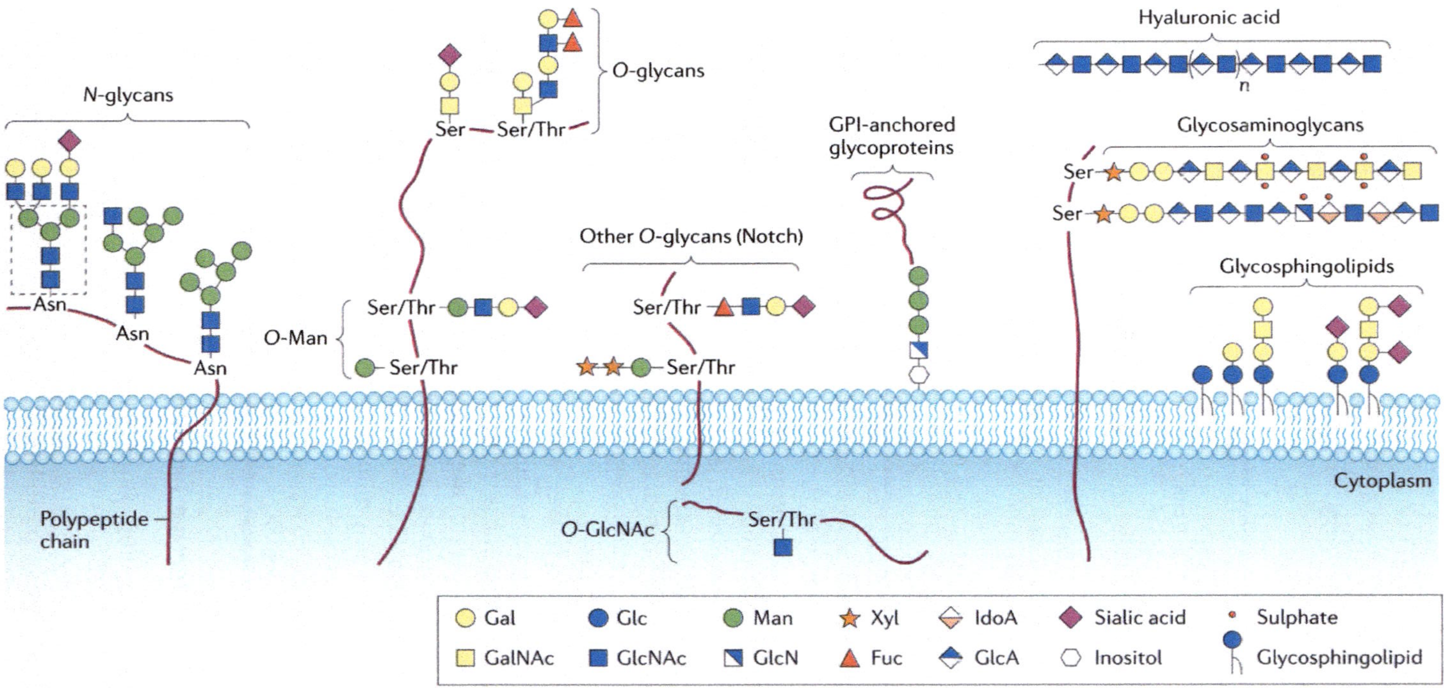

Figure 2.1 Common classes of glycoconjugates in mammalian cells.[8]
Reproduced from ref. 8 with permission from Springer Nature, Copyright 2015.

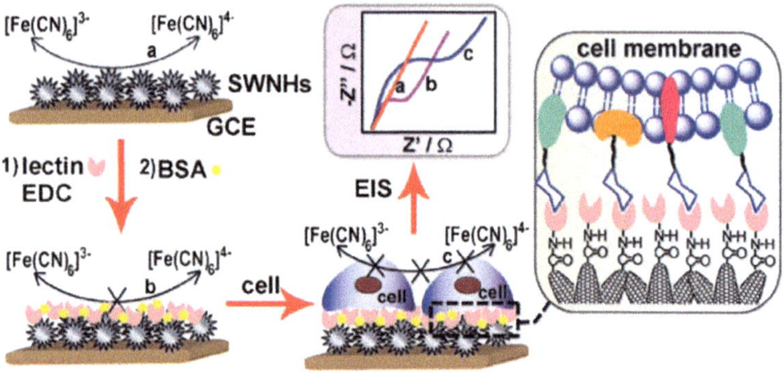

Figure 2.2 Scheme of the electrochemical label-free strategy for the analysis of glycan expression on cell surfaces.[28]
Reproduced from ref. 28 with permission from the Royal Society of Chemistry.

Craig *et al.* fabricated lectin-labelled silver nanoparticles to detect glycosylation expression on the cell surface by confocal Raman imaging, making use of the surface-enhanced Raman scattering (SERS) effect of silver nanoparticles.[27] The binding of three different lectins on three types of cells and the difference in glycosylation between non-cancerous and cancerous prostate cells were studied using this method. However, it should be noted that the specificity and avidity of lectins towards glycans are commonly weaker than those for antibody–antigen interactions, which limits their analytical applications to some extent. A promising solution to this problem is to conjugate multiple lectins on nanomaterials to enhance the binding avidity. Ding *et al.* immobilized the lectins on single-walled carbon nanohorn (SWNH)-modified electrodes, which could recognize the glycans on the cell surface effectively, thereby achieving cell capture.[28] The bound cells acted as an inert electron- and mass-transfer blocking layer at the electrode surface, leading to an increase in impedance, which reflected the corresponding extent of glycan expression on the cell surface (Figure 2.2). In a similar way, Han *et al.* designed an electrochemiluminescence (ECL) sensing strategy for the sensitive and dynamic monitoring of surface carbohydrate expression on living cells using a lectin-functionalized quantum dots (QDs)/chitosan–gold nanoparticles (AuNPs) composite membrane to capture cells.[29] In this case, the decrease in ECL emission reflected the glycan abundance on living cells.

2.2.2 Chemoselective Labelling

Chemoselective labelling strategies are based on specific chemical remodelling or conjugation of some types of glycans. For cell-surface glycan labelling, there are two commonly used reactions. The first relies on mild, controlled oxidation by periodate to transform selectively hydroxyl groups at specific positions of carbohydrates into aldehyde groups, which can then

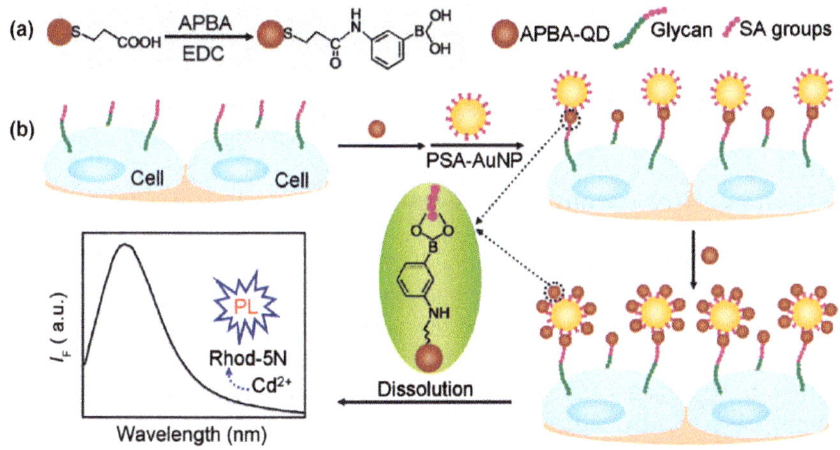

Figure 2.3 Schematic illustration of (a) functionalization of CdS QDs with APBA and (b) the fluorescent monitoring procedure for Sia on living cells.[31] Reproduced from ref. 31 with permission from John Wiley and Sons, © 2015 WILEY-VCH Verlag GmbH & Co. KGaA, Weinheim.

react bio-orthogonally with aminooxy or hydrazine functions to achieve labelling. A low concentration of aniline is also needed to catalyse this reaction. Han *et al.* used periodic acid and galactose oxidase (GO) to oxidize selectively hydroxyl groups of Sia and Gal on the cell surface to aldehyde groups, respectively.[30] Through a biotin–hydrazide linkage and avidin-mediated assembly of a gold nanoprobe, signal-amplified chemiluminescence imaging of these two types of carbohydrates simultaneously was achieved.

The second reaction is based on the specific binding of phenylboronic acid (PBA) to Sia under physiological pH conditions. Han *et al.* fabricated a 3-aminophenylboronic acid (APBA)-functionalized QD nanoprobe for specifically binding cell-surface Sia and further used the APBA–Sia reaction to recruit more QD nanoprobes through the bridging function of polysialic acid-stabilized AuNPs, thus developing a QD-based fluorescence method for monitoring Sia variation on the cell surface (Figure 2.3).[31]

2.2.3 Metabolic Oligosaccharide Engineering

The use of an MOE strategy for the imaging and identification of glycan-modified proteins was first pioneered by Bertozzi's group,[32] and this approach has grown into a powerful tool for glycan research.[33,34] In this strategy, the monosaccharide substrate is first linked with a bio-orthogonal chemical reporter (azide, alkyne, *etc.*). This non-natural precursor can be tolerated by cellular enzymes and installed on proteins by replacing the corresponding natural counterparts. The bio-orthogonal chemical reporter provides a convenient handle for visualization of the sugars. For example,

Lin *et al.* modified mannosamine with different bio-orthogonal groups (including alkyne, azide, cyano, *etc.*) to permit their installation on the cell surface.[35] The characteristic Raman peaks of these bio-orthogonal groups were obtained by using AuNPs as Raman-enhanced substrates and *in situ* Raman imaging of cell-surface Sia was realized using a confocal Raman microscope.

Importantly, the chemical reporter group allows labelling of cell-surface glycans by a bio-orthogonal reaction with the secondary probe. Commonly used bio-orthogonal reactions are the Staudinger ligation reaction, copper-catalysed azide–alkyne cycloaddition (CuAAC) and strain-promoted azide–alkyne cycloaddition (SPAAC or copper-free click chemistry). These bio-orthogonal reactions facilitate the application of metabolic labelling. Yuan *et al.* reported a multi-fluorescence resonance energy transfer (FRET) strategy based on hybridization chain reaction (HCR) amplification for the specific imaging of cell-surface glycosylation by MOE (Figure 2.4).[36] GalNAc was labelled on the target cells by MOE and linked with FRET donors by a copper-free click reaction and the protein of interest (POI) was labelled with an aptamer-functionalized HCR nanoassembly carrying multiple FRET acceptors. Intramolecular FRET occurred to indicate the existence of GalNAc on the given protein, thus achieving enhanced multi-FRET imaging of protein-specific glycosylation due to the amplification effect and spatial flexibility of HCR nanocomponents.

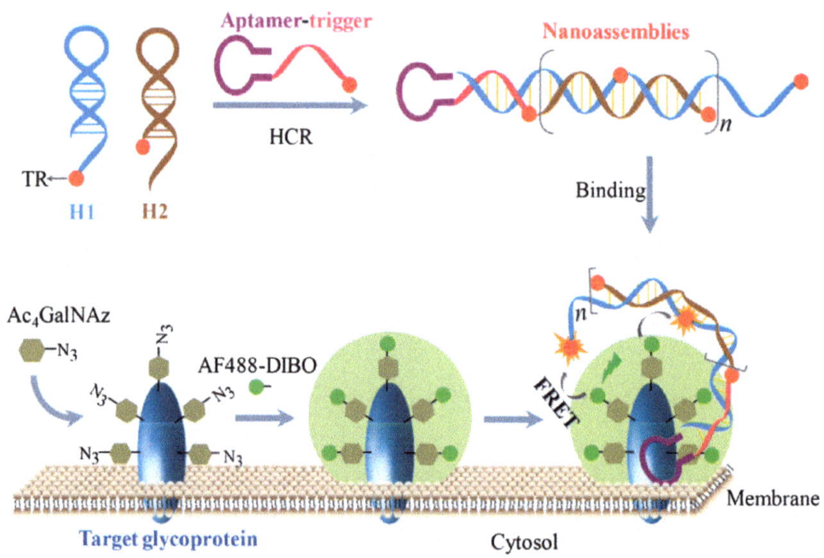

Figure 2.4 HCR amplification-based multi-FRET strategy for specific imaging of cell-surface glycosylation by employing MOE.[36]
Reproduced from ref. 36 with permission from American Chemical Society, Copyright 2018.

2.2.4 Chemoenzymatic Glycan Labelling

Despite its powerful labelling capability, MOE cannot be used to track the structures of higher-order glycans (disaccharides, trisaccharides, *etc.*). However, cell-surface glycoproteins often display glycans composed of complex sugar chains. To resolve this issue, chemoenzymatic glycan labelling (CeGL) strategies offer an elegant solution, utilizing glycosyltransferases to transfer monosaccharide analogues with chemically modified groups to corresponding glycan acceptors based on the recognition of the glycosyltransferase towards higher-order glycan substrate structures and the generation of the specific glycosidic linkages. Upon anchoring the analogues, the second labelling step in CeGL is similar to that in MOE: the secondary probes can be introduced to the remodelled glycan sites by reaction with other functional groups.[37–40]

Wu's group prepared a variety of glycosyltransferases, including sialyltransferase and fucosyltransferase, to assemble non-natural monosaccharide analogues on cell surfaces to develop a cell-based assay method (Figure 2.5).[41] In the same year, they used CeGL for the *in situ* incorporation of monosaccharides with different linkage modes into the receptor using recombinant glycosyltransferases and found that the *in situ* addition of α-2,3-linked Sia or α-1,3-linked Fuc could inhibit signalling through the epidermal growth factor receptor and fibroblast growth factor receptor.[42]

2.3 Global Analysis of Cell-surface Glycans

The detection of the extents of glycan expression on the cell surface can contribute to revealing the relationship between global glycosylation and cell states. To achieve this, one needs to design a sensing system composed of a glycan-recognition motif, a recognition signal transduction unit and a signal output module. The commonly used recognition motifs were introduced in the previous section; this section focuses on how to design a signal generation pathway to report the glycan recognition event, how to amplify the signals and how to implement multi-glycan analysis.

To transduce the signals from glycan recognition events, there are two main types of detection schemes: direct[25,43,44] and indirect.[45,46] The former involves the direct collection of signals from the recognition motifs bound on the target cells and the latter detects the residual non-binding recognition motifs after an incubation process. Compared with the direct format, the indirect method can be performed in a cell-free experimental setting, thus simplifying the operation and eliminating the influence of complex cellular components.

As a demonstration of the direct format, Cheng *et al.* prepared arginine–glycine–aspartic acid–serine (RGDS)-functionalized single-walled carbon nanotubes (SWNTs) to modify glassy carbon electrodes, which served as a platform for the efficient capture of cells through peptide–integrin

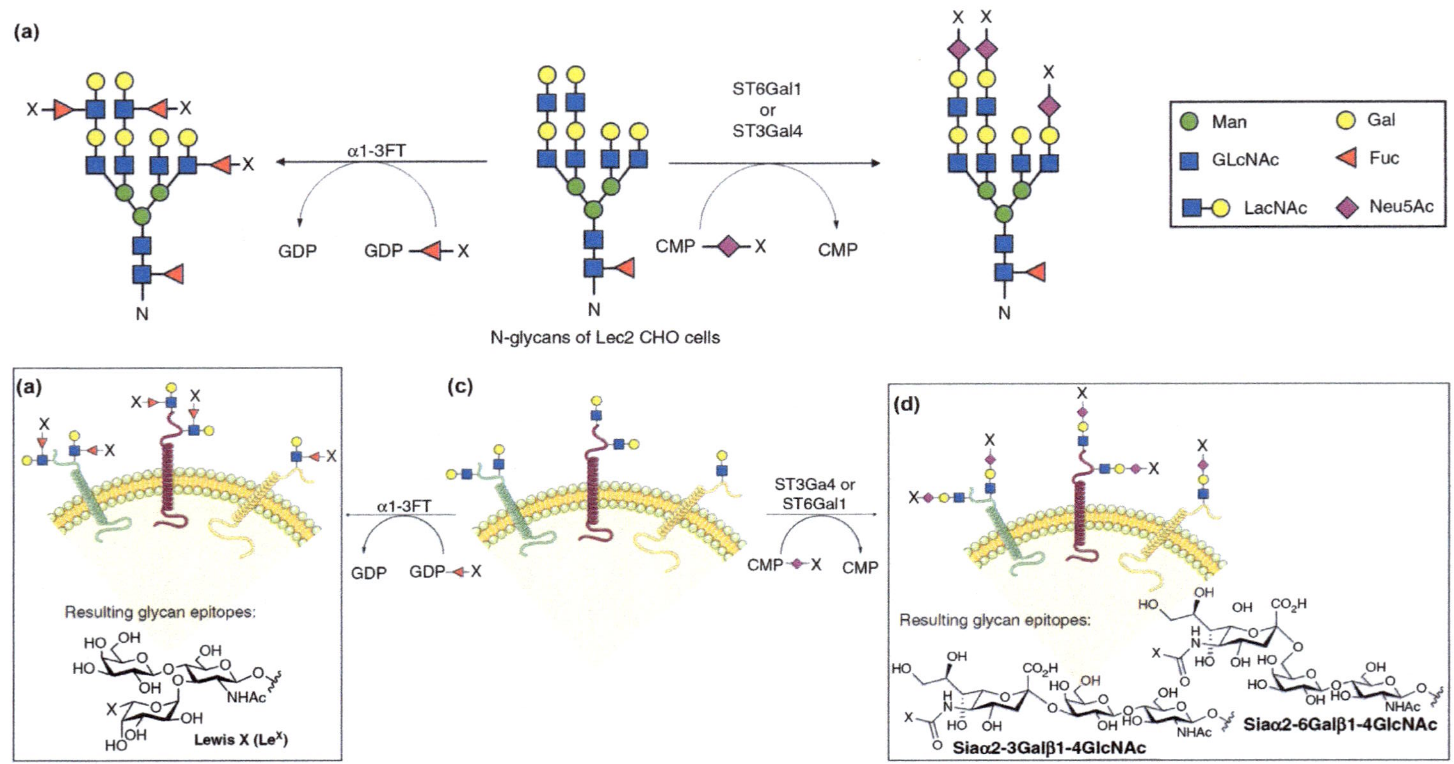

Figure 2.5 The workflow for constructing cell-based glycan arrays. (a) Distinct glycan epitopes are installed on *N*-glycans of Lec2 CHO cells *via* sialyltransferase-mediated *in situ* sialylation using the CMP-Neu5Ac donor (or its analogues) or fucosyltransferase-mediated *in situ* fucosylation using the GDP-Fuc donor (or its analogues). (b) Cell-based glycan arrays displaying Le^X or its derivatives. (c) *N*-Glycans of Lec2 CHO cells. (d) Cell-based glycan arrays displaying α-2,3- or α-2,6-linked sialosides or their derivatives. X represents unnatural substituents introduced at the fucose C5 or NeuNAc acetyl position.[41] Reproduced from ref. 41 with permission from Springer Nature, Copyright 2018.

interaction.[25] The mannose moieties on cells were then specifically bound with horseradish peroxidase (HRP)-labelled concanavalin A (Con A), a mannose-specific lectin, which endowed the system with signal amplification capability based on the catalysis by HRP of the oxidation of *o*-phenylenediamine (*o*-PD) by H_2O_2. The electrochemical signals directly represented the extent of mannose expression on the cell surface, which was further used for the quantification of mannose moieties per cell.

Considering that the above method could provide only average expression information on a cell population, the same group went on to develop electrochemical strategies to reflect the individual global glycosylation on a single cell. In this context, scanning electrochemical microscopy (SECM), which can quantitatively monitor local chemical reactions in a non-invasive manner under physiological conditions, offers a promising solution. Xue *et al.* combined HRP-labelled lectins with micropatterning technology and proposed an SECM strategy for the *in situ* imaging of four types of membrane glycan motifs on single adherent cells.[43] After the recognition of HRP-labelled lectins with target glycans, the reduction current at the SECM tip, coming from the enzymatic oxidation of ferrocenylmethanol (FMA) to generate FMA^+, was measured to indicate the glycan expression level.

As an alternative, SERS imaging, which possesses advantages of high sensitivity and a non-invasive nature, can also be used for the *in situ* evaluation of the extent of global expression on individual cells. Song *et al.* reported the use of a 3-mercaptophenylboronic acid (3-MPBA)-functionalized gold nanoflower as a bridge for connecting cell-surface Sia and polysialic acid-coated poly(amidoamine)-encapsulated AuNPs, to form a single-nuclear–multi-satellite nanostructure.[44] The imaging sensitivity of cell-surface Sia was enhanced by the plasmonic coupling of the two kinds of gold nanoprobes in the nanocomposite (Figure 2.6).

Regarding an indirect detection scheme, Xue *et al.* prepared an electrochemical lectin probe, ferrocene–concanavalin A (Fc–Con A), for the recognition of cell-surface mannose and detected the amount of non-binding probes electrochemically after cell-probe incubation using the signal of the ferrocenyl group.[45] The decrease in current (compared with the case when no cells were added) corresponded to the quantity of probes bound on the cell surface, and thus represented the extent of global expression of mannose. In contrast to the previous two electrochemical methods, this approach used non-adhered cells as the target, thus allowing authentic evaluation of the global expression.

To amplify the signal for the indirect format and simplify the detection equipment, Ding *et al.* proposed a scanometric method for *in situ* mannose analysis based on glyconanoparticle aggregation-regulated silver enhancement.[46] As shown in Figure 2.7, the mannan-functionalized AuNPs could aggregate in the presence of lectin Con A due to the four binding sites of Con A for sugar epitopes. After incubating Con A with cells, the residual Con A was collected by centrifugation and subjected to mixing with the

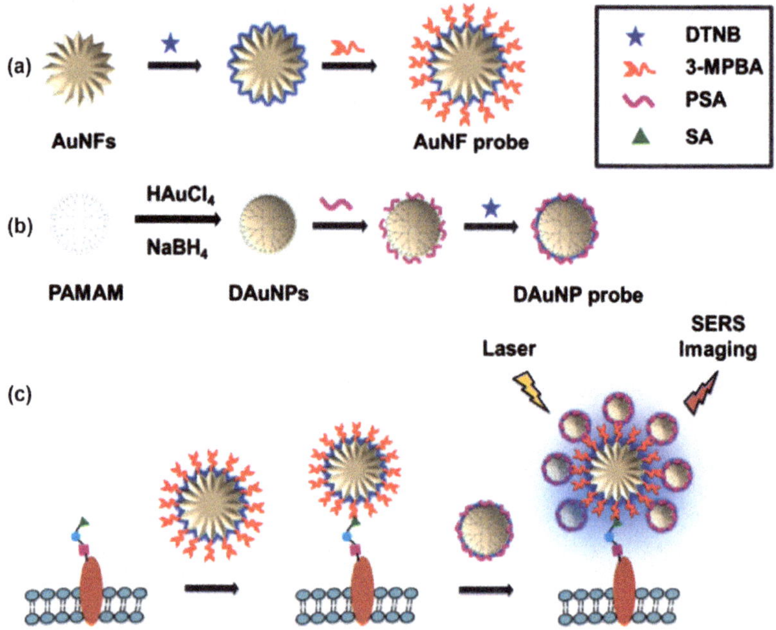

Figure 2.6 Schematic illustration of (a) the preparation of the AuNF probe, (b) the preparation of the DAuNP probe and (c) SERS imaging of the cell-surface Sia based on plasmonic coupling of the core AuNF with the satellite AuNPs. 3-MPBA represents 3-mercaptophenylboronic acid.[44]
Reproduced from ref. 44 with permission from the Royal Society of Chemistry.

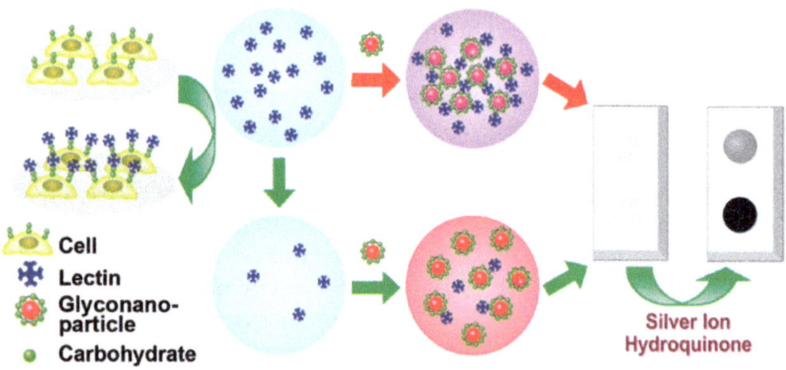

Figure 2.7 Scheme of the scanometric strategy for the *in situ* detection of mannose groups on living cells.[46]
Reproduced from ref. 46 with permission from American Chemical Society, Copyright 2010.

AuNP probe. The resulting AuNP aggregation could be sensitively and conveniently monitored by combining silver enhancement and scanometric detection. Hence the method could be used to analyse the expression level of mannose on the cell surface, with the advantages of a simple operation process, no need for pretreatment of cells and high sensitivity.

Considering that there are multiple types of glycans on the cell surface, the glycan expression profiles can provide the glycan signatures of the target cells, which may add a new dimension for the development of glycoprotein-based biomarkers and therapeutic targets. The commonly used multi-channel detection methods for glycans on living cells are based on the fabrication of lectin arrays,[47,48] on which multiple types of lectins with distinct sugar specificities are immobilized and subjected to incubation with fluorophore-labelled cells. The amount of captured cells, as reflected by the fluorescence intensity (FI), indicates the extent of glycan expression. Thus, by comparing the FIs of different spots, the glycan expression profiles can be obtained.

In spite of their convenience, these methods involve the surface confinement of the lectins, which often results in denaturation or inactivation of proteins due to chemical modifications or surface asymmetry of protein spots,[49] and the high density of protein molecules can also cause inaccessibility of active sites.[50] More importantly, these array-based approaches are based on an oversimplified comparison of the amount of cells bound on a lectin spot and therefore fail to provide *direct* quantitative information on glycan expression.

To resolve these issues, Cheng *et al.* changed the array fabrication format and developed a disposable sensor array for cell capture, which was then subjected to recognition by HRP-labelled lectins, with different electrode spots corresponding to different HRP-labelled lectins.[51] The differences in the electrochemical signals among different electrode spots directly reflected the characteristic glycan profile of the cells of interest, which was used for dynamic monitoring of the changes in the glycan pattern in response to drug treatment and during erythroid differentiation.

On the basis of this work, the same group continued to introduce coding-based differentiation into the multi-channel analysis of cell-surface glycans. Chen *et al.* combined DNA microarray technology with DNA coding technology to develop a quantitative detection strategy for multiple glycans on cell surfaces (Figure 2.8).[52] They designed four types of DNA-coded lectin probes, with each DNA strand containing an enzymatic cleavage site for releasing the DNA codes. After lectin-directed probe recognition on the cell surface, the DNA codes, liberated by endonuclease cleavage, could be captured by corresponding complementary strands on the DNA microarray. Then a pair of Cy5-labelled hairpin DNAs were used to amplify the fluorescence detection signals on the microarray by hybridization chain reaction (HCR), thus achieving the quantification of four types of glycans on BGC-823 cells.

In addition to multi-channel glycan detection, it is also beneficial to develop *in situ* analytical methods with the capability of multi-modal detection, to combine the distinct advantages of different detection modes and acquire wider information on the target glycans. Xue *et al.* prepared

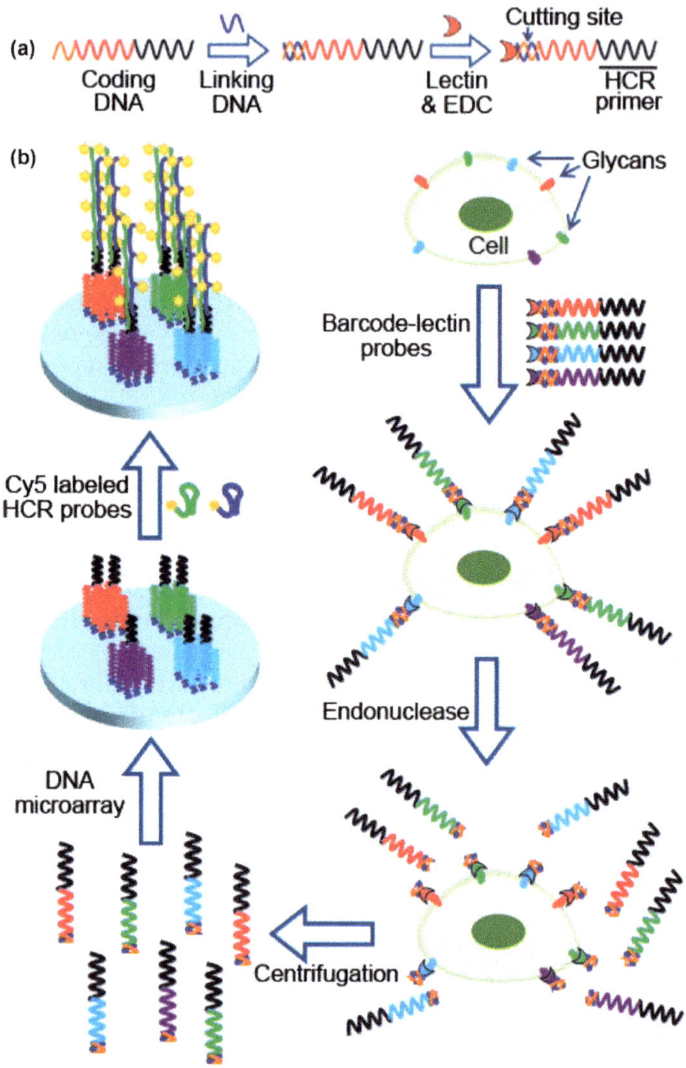

Figure 2.8 Schematic illustration of (a) barcode-lectin probe preparation and (b) assay procedure.[52]
Reproduced from ref. 52 with permission from American Chemical Society, Copyright 2013.

lectin-functionalized multi-walled carbon nanotubes (MWNTs) by non-covalent adsorption to serve as a glycan recognition-directed cell immobilization platform with favourable electronic and mechanical properties.[53] Based on this platform, the two-mode (electrochemical and optical) detection of cell-surface glycans could be achieved, which also permitted the monitoring of dynamic glycan expression on living cells in response to drugs.

2.4 Competition Systems for Quantification of Cell-surface Glycans

The previously described global detection methods for cell-surface glycans are all based on the recognition events participated in by two components in the system, cell-surface glycans and the glycan recognition motifs. To enhance the detection sensitivity and make the whole cell surface accessible to recognition motifs, a type of "two-surface–one-probe competition" system was developed that shows wide applicability with various detection equipment.

This concept was first proposed by Ding *et al.* and a mannan-modified monolayer was fabricated to compete with cell-surface mannose moieties towards a Con A-functionalized QD probe (Figure 2.9).[50] The amount of QDs captured on the mannan monolayer could be sensitively detected by anodic stripping analysis owing to its favourable signal amplification capability, which was used for the estimation of the extent of mannose expression on the cell surface. This recognition format eliminates the requirement for lectin immobilization and thus overcomes the shortcomings of lectin array-based methods (active-site inaccessibility and lectin denaturation) and has the advantage of whole-cell-surface detection.

Moreover, through this two-surface–one-probe competition process, the expression information of interest on a complex biological interface can be specifically extracted and transferred to a secondary surface, thus allowing

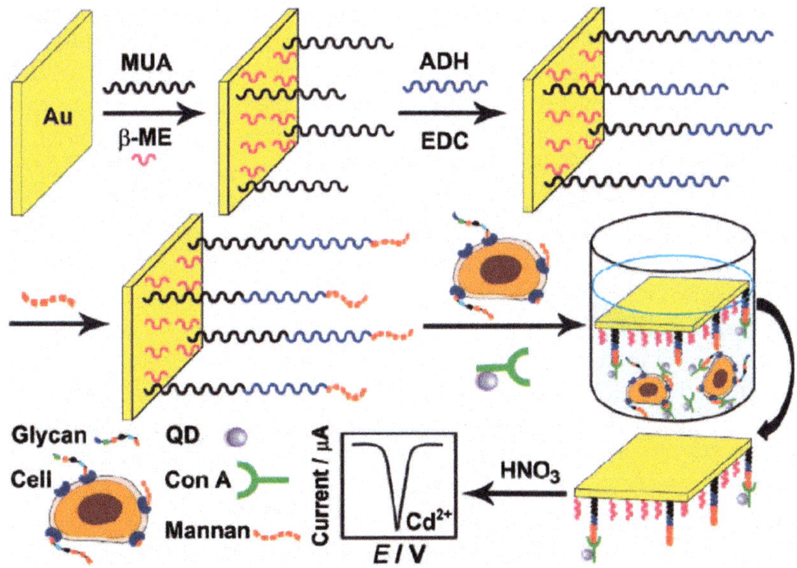

Figure 2.9 Schematic illustration of the monolayer fabrication and the competitive assay.[50]
Reproduced from ref. 50 with permission from American Chemical Society, Copyright 2008.

connection with a variety of detection techniques. As an illustrative demonstration, ECL, which combines the advantages of high sensitivity of chemiluminescence analysis and controllability of electrochemical potential, was employed by Han *et al.* to monitor the competition results.[54] In this work, CdS QDs, functionalized with mannan, were immobilized on the electrode, which could compete with cell-surface mannose towards Con A in solution. The binding of Con A on the electrode could lead to a decrease in ECL intensity due to the resistance of the lectin. The change reflected the amount of cells and the extent of expression of cell-surface carbohydrate. This protocol did not need labelling or immobilization of lectin and exhibited high sensitivity, good stability and acceptable reproducibility. The same group also reported a multi-channel glycan information transfer from a cell-adhered array to a mannan-modified array using Con A-functionalized AuNPs as the messenger.[55] After competition, the amount of AuNPs bound on the secondary array could be measured by silver enhancement followed by scanometric detection.

In the previously mentioned strategies, the secondary surfaces fabricated for competition with the primary cell surfaces are all on a millimetre scale, which is different from that of cells (micrometre scale). Hence the whole system can be regarded as a type of heterogeneous competition, which possesses the advantage of facile separation of the secondary surface for detection. To enhance the competition efficiency and at the same time maintain the capability of ease of separation, Chen *et al.* designed a micro-competition system that combined the two-surface–one-probe competition mechanism with SERS detection (Figure 2.10).[56] This system consisted of three components: Raman signal molecule and lectin dual-coded gold nanoprobes, silica bubbles functionalized by multiple-polysaccharide-coated gold nanostars (MGAuNS@B) and cells. The MGAuNS@B competed with the cells for binding with a mixture containing different types of gold nanoprobes and could be quickly collected by buoyancy. The amount of each type of bound nanoprobes, measured using the respective Raman characteristic peaks, represented the expression level of the corresponding type of glycan, thus achieving the simultaneous quantification of three types of glycans on the cell surface.

Inspired by these competition systems, displacement-based protocols were also developed for glycan imaging[57] and glycan density analysis[58] on living cells. For efficient recognition, proper spatial presentation of glycans on the cell surface is essential to match the conformation of native receptors. Hence the development of strategies for monitoring glycan density can contribute to the regulation of glycan binding avidity and selectivity and the interference of glycan-participating biological processes. To achieve this, Chen *et al.* designed an APBA-modified gold nanocluster probe, which was assembled on a density-adjustable dendrimer array by electrostatic adsorption [Figure 2.11(i)].[58] After incubating BGC-823 cells with the probe-adsorbed dendrimer array, the probe could be unloaded from the array by APBA–Sia binding, leading to a decrease in the FI of the probe on

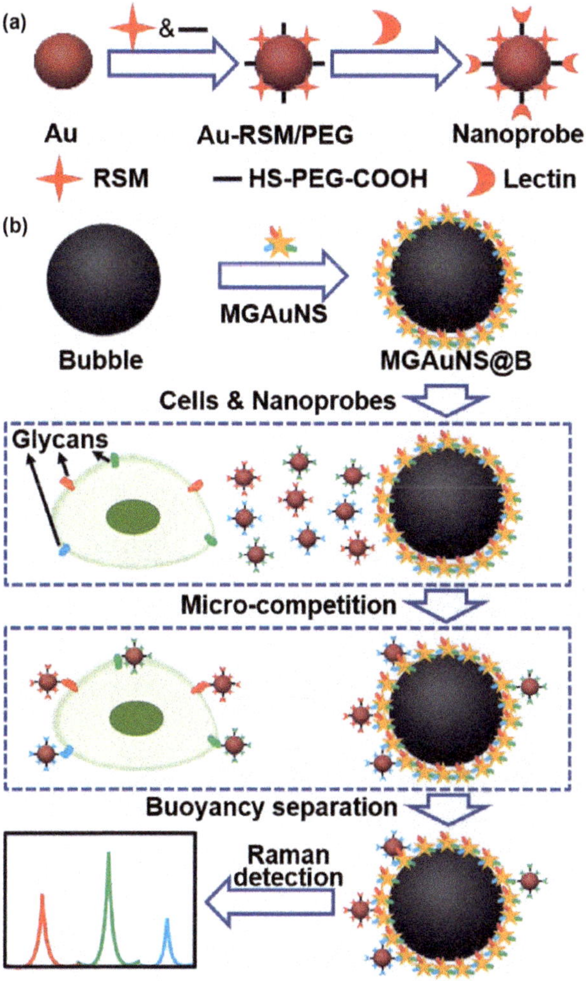

Figure 2.10 Schematic illustration of (a) synthesis of nanoprobes and (b) micro-competition system for multiple Raman detection of cell-surface glycans.[56] Reproduced from ref. 56 with permission from the Royal Society of Chemistry.

the slide. The unloading efficiency depended on the extent of matching between the spatial distribution of cell-surface Sia and the dendrimer density, and thus a method for estimating Sia density on the cell surface was developed [Figure 2.11(ii)].

2.5 Protein-specific Imaging of Cell-surface glycans

The focus of *in situ* cell-surface glycan analysis has gradually shifted from global to protein-specific glycan expression. From an in-depth pathological study of mouse models, it was found that their aetiology is not caused by the

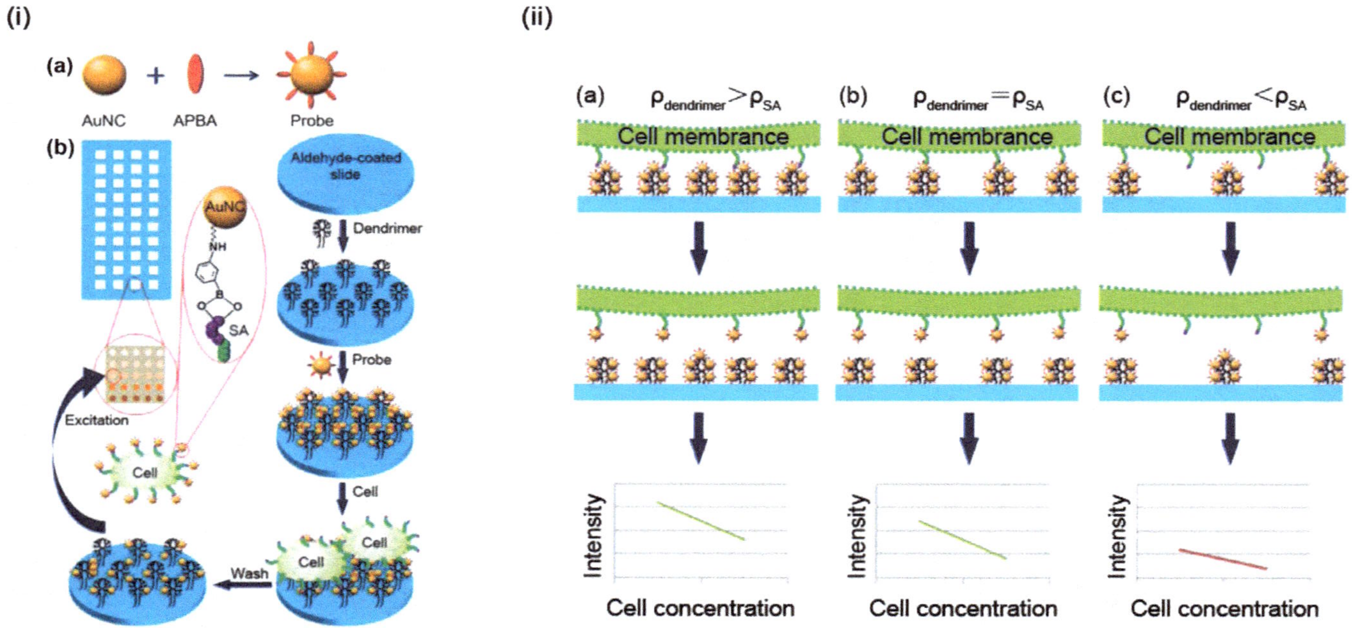

Figure 2.11 (i) Schematic illustration of (a) probe fabrication and (b) assay process. (ii) Signal change upon increasing the cell concentration at three types of arrays with dendrimer density (a) greater than, (b) just equal to and (c) smaller than the cell-surface Sia density within the same unloading time.[58]
Reproduced from ref. 58 with permission from the Royal Society of Chemistry.

change in the global glycosylation, but by the glycosylation of certain types of glycoproteins. Moreover, with the onset and development of diseases, the glycosylation of different glycoproteins will display distinct changes. Therefore, protein-specific imaging of glycans can visually display changes in glycosylation along with physiological/pathological processes, thus contributing to the exploration of disease mechanisms and the development of diagnostic tools.

The detection principle for protein-specific glycans is based primarily on separately labelling the target protein and the target glycan and designing distance-dependent off–on signal switches to visualize the glycans on the target protein. As a typical representative of this concept, FRET donors and acceptors are labelled on proteins and glycans, respectively. Only fluorophores on the glycans of the POI can be lit up by the protein-anchored donor owing to the distance limitation of FRET, thereby permitting the imaging of protein-specific cell-surface glycans. Haga *et al.* labelled the target transmembrane protein with green fluorescent protein (GFP) as the FRET fluorescence donor and conjugated fluorophore molecules on metabolically labelled SiaNAz as FRET acceptors, thus fabricating an intramolecular proximity reporting system (Figure 2.12).[59] This is a transmembrane FRET system, because GFP has to be fused to the cytosolic tail of the protein while the glycans reside on the other side of the cell membrane. Therefore, this strategy suffered from limited FRET efficiency owing to the inherent transmembrane distance.

To address this problem, in a study by Lin *et al.* the POI was fused with LAP (LplA acceptor peptide) at the N-termini and labelled using an enzyme-catalysed probe ligation method, and SiaNAl was introduced into cell-surface sialylated glycans by incubating cells with $Ac_4ManNAl$; thus a *cis*-membrane FRET-based strategy, with both the donor and acceptor installed on the extracellular terminus of the protein, was developed for the protein-specific imaging of glycans.[60] This method was successfully used for visualization of the sialylation of several important cell-surface receptors, including integrin $\alpha_X\beta_2$, epidermal growth factor receptor and transforming growth factor-beta receptor type I.

Both of the above two methods are reliant on protein fusion technology, hence they cannot be used for imaging endogenous glycoproteins. An alternative method for locating the POI is to use small targeting motifs that

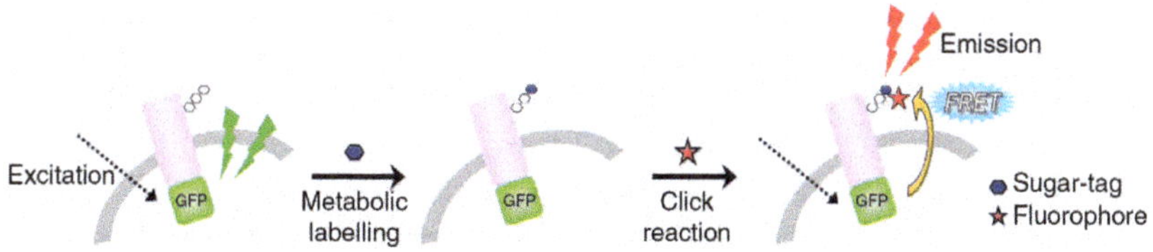

Figure 2.12 FRET-based imaging of the glycosylation pattern of a given glycoprotein.[59] Reproduced from ref. 59 with permission from Springer Nature, Copyright 2012.

can specifically recognize the POI and alleviate the influence on the energy transfer efficiency, such as antigen-binding fragment (Fab),[61] aptamer,[62–66] peptide, *etc.* In addition, these traditional FRET-based systems also suffer from the discrepancy between protein copy number and glycan abundance,[61] where the latter can be orders of magnitude greater than the former. This might lead to acceptor bleed-through due to the high acceptor concentration, which would result in a false-positive signal. In this regard, two-photon fluorescence lifetime imaging microscopy (FLIM) was employed by Belardi *et al.* to focus on the donor fluorophore emission. The fluorescence lifetime of the donor associated with FRET between fluorophores located on the POI and its Sia moieties was recorded to monitor the sialylation state of a given glycoprotein.[61]

In addition to FRET, SERS, an alternative distance-dependent signal generation mechanism, can also be used for imaging glycans on specific proteins. Chen *et al.* used an EpCAM-specific aptamer to guide the binding of an AuNP probe (30 or 40 nm) to the target protein EpCAM as the SERS substrate and anchored the Raman signal molecule-functionalized AuNP probe (10 nm) to the target glycans *via* MOE on the cell surface [Figure 2.13(i)].[62] The size of the AuNP substrate probe was adjusted to modulate the effective SERS zone to match the protein-specific glycan expression region. Three types of glycans on the cell-surface EpCAM were successfully visualized by SERS imaging [Figure 2.13(ii)]. This is the first example of the use of SERS to achieve protein-specific glycan imaging.

Despite the encouraging progress, the drawbacks of MOE as a glycan labelling tool are obvious: MOE is a non-selective, global chemical remodelling (GCM) technology that inevitably disturbs the whole cell, and thus fails to reflect the original states of the cells. In addition, owing to its dependence on intracellular enzymatic processing, the metabolic efficiency and percentage installation of MOE are unpredictable among cells of different types/states. To resolve these issues, an ingenious approach, called localized chemical remodelling (LCM), was developed by Hui *et al.*, which allowed protein-specific labelling and imaging of glycans on living cells (Figure 2.14).[63] In this work, a remodelling probe was prepared by conjugating an aptamer sequence with GO *via* a DNA linker. This probe, with enzymatic activity first inactivated by $K_4[Fe(CN)_6]$, could specifically bind the target protein by aptamer recognition. After removing non-binding probes, the GO, anchored to the target protein, could be activated by $K_3[Fe(CN)_6]$ and oxidize the C6-hydroxy group of terminal Gal/GalNAc to an aldehyde group, which provided a reactive handle for bio-orthogonal labelling and imaging. The protein specificity was ensured owing to the confinement of the GO in the region of the POI. This strategy essentially provides a molecular scale that can be utilized to report the intramolecular distance by using chemical reactions as a reporting system rather than relying on physical processes (such as FRET or SERS). Compared with MOE, LCM possesses several advantageous features, including minimal target cell perturbations, short reaction time ranges and quantitative analytical capabilities.

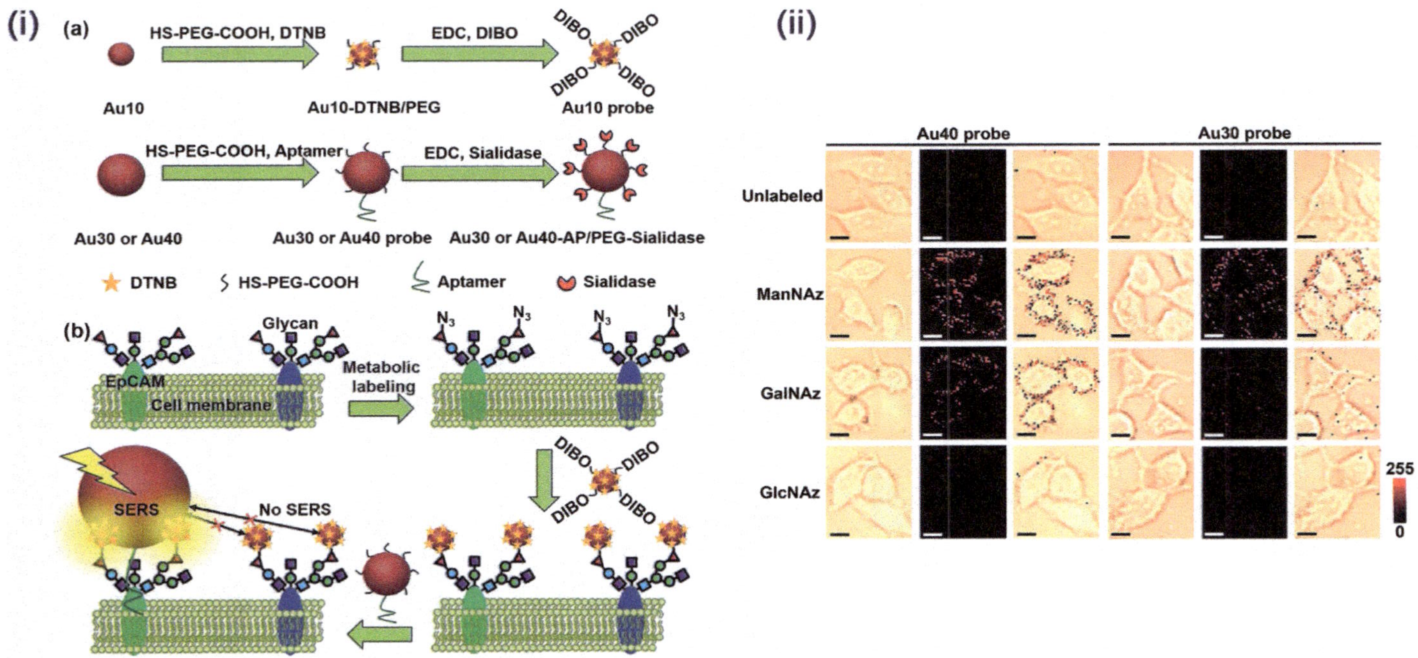

Figure 2.13 (i) Schematic illustration of (a) the synthesis of two types of Au nanoprobes and (b) the zone-controllable SERS effect for imaging of protein-specific glycans on the cell surface. (ii) Bright field, Raman and overlay images of unlabelled and ManNAz, GalNAz and GlcNAz metabolically labelled MCF-7 cells after being treated with different substrate probes. Scale bar: 10 μm.[62] Reproduced from ref. 62 with permission from the Royal Society of Chemistry.

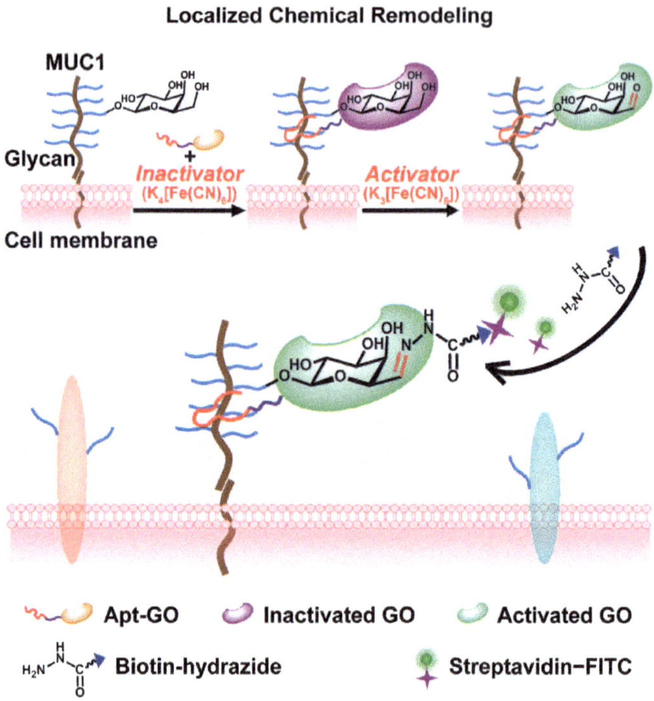

Figure 2.14 Scheme of the localized chemical remodelling (LCM) strategy for live cell imaging of protein-specific glycoforms.[63]
Reproduced from ref. 63 with permission from John Wiley and Sons, © 2017 Wiley-VCH Verlag GmbH & Co. KGaA, Weinheim.

Considering the diversity and complexity of the glycoforms, simultaneous imaging of multiple monosaccharides on a given cell-surface protein is beneficial for revealing complex glycan-involved signal pathway machinery. To achieve the differentiation of glycans, Wu *et al.* used different fluorophores to label different monosaccharides on the cell surface *via* MOE and installed an upconversion luminescent nanoparticle (UNP) on the POI *via* aptamer recognition [Figure 2.15(i)].[65] The POI-located UNP can be excited at 980 nm and exhibits two energy emission bands (EEB1 and EEB2), which respectively matched to light up the fluorophores on glycans in close proximity through luminescence resonance energy transfer (LRET), thus constructing a novel single excitation-duplexed imaging system for protein-specific glycoforms. Using this method, MUC1-specific imaging of two types of tumour-associated terminal glycans was achieved on three types of mammary epithelial cells, namely cancerous cells MCF-7 and T47D and normal cells MCF-10A [Figure 2.15(ii)].

Considering the limited resolution capability of fluorophore labelling, Li *et al.* made use of the powerful coding function of DNA sequences to construct a hierarchical coding (HieCo) strategy.[66] Combining aptamer recognition and MOE, the two different levels of structural units, protein and

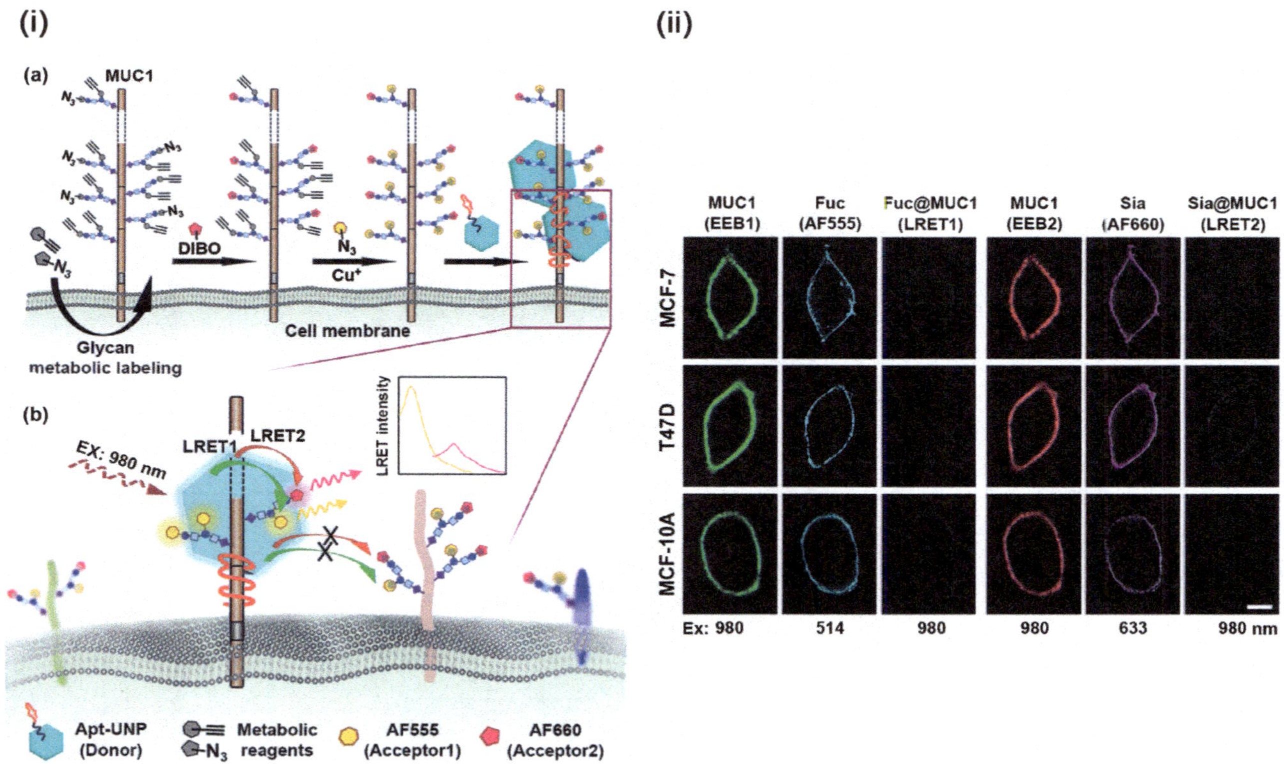

Figure 2.15 (i) The site-specific D-LRET method for duplexed imaging and dynamic monitoring of protein-specific monosaccharides on an intact cell surface. (a) Two types of monosaccharides and the underlying protein were sequentially tagged with fluorescence acceptor dyes and aptamer-functionalized UNP (Apt-UNP) as a donor by dual metabolic labelling and aptamer recognition, respectively. (b) Upon excitation at 980 nm, two non-interfering parallel energy transfer processes from UNP to two acceptors displayed the existence of corresponding monosaccharides. (ii) Duplexed protein-specific imaging of metabolically labelled Fuc and Sia of MUC1 on MCF-7, T47D and MCF-10A cells. Scale bar: 10 µm.[65]
Reproduced from ref. 65 with permission from John Wiley and Sons, © 2016 WILEY-VCH Verlag GmbH & Co. KGaA, Weinheim.

glycans, were separately encoded and masked. The decoding process was initiated by the hybridization of the trigger sequence and the time code, and sequential decoding from the high level to the low level was realized. Compared with the existing protein-specific glycoform imaging strategy, this method can reflect the true hierarchical structure of the target glycoprotein and provide any extended monosaccharide detection channel. Suzuki

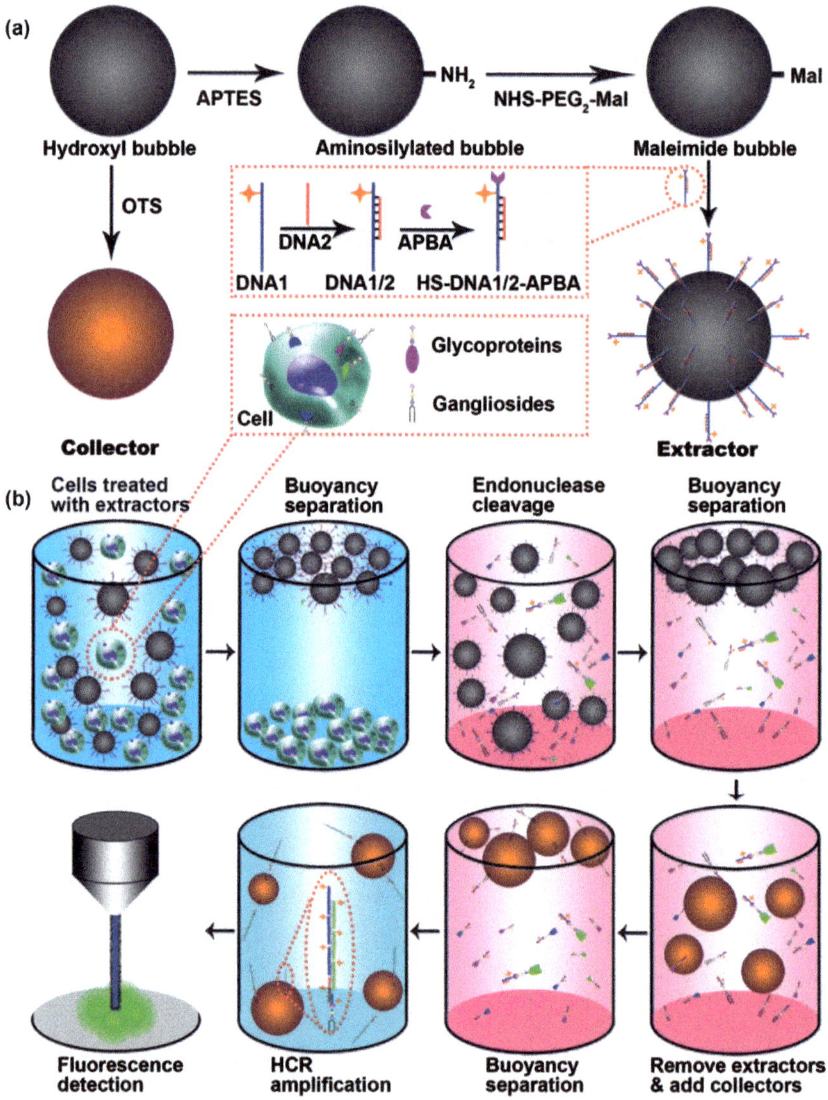

Figure 2.16 Schematic of (a) the preparation of the extractor and collector and (b) quantification of cell-surface gangliosides.[68]
Reproduced from ref. 68 with permission from John Wiley and Sons,

highlighted this work in *Nature*'s News and Views column and pointed out that "this work shows great potential and could be an important step in the development of a system analogous to the use of green fluorescent proteins for protein tagging".[67]

In addition to glycoproteins, glycolipids, another type of glycoconjugate, also play essential roles in the regulation of cell processes. Among various glycolipids, gangliosides, a kind of glycosphingolipid terminated with one or more Sia residues, have been found to mediate cell structure, adhesion and signal transduction. A quantitative screening strategy based on non-destructive extraction and hydrophobic collection for cell-surface gangliosides was developed by Chen *et al.* (Figure 2.16).[68] In this work, two probes, an extractor probe and a collector probe, were designed. The extractor was constructed by conjugating an endonuclease cleavage site-carrying, APBA-capped DNA sequence on silica bubbles. The collector was a trichloro(octadecyl)silane (OTS) silylated microbubble. After incubating the extractors with the cells, Sia-containing compounds could bind to APBA on extractors and be liberated from the cells through buoyancy. The extractors were then applied to endonuclease cleavage to release the Sia-containing compounds, among which oligonucleotide gangliosides could subsequently be collected by the collectors through hydrophobic interaction and sensitively detected with HCR. Using this method, the profiling of ganglioside subtypes was achieved through sialidase cleavage of Sia followed by recognition of fluorescein-labelled lectins. The regenerative process of gangliosides on the cell surface was also monitored.

2.6 Perspective

Considering that protein glycosylation is under elaborate cellular synthetic control and plays a dominant role in defining the properties of the protein and residing cells, live cell imaging of glycans, especially in a protein-specific fashion, can not only provide a deep insight into glycosylation regulation mechanisms but also contribute to the development of diagnostic markers and therapeutic targets. Despite the encouraging progress in this field, there are still several important aspects that remain to be explored. For example, considering the close relationship between some glycoforms (T antigen, SLe^a, SLe^x, *etc.*) and disease progression, it would be useful to develop protein-specific methods to visualize these complex glycan structures on living cells. Moreover, live cell glycan imaging is the basis for live cell glycan remodelling, hence a promising developing trend is to achieve the versatile editing of glycan chains at different specificity levels, such as protein-specific, cell-specific or even tissue-specific levels, which may contribute to the artificial control or regulation of the functions of proteins and cells. Further, most current protein-specific glycan imaging methods use cell lines as the model, hence the development of strategies that fit with *in vivo* applications is an urgent need.

References

1. M. Kleinert, N. Röckendorf and T. K. Lindhorst, *Eur. J. Org. Chem.*, 2004, 3931–3940.
2. M. J. Paszek, C. C. DuFort, O. Rossier, R. Bainer, J. K. Mouw, K. Godula, J. E. Hudak, J. N. Lakins, A. C. Wijekoon, L. Cassereau, M. G. Rubashkin, M. J. Magbanua, K. S. Thorn, M. W. Davidson, H. S. Rugo, J. W. Park, D. A. Hammer, G. Giannone, C. R. Bertozzi and V. M. Weaver, *Nature*, 2014, **511**, 319–325.
3. J. D. Marth and P. K. Grewal, *Nat. Rev. Immunol.*, 2008, **8**, 874–887.
4. L. Krishnamoorthy and L. K. Mahal, *ACS Chem. Biol.*, 2009, **4**, 715–732.
5. A. Varki, *Nature*, 2007, **446**, 1023–1029.
6. D. H. Dube and C. R. Bertozzi, *Nat. Rev. Drug Discovery*, 2005, **4**, 477–488.
7. M. M. Fuster and J. D. Esko, *Nat. Rev. Cancer*, 2005, **5**, 526–542.
8. S. P. Salomé and A. R. Celso, *Nat. Rev. Cancer*, 2015, **15**, 540–555.
9. P. M. Rudd, T. Elliott, P. Cresswell, I. A. Wilson and R. A. Dwek, *Science*, 2001, **291**, 2370–2376.
10. J. J. Caramelo and A. J. Parodi, *Semin. Cell Dev. Biol.*, 2007, **18**, 732–742.
11. R. Raman, S. Raguram, G. Venkataraman, J. C. Paulson and R. Sasisekharan, *Nat. Methods*, 2005, **2**, 817–824.
12. K. Marino, J. Bones, J. J. Kattla and P. M. Rudd, *Nat. Chem. Biol.*, 2010, **6**, 713–723.
13. A. Helenius and M. Aebi, *Science*, 2001, **291**, 2364–2369.
14. K. W. Moremen, M. Tiemeyer and A. V. Nairn, *Nat. Rev. Mol. Cell Biol.*, 2012, **13**, 448–462.
15. A. Varki and M. J. Chrispeels, *Glycobiology*, 1993, **3**, 97–130.
16. E. Meezan, H. C. Wu, P. H. Black and P. W. Robbins, *Biochemistry*, 1969, **8**, 2518–2524.
17. J. S. Axford, *Biochim. Biophys. Acta, Mol. Basis Dis.*, 1999, **1455**, 219–229.
18. G. A. Turner, *Clin. Chim. Acta*, 1992, **208**, 149–171.
19. A. Mackiewicz and K. Mackiewicz, *Glycoconjugate J.*, 1995, **12**, 241–247.
20. S. Hakomori, *Adv. Exp. Med. Biol.*, 2001, **491**, 369–402.
21. L. Krasnova and C. H. Wong, *Annu. Rev. Biochem.*, 2016, **85**, 599–630.
22. H. Xiao, E. C. Woods, P. Vukojicic and C. R. Bertozzi, *Proc. Natl. Acad. Sci. U. S. A.*, 2016, **113**, 10304–10309.
23. S. J. Danishefsky, Y. K. Shue, M. N. Chang and C. H. Woog, *Acc. Chem. Res.*, 2015, **48**, 643–652.
24. H. Lis and N. Sharon, *Chem. Rev.*, 1998, **98**, 637–642.
25. W. Cheng, L. Ding, J. P. Lei, S. J. Ding and H. X. Ju, *Anal. Chem.*, 2008, **80**, 3867–3872.
26. Y. D. Xue, L. Ding, J. P. Lei, F. Yan and H. X. Ju, *Anal. Chem.*, 2010, **82**, 7112–7118.
27. D. Craig, S. McAughtrie, J. Simpson, C. McCraw, K. Faulds and D. Graham, *Anal. Chem.*, 2014, **86**, 4775–4782.
28. L. Ding, W. Cheng, X. J. Wang, Y. D. Xue, J. P. Lei, Y. B. Yin and H. X. Ju, *Chem. Commun.*, 2009, **46**, 7161–7163.

29. E. Han, L. Ding, H. Z. Lian and H. X. Ju, *Chem. Commun.*, 2010, **46**, 5446–5448.
30. E. Han, L. Ding, R. C. Qian, L. Bao and H. X. Ju, *Anal. Chem.*, 2012, **84**, 1452–1458.
31. E. Han, L. Ding and H. X. Ju, *Anal. Chem.*, 2011, **83**, 7006–7012.
32. E. Saxon and C. R. Bertozzi, *Science*, 2000, **287**, 2007–2010.
33. S. H. Rouhanifard, L. U. Nordstrom, T. Zheng and P. Wu, *Chem. Soc. Rev.*, 2013, **42**, 4284–4296.
34. P. V. Robinson, G. de Almeida Escobcdo, A. E. de Groot, J. L. McKechnie and C. R. Bertozzi, *J. Am. Chem. Soc.*, 2015, **137**, 10452–10455.
35. L. Lin, X. D. Tian, S. L. Hong, P. Dai, Q. H. You, R. Y. Wang, L. S. Feng, C. Xie, Z. Q. Tian and X. Chen, *Angew. Chem., Int. Ed.*, 2013, **52**, 7266–7271.
36. B. Y. Yuan, Y. Y. Chen, Y. Q. Sun, Q. P. Guo, J. Huang, J. B. Liu, X. X. Meng, X. H. Yang, X. H. Wen, Z. H. Li, L. Li and K. M. Wang, *Anal. Chem.*, 2018, **90**, 6131–6137.
37. P. R. Wratil, R. Horstkorte and W. Reutter, *Angew. Chem., Int. Ed.*, 2016, **55**, 9482–9512.
38. S. L. Hong, L. Lin, M. Xiao and X. Chen, *Curr. Opin. Chem. Biol.*, 2015, **24**, 91–96.
39. N. D. Pham, R. B. Parker and J. J. Kohler, *Curr. Opin. Chem. Biol.*, 2013, **17**, 90–101.
40. N. D. Pham, C. S. Fermaintt, A. C. Rodriguez, J. E. McCombs, N. Nischan and J. J. Kohler, *Glycoconjugate J.*, 2015, **32**, 515–529.
41. J. G. Briard, H. Jiang, K. W. Moremen, M. S. Macauley and P. Wu, *Nat. Commun.*, 2018, **9**, 880.
42. H. Jiang, A. Llpez-Aguilar, L. Meng, Z. W. Gao, Y. N. Liu, X. Tian, G. L. Yu, B. Ovryn, K. W. Moremen and P. Wu, *Angew. Chem., Int. Ed.*, 2018, **57**, 967–971.
43. Y. D. Xue, L. Ding, J. P. Lei, F. Yan and H. X. Ju, *Anal. Chem.*, 2010, **82**, 7112–7118.
44. W. Y. Song, L. Ding, Y. L. Chen and H. X. Ju, *Chem. Commun.*, 2016, **52**, 10640–10643.
45. Y. D. Xue, L. Ding, J. P. Lei and H. X. Ju, *Biosens. Bioelectron.*, 2010, **26**, 169–174.
46. L. Ding, R. C. Qian, Y. D. Xue, W. Cheng and H. X. Ju, *Anal. Chem.*, 2010, **82**, 5804–5809.
47. K. Hsu, K. T. Pilobello and L. K. Mahal, *Nat. Chem. Biol.*, 2006, **2**, 153–157.
48. K. T. Pilobello, D. E. Slawek and L. K. Mahal, *Proc. Natl. Acad. Sci. U. S. A.*, 2007, **104**, 11534–11539.
49. P. Mitchell, *Nat. Biotechnol.*, 2002, **20**, 225–229.
50. L. Ding, W. Cheng, X. J. Wang, S. J. Ding and H. X. Ju, *J. Am. Chem. Soc.*, 2008, **130**, 7224–7225.
51. W. Cheng, L. Ding, S. J. Ding, Y. B. Yin and H. X. Ju, *Angew. Chem., Int. Ed.*, 2009, **48**, 6465–6468.

52. Y. L. Chen, L. Ding, T. T. Liu and H. X. Ju, *Anal. Chem.*, 2013, **85**, 11153–11158.
53. Y. D. Xue, L. Bao, X. R. Xiao, L. Ding, J. P. Lei and H. X. Ju, *Anal. Biochem.*, 2011, **410**, 92–97.
54. E. Han, L. Ding, S. Jin and H. X. Ju, *Biosens. Bioelectron.*, 2011, **26**, 2500–2505.
55. L. Ding, X. R. Xiao, Y. L. Chen, R. C. Qian, L. Bao and H. X. Ju, *Chem. Commun.*, 2011, **47**, 3742–3744.
56. Y. L. Chen, L. Ding, J. Q. Xu, W. Y. Song, M. Yang, J. J. Hua and H. X. Ju, *Chem. Sci.*, 2015, **6**, 3769–3774.
57. R. C. Qian, L. Ding, L. W. Yan and H. X. Ju, *Anal. Chim. Acta*, 2015, **894**, 85–90.
58. Y. L. Chen, L. Ding and H. X. Ju, *Chem. Commun.*, 2013, **49**, 862–864.
59. Y. Haga, K. Ishii, K. Hibino, Y. Sako, Y. Ito, N. Taniguchi and T. Suzuki, *Nat. Commun.*, 2012, **3**, 907.
60. W. Lin, Y. F. Du, Y. T. Zhu and X. Chen, *J. Am. Chem. Soc.*, 2014, **136**, 679–687.
61. B. Belardi, A. de la Zerda, D. R. Spiciarich, S. L. Maund, D. M. Peehl and C. R. Bertozzi, *Angew. Chem., Int. Ed.*, 2013, **52**, 14045–14049.
62. Y. L. Chen, L. Ding, W. Y. Song, M. Yang and H. X. Ju, *Chem. Sci.*, 2016, **7**, 569–574.
63. J. J. Hui, L. Bao, S. Q. Li, Y. Zhang, Y. M. Feng, L. Ding and H. X. Ju, *Angew. Chem., Int. Ed.*, 2017, **56**, 8139–8143.
64. Y. R. Liu, L. Liu, S. Q. Li, G. Y. Wang, H. X. Ju and L. Ding, *Anal. Chem.*, 2019, **91**, 6027–6034.
65. N. Wu, L. Bao, L. Ding and H. X. Ju, *Angew. Chem., Int. Ed.*, 2016, **55**, 5220–5224.
66. S. Q. Li, Y. R. Liu, L. Liu, Y. M. Feng, L. Ding and H. X. Ju, *Angew. Chem., Int. Ed.*, 2018, **57**, 12007–12011.
67. T. Suzuki, *Nature*, 2018, **561**, 38–40.
68. Y. L. Chen, H. P. Liu, Y. Y. Xiong and H. X. Ju, *Angew. Chem., Int. Ed.*, 2018, **57**, 785–789.

CHAPTER 3

In Situ *Detection of Intracellular Messenger RNA and MicroRNA*

3.1 Introduction

Both mRNA and miRNA are emerging biomarker candidates for early disease diagnosis and prognosis and also therapy targeting. The detection of mRNA/miRNA abundance is significant for deciphering cellular signalling pathways, understanding cellular behaviour and predicting the malignant progression of tumours.[1] Conventional techniques for RNA analysis such as northern blotting, western blotting, reverse transcription polymerase chain reaction (RT-PCR) and DNA microarrays rely on the fixation or lysis of cells. However, these techniques are labour intensive and time consuming. More importantly, the information about the dynamics of RNA inside cells and cell-to-cell heterogeneity is often lost in such cases owing to ensemble averaging.[2] The accurate analysis of the abundance of miRNA in real time in living cells is needed for monitoring the amount and localization of miRNA. However, monitoring of intracellular mRNA and miRNA levels in living cells remains a great challenge owing to the low gene probe transfection efficiency and the low abundance of mRNA and miRNA. This chapter first introduces the biofunction of mRNA and miRNA, then various gene probe vectors are summarized and the advanced techniques involved in intracellular mRNA and miRNA detection are discussed.

Detection Science Series No. 16
In Situ Analysis of Cellular Functional Molecules
By Huangxian Ju, Bo Tang and Lin Ding

Published by the Royal Society of Chemistry, www.rsc.org

3.2 Biofunction of mRNA and miRNA

3.2.1 Biofunction of mRNA

RNA is not only responsible for the storage and transfer of genetic information, but also plays a role as a ribozyme in cell metabolism. In the process of transferring genetic information from DNA to protein, RNA does not simply transfer genetic information. RNA can regulate the gene expression and genetic information through splicing, editing and recoding, so as to achieve the regulation of biological development and differentiation.[3] As the messenger of DNA and template of protein synthesis, messenger RNA (mRNA) plays an important role in genetic activities. It is present in the cytoplasm of prokaryotes and eukaryotes and in some organelles of eukaryotes, such as mitochondria and chloroplasts. Generally, biological information stored in DNA must be transcribed into RNA and the transcribed RNA cannot immediately and directly serve as a template to guide the synthesis of peptides. Mature mRNA must be formed through cutting, splicing and other editing processes. mRNA plays important roles in the cell cycle, proliferation and cell apoptosis. Therefore, the study of mRNA is not only of great significance for understanding the structure and function of the genome, but is also of great value for studying the mechanism, prevention and treatment of diseases. The systematic and effective analysis of mRNA is crucial to understanding genes, providing the possibility of exploring potential genetic applications.

Many genetic diseases are closely related to the structural changes of specific endogenic genes and the ectogenic genes. Fragile X mental retardation protein (FMRP) is one of the most common hereditary mental retardations throughout the world. It has been reported that the commonly expressed mRNA nuclear output factor NXF1 is the regulatory target of neuronal FMRP *in vivo*.[4] Ichthyosis is a group of hereditary keratosis skin diseases considered to be caused by the instability of mRNA and a defect of the post-transcriptional control mechanism.[5] Furthermore, the abnormal expression of tumour-related mRNAs is associated with tumour progression and tumour-related mRNA has been widely used as a specific marker to evaluate the migration of tumour cells in local blood flow. Therefore, tumour-related mRNA analysis in cancer cells provides a powerful tool for the identification of cancer cells. Importantly, mRNA can also be used in the evaluation of drug effectiveness. When cells are stimulated by external stimuli or drugs, the expression level of some genes often changes. By analysing the mRNA expression change in response to a drug, the time needed to establish the target gene of the drug could be effectively shortened, providing a more objective and accurate approach for the evaluation of drug effectiveness and reducing drug development costs.[6] Moreover, the recent advances in mRNA synthesis, modification and delivery technology make it possible to develop mRNA-related gene therapy. mRNA vaccine exhibits an intriguing validity for many new viral diseases,[7,8] and is safer than traditional vaccines because mRNA is not integrated into the genome and is degraded by normal cells with a reduced cellular immune response.

3.2.2 Biofunction of miRNA

MicroRNA (miRNA) is a type of no-coding and small-sized RNA (18–22 nt), which plays significant post-transcriptional regulatory roles in gene expression. The partially or fully complementary hybridization between miRNA and the 3′ non-coding region (3′ UTRs) of the target mRNA induce protein translation suppression or mRNA degradation. The bio-function of miRNA is mainly derived from inhibiting the expression of downstream genes, weakening or eliminating the function of downstream genes and regulating the physiological and pathological state. The miRNA regulation is a complex network process in which each miRNA has multiple target genes and several miRNAs also jointly regulate the expression of a single gene.[9,10] The miRBase database (miRBase 20: http://www.mirbase.org/) has reported more than 1800 precursors and 2500 mature miRNA sequences for the human species, which regulate more than two-thirds of functional protein genes, demonstrating the importance of miRNA.

miRNAs are involved in a variety of physiological processes, including regulation of development, cellular differentiation, proliferation and apoptosis. The miRNAs lin-4 and let-7 were first discovered in nematodes, and both can partially bind to the 3′ non-coding region (3′ UTRs) of the target mRNA to regulate the development of nematodes by inhibiting the translation of key proteins.[11–13] miRNA-14 and miRNA-23 are involved in the post-transcriptional regulation of genes that play an important role in cell differentiation and tissue development. The mature miRNAs are cleaved from pre-miRNA by the RNase-III enzyme Dicer and it has been reported that Dicer knockout mice died after 7.5 days, suggesting that miRNA plays an important role in the development of mice.[14] The miRNA-150 level changes dynamically during the development of lymphocytes, which will lead to an inverse change of the target gene c-Myb, which is an important transcription regulator of cell differentiation. Dicer also plays an important role in the development of T cells.[15]

The aberrant expression of miRNAs is associated with various pathological processes such as regulation of animal immune function, resistance to virus invasion, diseases of organs or systems and tumours. Hence miRNAs are emerging biomarkers for disease diagnosis and prognosis and are promising gene therapy targets. Organisms have developed the function of using miRNA to resist virus invasion during evolution. miRNA has the potential to resist HIV virus infection by binding non-specifically to HIV-1 Gag protein.[16] The formation and transmission of HIV virus are suppressed by miRNA-mediated interference with the combination of Gag and target skeleton RNA. miRNA-122 is specifically expressed in the liver and is able to recognize directly the pre-genomic RNA of hepatitis B virus (HBV) to inhibit the replication and translation of HBV. However, miRNA-122 can also bind to the 5′ UTRs of HCV RNA and promote the proliferation of HCV.[17] miRNA not only regulates nervous system growth, development, synapse formation and memory, but also is related to the occurrence and development of nervous system diseases. In Alzheimer's disease (AD), amounts of aberrant expression of miRNAs regulating the activity of β-starch precursor protein lyase (BACE1) are observed. For example, miRNA-107 is significantly reduced in the early stage of AD, which binds to the 3′ UTR region of BACE1 to increase the level of BACE1 mRNA.[18]

miRNAs are also associated with human metabolic life activities. PRDM16 is a zinc finger protein that is related to muscle and brown fat differentiation and miRNA-149-3p was found to inhibit the expression of PRDM16.[19] Especially, dysfunction of miRNA was associated with diverse tumours and it regulates tumour proliferation, tumour autonomic growth, apoptosis, angiogenesis and metastasis. For example, miRNA-21 is highly expressed in breast, colorectal, lung, prostate, pancreatic and gastric cancer, and inhibits tumour apoptosis. It has been identified that the targets of miRNA-21 are PTEN and PDCD4, both of which significantly regulate apoptosis.[20] miRNA-10b is highly expressed in migratory breast cancer cells and it is positively correlated with tumour migration and invasiveness.[21] Overexpression of miRNA-10b in non-migratory breast cancer cells can induce migration. It was found that that miRNA-10b promotes tumour metastasis by inhibiting the expression of HOXD10 gene, which leads to high expression of migration protein RhoC.

3.3 Viral Gene Vectors

With rapid advances in understanding mRNA- and miRNA-related genomic pathogenesis of human diseases, it was anticipated that increasing numbers of genes or targets as biomarkers would become available for the accurate diagnosis of and targeted therapies for cancer. However, the poor efficiency of the gene delivery system is the main limiting factor for the detection of most biomarkers or gene therapies. Gene vectors are generally divided into two main types: viral vectors and non-viral vectors. In comparison with non-viral vectors, the viral delivery systems show various advantages, including high targeting ability, good gene loading capacity and high transfection efficiency, which inspired researchers to focus on developing various virus-derived gene vector systems for gene delivery, performing well with respect to both diagnostic accuracy and therapeutic efficacy. Here, we review the applications of different types of viral vectors to carry out gene delivery and target mammalian cells or cancer cells, including lenti-/retrovirus, adenovirus (Ad), adeno-associated virus (AAV) and some plant viruses (*e.g.* tobacco mosaic virus, beet necrotic yellow vein virus).

3.3.1 Lenti-/Retrovirus-mediated Vectors for Gene Delivery

As a retrovirus, lentivirus-mediated vectors containing RNA gene must undergo a process with reverse transcriptase enzymes to generate double-stranded DNA proviral transgenes. In comparison with other viral delivery systems, the lenti-/retroviruses show some drawbacks, including inefficiency and poor fidelity of gene targeting. However, when these problems are overcome, the lentivirus-mediated vectors could be applied for gene targeting to human stem cells and provide treatment for human diseases. Guo *et al.* developed lentivirus-mediated miRNA-26a-modified neural stem cells (NSCs) to treat brain injury in rats with cerebral palsy (CP), demonstrating that miRNA-26a-modified NSCs mediated by lentivirus could improve brain injury, inhibiting apoptosis of brain cells and activation of astrocytes in CP rats.[22]

3.3.2 Adenovirus-mediated Vectors for Gene Delivery

Adenovirus (Ad), a type of non-enveloped, linear, double-stranded DNA virus with 57 identified human serotypes, has been developed as effective gene delivery vectors for constructing multifunctional theranostic nanoplatforms. The Ad-derived vectors show several advantages compared with other viral gene delivery systems. First, Ad-mediated vectors could specifically deliver genes to most human cells that express the primary Ad receptors or the secondary integrin receptors with high efficiency. Furthermore, the rapid development of gutless adenoviral vectors allows us to circumvent anti-adenoviral vector immunity and reduce immune clearance. In fact, Ad-mediated gene delivery systems offer versatile platforms for cancer theranostic strategies *via* modifying viral capsids to enhance the therapeutic properties and improve the targeting specificity of the virus. Recently, Huang *et al.* developed an Ad-mediated ABCC6 gene delivery system and offered a promising prevention therapy for pseudoxanthoma elasticum and generalized arterial calcification of infancy, which currently lack preventive or therapeutic options.[23]

3.3.3 Adeno-associated Virus Vectors for Gene Delivery

Adeno-associated virus (AAV) is a small (25 nm) and helper-dependent virus derived from the Parvoviridae family, and is composed of a non-enveloped icosahedral capsid (protein shell) that contains a linear single-stranded DNA genome.[24] Compared with Ad-mediated vectors, the AAV gene delivery systems represent a safer option for transfection, given their naturally attenuated pathogenicity and immunogenicity in the human body. Importantly, the duration of human exposure to AAV has a direct impact on the immunological response of the body to the recombinant adeno-associated viral (rAAV) vectors. In recent years, rAAV gene delivery systems have displayed great potential for the treatment of various human diseases. Pinto *et al.*

explored the possibility of using the rAAV-mediated gene vector of short hairpin RNAs (shRNAs) for targeting basal-like breast cancer genetic vulnerabilities.[25] The prepared rAAV–shRNA vectors showed a good capacity for efficiently transducing the BLBC cell lines MDA-MB-468 and HCC1954. In MDA-MB-468, transduction of rAAV–shRNA vector targeting PSMA2 was associated with significant decreases in cell viability and apoptosis induction, thus presenting a potential approach for this cancer therapy. Additionally, rAAV vectors have also shown potential in several human diseases, such as employing the matrix metalloproteinases to activate the AAV vector for epithelial ovarian cancer or various heart diseases,[26] or applying DNA-encoded small RNA molecules for the post-transcriptional regulation of oncogenes, immunogenic cell surface molecules and tumour antigens.[26]

In addition to the above three major viral vector categories in the gene delivery "tool box", there are many other virus-mediated gene vectors.[27] For example, Leber *et al.* described a systematic and comprehensive analysis of a post-entry targeting strategy based on the incorporation of miRNA target sites into the oncolytic measles virus genome for developing a safe and efficient strategy for cancer viro-therapy.[27] Further, some plant viruses have also been employed for constructing gene delivery systems.[28,29] Tobacco mosaic virus (TMV), a rod-like plant virus with a high aspect ratio, exhibited promising potential for gene delivery. As shown in Figure 3.1, Tian *et al.* developed a transacting activator of transduction (TAT)-modified and silencer green fluorescent protein (GFP) siRNA-loaded TMV vector for interfering GFP expression in GFP-expressing cells

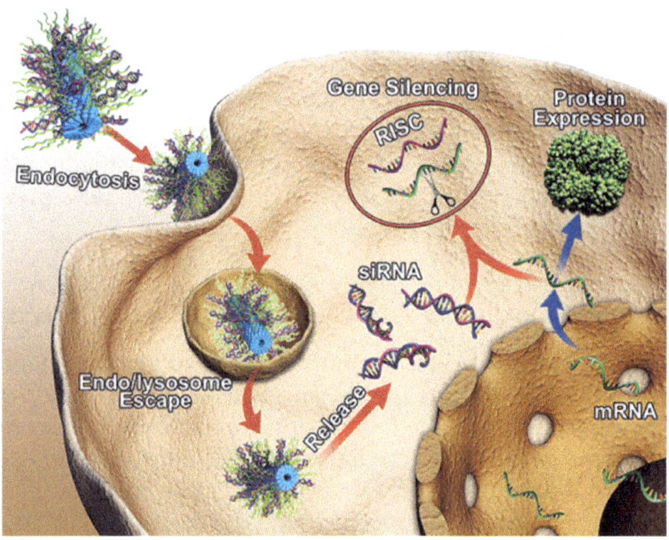

Figure 3.1 Schematic illustration of the gene-silencing process based on a tobacco mosaic virus gene delivery system.[30]
Reproduced from ref. 30 with permission from American Chemical Society, Copyright 2018.

(metastatic hepatocellular carcinoma).[30] This work combined the high efficiency of viral vectors and the safety of non-viral vectors, providing a promising strategy for gene-silencing technology.

3.4 Organic Gene Vectors

Organic gene vectors have been extensively explored to protect fragile nucleic acids from degradation and deliver them to the targeted cells owing to their biodegradability, biocompatibility and flexibility for functionalization. Generally, organic gene vectors can be categorized as cationic polymers and lipid-based cationic (cationic lipids) gene vectors.

3.4.1 Cationic Polymers

Polylysine (PLL), polyethylenimine (PEI), polyamidoamine (PAMAM) and chitosan have been widely developed as cationic polymer gene vectors.[31] They can conjugate with negatively charged nucleic acids to form nanosized polyplexes and thus protect the payload gene probe against enzymatic degradation and enhance their cellular endocytosis.

PLL is one of the most commonly used polycations for gene vectors owing to the biodegradable peptide structure. However, the gene transfection efficiency of PLL polyplexes is poor because they can become trapped in endosomes and lysosomes after cellular endocytosis.[32] Therefore, PLL is usually modified with endosomolytic groups such as histidyl and imidazole to facilitate its lysosomal escape and improve the transfection activity.

PEI with a molecular weight (MW) in the range 20–30 kDa, particularly 25 kDa, exhibits a high transfection efficiency.[33] Controversially, the high transfection efficiency stems from its ability to rupture endosomes through the proton-sponge mechanism.[34,35] The gene transfection efficiency and cytotoxicity of PEI are closely associated with its MW. Otherwise, high-MW PEI usually has a favourable transfection efficiency but possesses high cytotoxicity that generates deficiencies in the cell surface and mitochondria or nuclear membranes, triggering apoptosis and inhibition of mitochondrial ATP synthesis.[36,37]

PAMAM, a cationic dendrimer, is the most studied dendrimer as a gene vector, primarily owing to its good monodispersity, three-dimensional structure, well-controlled size, abundant surface groups for functionalization, high gene transfection efficacy and commercial availability.[38,39] PAMAM also possesses significant generation-dependent cytotoxicity.[40] It was reported that linear 10 kDa amphoteric PAMAM is optimal for its good biocompatibility and favourable gene transfection efficiency.[41]

Chitosan (poly-D-glucosamine, generated by deacetylation of chitin) is a biodegradable natural polysaccharide. Chitosan and its derivatives have been successfully applied for gene delivery, resulting from its cationic nature, biocompatibility, low immunogenicity, non-toxicity and low cost.[42–44] The encapsulation and transfection of chitosan are associated with its MW and degree of deacetylation (number of charges).[45,46] Chitosan can be functionalized by methylation and PEGylation to improve the formation and colloidal stabilization of gene polyplexes.[42,43] Chitosan has also been functionalized by thiolation, targeting ligands for enhanced cellular uptake or improved endosomal escape to enhance the gene transfection efficiency.[44,47–49] Additionally, PEI-grafted chitosan shows a comparable gene transfection efficiency to 25 kDa PEI, while maintaining the favourable biocompatibility and low toxicity.[50,51]

3.4.2 Cationic Lipids

Cationic lipids have been widely used for gene delivery,[52–54] and some of them have been applied in clinical trials, such as Lipofectin and Lipofectamine. Lipofectin is a commercialized gene vector consisting of 1 : 1 (w/w) polyplexes of a neutral lipid, 1,2-dioleoylphosphatidylethanolamine (DOPE), and a cationic lipid, *N*-[1-(2,3-dioleyloxy)propyl]-*N*,*N*,*N*-trimethylammonium chloride (DOTMA). Lipofectamine is a commercialized polyplex composed of 3 : 1 (w/w) DOPE and a polycationic lipid, 2,3-dioleyloxy-*N*-[2-(sperminecarboxamido)ethyl]-*N*,*N*-dimethyl-1-propanaminium trifluoroacetate (DOSPA).[55] DOPE or cholesterol and helper lipids are usually incorporated into cationic lipids to promote the condensation of nucleic acids into a stable hexagonal-phase structure, which is beneficial to transfection efficiency.[56,57] In addition to forming a stable hexagonal phase structure, DOPE also facilitates the escape of nucleic acids from endosomes and release from the polyplexes.[58] Cholesterol not only improves the stability of polyplexes but also prolongs its blood circulation.[59]

The relationships between the structure of cationic lipids, genes and transfection efficiency have been extensively studied by investigating the roles of three functional domains: (1) the polar hydrophilic cationic head group, which is used to conjugate with the nucleic acids;[60] (2) the hydrophobic anchor group, which is usually composed of two alkyl chains or cholesteryl groups for forming liposomes and interacting with the cell membrane;[61] and (3) the spacer group, which connects the head and anchor groups and optimizes the interactions of the polyplexes with nucleic acids and cell membranes. The head group, such as amines, quaternary ammonium compounds, guanidinium compounds or amino acids, carries a positive charge and its amount determines the interaction strength of the lipids with phosphate groups in nucleic acids.[62] The chain length and chemical structure of the anchor group also have a great influence on the gene transfection activity.[63]

Lipid-based hybrids have also been used for systemic gene delivery, including the multifunctional envelope-type nanodevice (MEND) (Figure 3.2),[64] stable nucleic acid–lipid particle (SNALP)[52] and stabilized plasmid–lipid particle (SPLP).[65] In particular, lipopolyplexes, polyplexes encapsulated in a PEGylated liposome,[66] take advantage of both the polymer and lipid carrier to create effective vectors for systemic gene delivery.[67]

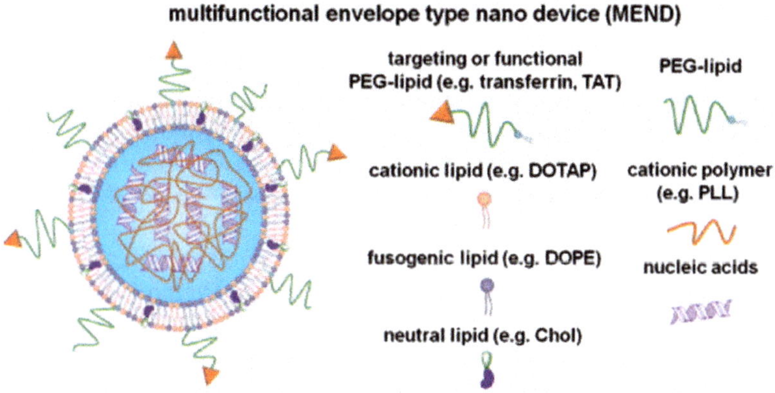

Figure 3.2 Schematic representation of a multifunctional envelope-type nano device (MEND).[64]
Reproduced from ref. 64 with permission from Elsevier, Copyright 2017.

3.5 Inorganic Gene Vectors

In comparison with viral delivery systems with deficiencies such as high cytotoxicity and immunogenicity and high cost, non-viral vectors exhibit unique advantages.[68,69] Recently, inorganic nanomaterial-based gene vectors have emerged as promising non-viral vectors for gene delivery owing to their unique physicochemical properties, including ultra-small size, large surface area-to-volume ratio and high reactivity.[70] A variety of advanced inorganic nanomaterials have been developed for gene delivery, particularly gold nanomaterials such as gold nanoparticles (AuNPs) and gold nanorods (AuNRs), graphene oxide (GO), quantum dots (QDs) and transition metal dichalcogenide (TMD) nanosheets, including MoS_2 and WS_2.

3.5.1 Gold-based Nanomaterials as Gene Vectors

Gold nanoparticles (AuNPs) have unique optical and physicochemical properties, excellent biocompatibility and a flexible surface for functionalization, and are easily modified by DNA or RNA strands and other cationic polymers for enhancing the transfection efficiency.[71,72] Zhong and co-workers developed a PEI 800 Da-modified AuNP, called a PEI–Au gene vector, which exhibited highly stable binding properties with respect to DNA molecules and a high transfection efficiency.[73] The fluorescence quench properties of AuNPs combined with a high affinity and loading capacity with respect to nucleic acid strands make them promising for endogenous mRNA and miRNA detection in living cells. For example, Degliangeli *et al.* developed a fluorescently labelled DNA probe immobilized on a PEGylated AuNP nanoplatform for intracellular miRNA detection.[74] Dong and co-workers developed 3D DNA tetrahedron-structured probes (TSPs) combing duplex-specific nuclease for highly sensitive and multiple miRNA detection *in vitro*.[75]

3.5.2 Two-dimensional GO and TMD Nanosheets as Gene Vectors

The large specific surface area, unique optical, electrical and physicochemical properties and adaptability for surface functionalization promote graphene-based nanomaterials as significant gene vectors.[76] The π–π stacking interactions between GO and nucleic acid strands (non-covalent interactions between aromatic rings) combined with the large available surface area provide GO with a high gene loading capacity and controlled release ability (Figure 3.3).[77] In fact, the negatively charged GO surface and the electrostatic repulsion between the negatively charged phosphate backbone of DNA/RNA molecules and GO limit the biological applications of GO. Hence GO needs to be functionalized *via* cationic polymers such as polyethylenimine (PEI) and poly(ethylene glycol) (PEG). The interaction between endogenous biomolecules and GO *in vivo* could form a protein corona on the surface of the nanomaterials, inducing an immune response and leading to a decrease in cellular uptake efficiency and low biocompatibility. Hence researchers have generally employed PEG or bovine serum albumin to modify GO nanomaterials for low cytotoxicity and to enhance biosecurity.[78,79]

Two-dimensional (2D) TMDs possess unique optical and physicochemical properties, good stability and a large specific surface area, and have also attracted considerable attention for constructing gene delivery.[80,81]

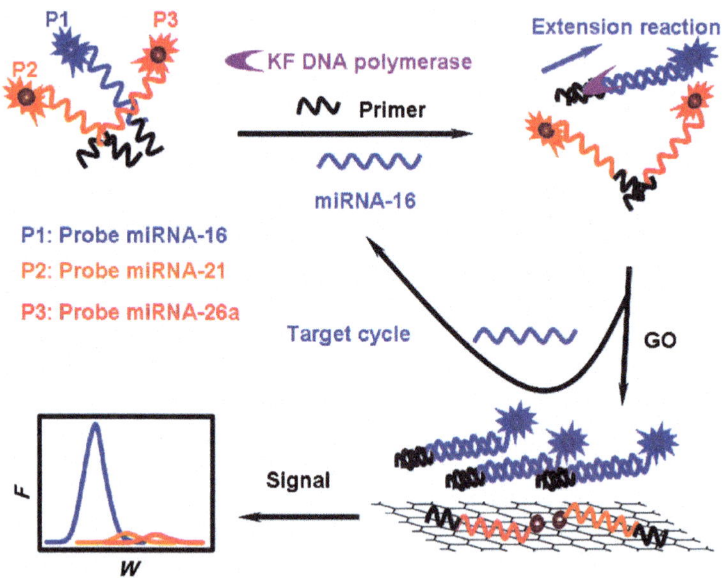

Figure 3.3 Illustration of GO fluorescence quenching and isothermal strand-displacement polymerase reaction-based multiple miRNA analysis.[77] Reproduced from ref. 77 with permission from American Chemical Society, Copyright 2012.

MoS_2 nanosheets have attracted intense interest for gene cargo delivery for imaging of biomarkers.[82,83] Dong's group[84] developed a facile approach for preparing size-tuneable blue-luminescent MoS_2 QDs with good compatibility and physiological stability, showing a good capacity for intracellular miRNA imaging analysis. In addition to the aforementioned inorganic gene vectors, various other nanomaterials, including C_3N_4, WS_2, Mo_2C, black phosphorus and MnO_2 nanosheets,[85–87] have also been widely explored to construct gene delivery systems.

3.6 Intracellular mRNA Detection

Owing to the low abundance and complex environment, it is still a major challenge to detect mRNA in living cells. The conventional nucleic acid probes, such as linear hybridization probes and molecular beacons (MBs), follow a "one-to-one" pattern to generate a signal for detection, which is insufficient for low-abundance mRNA detection. "One-to-more" signal amplification where an mRNA target can trigger multiple signal outputs is highly desired for intracellular mRNA imaging detection, allowing a very small amount of mRNA to produce a signal that reaches a detectable range in a short period of time. Generally, the signal amplification strategy can be divided into two main types: a molecular biology-based strategy and a nanomaterials-related approach.

Enzyme-free molecular biology-based signal amplification allowed the sensitive *in vitro* detection of biomolecules, including proteins and nucleic acids. However, monitoring or imaging targets of interest in living cells *via* enzyme-free amplification remains a challenge, especially for analytes with low concentrations. Various strand displacement isothermal signal amplification strategies have been explored for intracellular mRNA detection. For example, Jiang and co-workers developed an electrostatic DNA nanoassembly realizing a hybridization chain reaction (HCR), a target-initiated alternating hybridization reaction between two hairpin probes, for signal amplification in living cells.[88] The DNA nanoassembly consisted of a core gold nanoparticle, a cationic peptide interlayer and an electrostatically assembled outer layer of fluorophore-labelled hairpin DNA probes. This approach showed a high efficiency for cellular delivery of DNA probes through an endocytosis-independent mechanism that overcomes endosomal entrapment and permitted the ultrasensitive fluorescence imaging of mRNA expression with a picomolar detection limit due to the HCR signal amplification.

Cell permeation and fragility are poor in complex cytoplasmic milieu owing to the well-established character of nucleic acid molecules. To circumvent these problems, Tan and co-workers reported a molecularly engineered, entropy-driven, three-dimensional DNA amplifier (EDTD), which can respond to a specific mRNA target in living cells (Figure 3.4).[89] The EDTD shows enhanced biostability and cellular uptake efficiency and the mRNA target–EDTD interaction can specifically activate an autonomous DNA circuit inside living cells through the exclusive entropy-driven force, providing enormous signal amplification for ultrasensitive detection of the mRNA. Tan and co-workers reported an mRNA imaging strategy inside live cells using a non-enzymatic hairpin DNA cascade reaction termed a hairpin DNA cascade amplifier (HDCA) for signal amplification.[90] The recycled mRNA target in the HDCA serves as a catalyst for the assembly of multiple DNA duplexes, generating an amplified fluorescent signal of the target mRNA for intracellular imaging.

Many nanomaterials, such as graphene,[91] MnO_2,[92] AuNPs[93,94] and Fe_3O_4,[95] are widely used in the detection of nucleic acid in cells or *in vivo* owing to their advantageous properties. For example, Lin *et al.* synthesized Fe_3O_4@polydopamine core–shell nanocomposites (Fe_3O_4@PDA NCs) through an *in situ* self-polymerization route.[95] The Fe_3O_4@PDA NCs demonstrated good performance for intracellular mRNA detection and multimodal imaging-guided photothermal therapy.

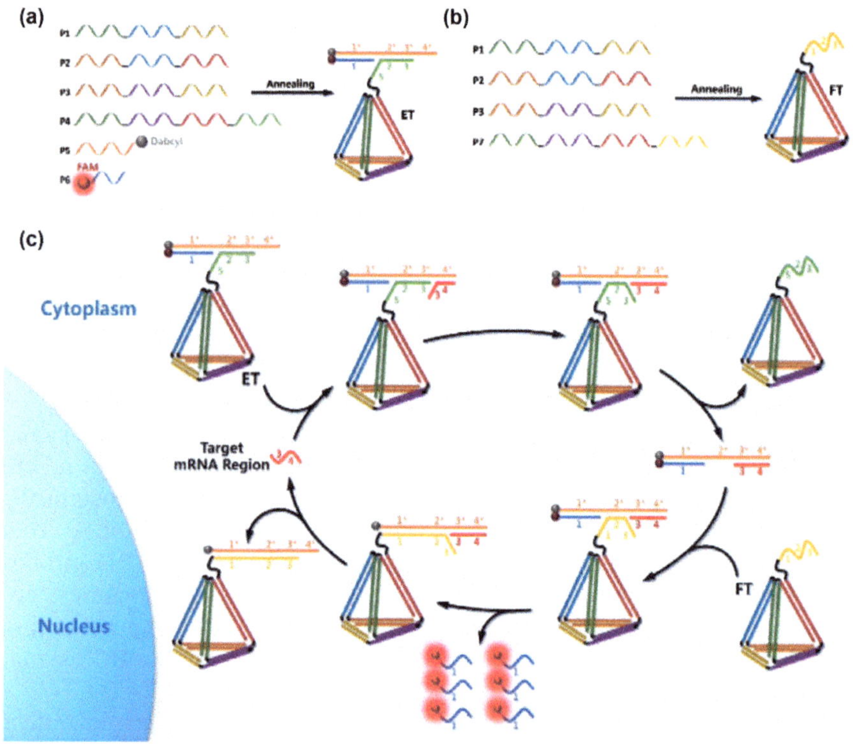

Figure 3.4 Assembly of (a) the ET module and (b) the FT module. (c) Mechanism of EDTD for catalytic signal enhancement of specific mRNA expression in living cells.[89]
Reproduced from ref. 89 with permission from American Chemical Society, Copyright 2018.

3.7 Intracellular miRNA Detection

3.7.1 "One-to-One" Signal Detection

Molecular beacons (MBs) are designed based on a one-to-one pattern where one target would bind to and activate one MB for signal generation. They show a stronger specificity to distinguish homologous or mismatched miRNA than unstructured linear probes owing to their loop and stem structure. They have been widely used for single or multiple miRNA imaging in living cells.[96] For example, Choi *et al.* used fluorescently labelled hairpin DNA strands (hpDNAs) to detect two important osteogenic markers, miRNA-29b and miRNA-31, in living stem cells.[97] The corresponding fluorescence signals would recover only in differentiated stem cells where miRNA-29b and miRNA-31 were overexpressed.

The toehold-mediated strand displacement (TMSD) reaction has also been used for miRNA detection in living cells. It was demonstrated that the length and base composition of the toehold are crucial elements impacting the reaction rate. The kinetic reaction rate of TMSD showed an exponential increase when the base number increased from zero to six. Moreover, a higher GC content in the toehold led to a faster reaction rate. Hemphill and Deiters reported an engineered oligonucleotide AND gate to respond to specific miRNA inputs in living mammalian cells based on the TMSD reaction.[98] The single- and dual-sensing miRNA-based computation devices were designed for cell-specific identification of endogenous miRNA-21 and miRNA-122. Additionally, other nanoprobes such as nanoflares have also been used for miRNA detection.

3.7.2 Sensitive Detection Based on Signal Amplification

The "one-to-one" signal read-out pattern precludes low-abundance miRNA imaging in living cells and a variety of enzyme- and non-enzyme-catalysed signal amplification techniques had been explored for monitoring intracellular miRNA. There are mainly two approaches for outputting the amplified signals: in the first, one target binds to one probe and activates a series of cascade reactions along a programmed path and produces a "one-to-more" signal read-out, such as HCR. The other approach usually employs recycled miRNA as a catalyst in the detection system. The target binds to one probe and produces one signal, then it would be released from the probe through fuel entropy-driven progress and activates another recycle. The read-out signal is accumulated with continuous involvement of miRNA. The fuels could be DNA single strand,[99] DNA hairpin probe,[100] ATP, enzyme,[101] pH, *etc.*, which play crucial roles in the detection system.

These two approaches have been widely used for intracellular low-abundance miRNA analysis. For instance, Dong and co-workers designed two pairs of programmable hairpin oligonucleotide probes for the analysis of intracellular miRNA-155 and its pre-miRNA based on HCR.[102] The HCR

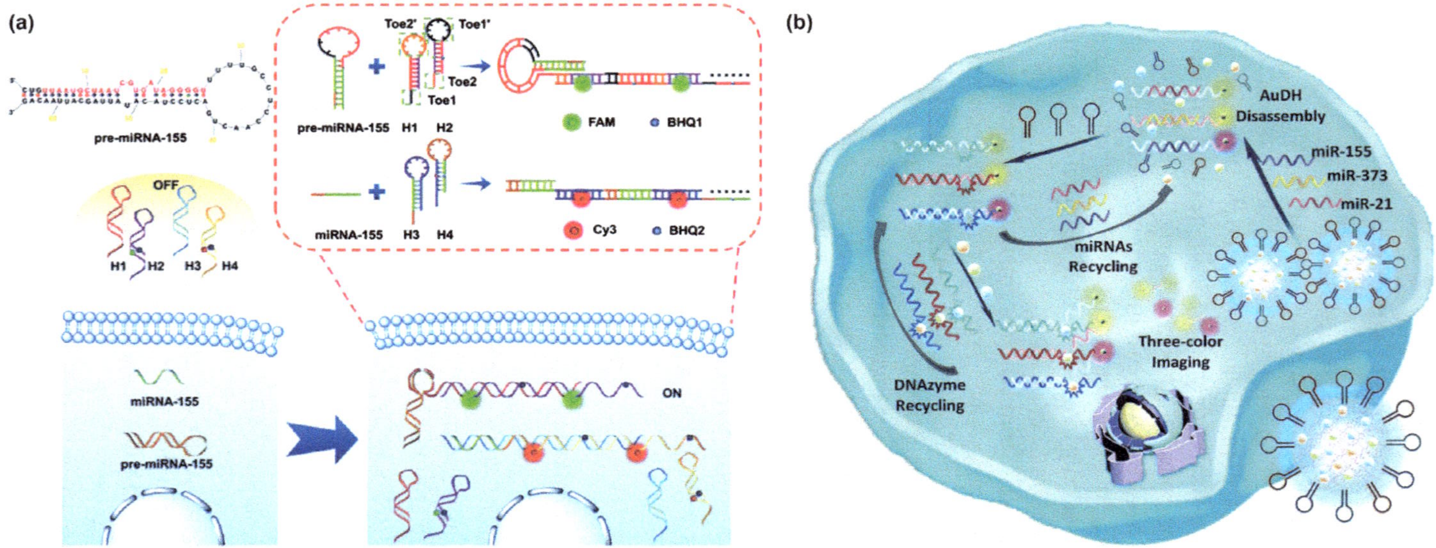

Figure 3.5 (a) Sensitively distinguishing intracellular precursor and mature miRNA abundance. (b) Multiplex miRNA imaging in living cells using DNA-capped–Au assembled hydrogels.[102,104]
Reproduced from ref. 102 and 104 with permission from the Royal Society of Chemistry.

system could discriminatorily and accurately assess the miRNA and pre-miRNA abundance *in vitro* compared with MBs (Figure 3.5a). The HCR system also showed a stronger fluorescence signal related to miRNA-155 and pre-miRNA-155 in three different cells than with MB-based detection, which was also proved by flow cytometry analysis. Chang and co-workers utilized DNA polymerase for the *in situ* growth of a three-way junction-based DNA structure for miRNA imaging.[103] In this system, the miRNA acted as the target for detection, which bound to the first hairpin probe (HPa) and triggered a series of DNA polymerization reactions to form Y-shaped DNA. The released miRNA further initiated the next round of the reaction cycle.

The two approaches have also been combined for intracellular miRNA analysis with a multiple magnification effect. Wang *et al.* combined a catalytic hairpin assembly (CHA) and HCR with a concatenated CHA–HCR system for miRNA-21 detection *in vivo*.[101] Dong and co-workers developed a toehold strand-displacement reaction and hairpin-locked DNAzyme-assisted miRNA recycling dual signal amplification system for imaging intracellular multiple miRNA (Figure 3.5b).[104] miRNA imaging in cancer cells could be realized even if the expression of related miRNA was downregulated due to the dual signal amplification.

3.8 Perspective and Challenges

Imaging of low-abundance intracellular mRNA/miRNA would provide valuable and useful information for monitoring pathological conditions and evaluating drug efficacy. Efficient gene vectors and intracellular signal amplification and signal read-out methods are the prerequisites for satisfactory monitoring of intracellular mRNA and miRNA levels in living cells. Especially for the intracellular or *in situ* detection of miRNA, when the sample consists of pre-miRNA and mature miRNA in a complex cellular environment, the non-specific recognition of pre-miRNA containing the same sequence as miRNA could induce a false-positive signal for expression levels of mature miRNA. Hence efficient non-invasive monitoring approaches are highly desirable to provide detailed spatial and temporal expression patterns for miRNAs with high specificity and selectivity. Overall, advanced gene cellular vectors and facile, highly selective and sensitive mRNA/miRNA imaging methods that can effectively implement intracellular expression profiles are still in demand.

Acknowledgement

Dr Haifeng Dong is thanked for expert assistance with the writing of this chapter.

References

1. A. Lujambio and S. W. Lowe, *Nature*, 2012, **482**, 347–355.
2. Z. Qing, J. Xu, J. Hu, J. Zheng, L. He, Z. Zou, S. Yang, W. Tan and R. Yang, *Angew. Chem., Int. Ed. Engl.*, 2019, DOI: 10.1002/anie.201812449.

3. M. A. Surani, *Cell*, 1998, **93**, 309–312.
4. E. Pasciuto and C. Bagni, *Cell*, 2014, **158**, 1446.
5. T. Takeichi, K. Sugiura, K. Matsuda, M. Kono and M. Akiyama, *J. Dermatol. Sci.*, 2013, **69**, 259–261.
6. D. J. Matthews and J. Kopczynski, *Drug Discovery Today*, 2001, **6**, 141–149.
7. J. Fueyo, C. Gomez-Manzano, W. K. Yung, G. L. Clayman, T. J. Liu, J. Bruner, V. A. Levin and A. P. Kyritsis, *Oncogene*, 1996, **12**, 103–110.
8. M. X. Wei, T. Tamiya, J. Robert, K. Hurford, E. J. Boviatsis, R. I. Tepper and E. A. Chiocca, *Hum. Gene. Ther.*, 1995, **6**, 437–443.
9. G. Obernosterer, P. J. F. Leuschner, M. Alenius and J. Martinez, *RNA*, 2006, **12**, 1161–1167.
10. R. I. Gregory and R. Shiekhattar, *Cancer Res.*, 2005, **65**, 3509–3512.
11. R. C. Lee, R. L. Feinbaum and V. Ambros, *Cell*, 1993, **75**, 843–854.
12. B. J. Reinhart, F. J. Slack, M. Basson, A. E. Pasquinelli, J. C. Bettinger, A. E. Rougvie, H. R. Horvitz and G. Ruvkun, *Nature*, 2000, **403**, 901.
13. A. E. Pasquinelli, B. J. Reinhart, F. Slack, M. Q. Martindale, M. I. Kuroda, B. Maller, D. C. Hayward, E. E. Ball, B. Degnan, P. Müller, J. Spring, A. Srinivasan, M. Fishman, J. Finnerty, J. Corbo, M. Levine, P. Leahy, E. Davidson and G. Ruvkun, *Nature*, 2000, **408**, 86.
14. E. Bernstein, S. Y. Kim, M. A. Carmell, E. P. Murchison, H. Alcorn, M. Z. Li, A. A. Mills, S. J. Elledge, K. V. Anderson and G. J. Hannon, *Nat. Genet.*, 2003, **35**, 215–217.
15. C. Xiao, D. P. Calado, G. Galler, T.-H. Thai, H. C. Patterson, J. Wang, N. Rajewsky, T. P. Bender and K. Rajewsky, *Cell*, 2007, **131**, 146–159.
16. A. K. Chen, P. Sengupta, K. Waki, S. B. Van Engelenburg, T. Ochiya, S. D. Ablan, E. O. Freed and J. Lippincott-Schwartz, *Proc. Natl. Acad. Sci. U. S. A.*, 2014, **111**, E2676–E2683.
17. C. L. Jopling, M. Yi, A. M. Lancaster, S. M. Lemon and P. Sarnow, *Science*, 2005, **309**, 1577–1581.
18. P. T. Nelson and W. X. Wang, *J. Alzheimer's Dis.*, 2010, **21**, 75–79.
19. H. Y. Ding, S. S. Zheng, D. Garcia-Ruiz, D. X. Hou, Z. Wei, Z. C. Liao, L. M. Li, Y. J. Zhang, X. Han, K. Zen, C. Y. Zhang, J. Li and X. H. Jiang, *Nat. Commun.*, 2016, **7**, 17.
20. A. W. Tong and J. Nemunaitis, *Cancer Gene Ther.*, 2008, **15**, 341.
21. L. Ma, J. Teruya-Feldstein and R. A. Weinberg, *Nature*, 2007, **449**, 682–688.
22. Q. Guo, J. Zhang, Z. Zheng, X. Li, F. Wang and S. Liu, *J. Cell. Physiol.*, 2019, DOI: 10.1002/jcp.29043.
23. J. Huang, A. E. Snook, J. Uitto and Q. Li, *J. Invest. Dermatol.*, 2019, **139**, 1254–1263.
24. G. R. Jayandharan and B. Balakrishnan, *Curr. Gene Ther.*, 2014, **14**, 86–100.
25. C. Pinto, G. Silva, A. S. Ribeiro, M. Oliveira, M. Garrido, V. S. Bandeira, A. Nascimento, A. S. Coroadinha, C. Peixoto, A. Barbas, J. Paredes, C. Brito and P. M. Alves, *J. Biotechnol.*, 2019, **300**, 70–77.
26. D. G. Miller, P. R. Wang, L. M. Petek, R. K. Hirata, M. S. Sands and D. W. Russell, *Nat. Biotechnol.*, 2006, **24**, 1022–1026.

27. M. F. Leber, M. A. Baertsch, S. C. Anker, L. Henkel, H. M. Singh, S. Bossow, C. E. Engeland, R. Barkley, B. Hoyler, J. Albert, C. Springfeld, D. Jager, C. von Kalle and G. Ungerechts, *Mol. Ther.-Oncolytics*, 2018, **9**, 30–40.
28. N. Jiang, C. Zhang, J. Y. Liu, Z. H. Guo, Z. Y. Zhang, C. G. Han and Y. Wang, *Plant Biotechnol. J.*, 2019, **17**, 1302–1315.
29. T. Mascia, F. Nigro, A. Abdallah, M. Ferrara, A. De Stradis, R. Faedda, P. Palukaitis and D. Gallitelli, *Proc. Natl. Acad. Sci. U. S. A.*, 2014, **111**, 4291–4296.
30. Y. Tian, M. Zhou, H. Shi, S. Gao, G. Xie, M. Zhu, M. Wu, J. Chen and Z. Niu, *Nano Lett.*, 2018, **18**, 5453–5460.
31. U. Lächelt and E. Wagner, *Chem. Rev.*, 2015, **115**, 11043–11078.
32. M. A. Mintzer and E. E. Simanek, *Chem. Rev.*, 2008, **109**, 259–302.
33. S. H. Huh, H. J. Do, H. Y. Lim, D. K. Kim, S. J. Choi, H. Song, N. H. Kim, J. K. Park, W. K. Chang, H. M. Chung and J. H. Kim, *Biologicals*, 2007, **35**, 165–171.
34. D. W. Pack, A. S. Hoffman, S. Pun and P. S. Stayton, *Nat. Rev. Drug Discovery*, 2005, **4**, 581–593.
35. J.-P. Behr, *CHIMIA Int. J. Chem.*, 1997, **51**, 34–36.
36. W. Godbey, K. K. Wu and A. G. Mikos, *J. Biomed. Mater. Res.*, 1999, **45**, 268–275.
37. S. M. Moghimi, P. Symonds, J. C. Murray, A. C. Hunter, G. Debska and A. Szewczyk, *Mol. Ther.*, 2005, **11**, 990–995.
38. C. Dufes, I. F. Uchegbu and A. G. Schätzlein, *Adv. Drug Deliverery Rev.*, 2005, **57**, 2177–2202.
39. Y. C. Zhou, Z. X. Zhou and Y. Q. Shen, *Acta Polym. Sin.*, 2017, **16295**, 359–366.
40. R. Duncan and L. Izzo, *Adv. Drug Delivery Rev.*, 2005, **57**, 2215–2237.
41. J. Haensler and F. C. Szoka Jr, *Bioconjugate Chem.*, 1993, **4**, 372–379.
42. M. Thanou, B. Florea, M. Geldof, H. Junginger and G. Borchard, *Biomaterials*, 2002, **23**, 153–159.
43. O. Germershaus, S. Mao, J. Sitterberg, U. Bakowsky and T. Kissel, *J. Controlled Release*, 2008, **125**, 145–154.
44. D. Lee, W. Zhang, S. A. Shirley, X. Kong, G. R. Hellermann, R. F. Lockey and S. S. Mohapatra, *Pharm. Res.*, 2007, **24**, 157–167.
45. M. Köping-Höggård, I. Tubulekas, H. Guan, K. Edwards, M. Nilsson, K. M. Vårum and P. Artursson, *Gene Ther.*, 2001, **8**, 1108–1121.
46. M. Huang, C.-W. Fong, E. Khor and L.-Y. Lim, *J. Controlled Release*, 2005, **106**, 391–406.
47. S. Mao, W. Sun and T. Kissel, *Adv. Drug Delivery Rev.*, 2010, **62**, 12–27.
48. I. Park, T. Kim, Y. Park, B. Shin, E. Choi, E. Chowdhury, T. Akaike and C. Cho, *J. Controlled Release*, 2001, **76**, 349–362.
49. K.-L. Chang, Y. Higuchi, S. Kawakami, F. Yamashita and M. Hashida, *Bioconjugate Chem.*, 2010, **21**, 1087–1095.
50. K. Wong, G. Sun, X. Zhang, H. Dai, Y. Liu, C. He and K. W. Leong, *Bioconjugate Chem.*, 2006, **17**, 152–158.

51. X. Zhao, Z. Li, H. Pan, W. Liu, M. Lv, F. Leung and W. W. Lu, *Acta Biomater.*, 2013, **9**, 6694–6703.
52. S. C. Semple, A. Akinc, J. Chen, A. P. Sandhu, B. L. Mui, C. K. Cho, D. W. Sah, D. Stebbing, E. J. Crosley and E. Yaworski, *Nat. Biotechnol.*, 2010, **28**, 172–176.
53. C. H. Jones, C.-K. Chen, A. Ravikrishnan, S. Rane and B. A. Pfeifer, *Mol. Pharmaceutics*, 2013, **10**, 4082–4098.
54. K. Zhang, Y. Zhang, X. Meng, H. Lu, H. Chang, H. Dong and X. Zhang, *Biomaterials*, 2018, **185**, 301–309.
55. B. Dalby, S. Cates, A. Harris, E. C. Ohki, M. L. Tilkins, P. J. Price and V. C. Ciccarone, *Methods*, 2004, **33**, 95–103.
56. D. Lasic and N. Templeton, *Adv. Drug Delivery Rev.*, 1996, **20**, 221–266.
57. X.-z. Gao and L.-b. Huang, *Gene Ther.*, 1995, **2**, 710–722.
58. D. Hoekstra, J. Rejman, L. Wasungu, F. Shi and I. Zuhorn, *Biochem. Soc. Trans.*, 2007, **35**, 68–71.
59. R. Koynova and B. Tenchov, *Nucleic Acid Transfection*, Springer, 2010, pp. 51–93.
60. D. Zhi, S. Zhang, S. Cui, Y. Zhao, Y. Wang and D. Zhao, *Bioconjugate Chem.*, 2013, **24**, 487–519.
61. R. Koynova, B. Tenchov, L. Wang and R. C. MacDonald, *Mol. Pharmaceutics*, 2009, **6**, 951–958.
62. S. Chesnoy and L. Huang, *Annu. Rev. Biophys. Biomol. Struct.*, 2000, **29**, 27–47.
63. J. H. Felgner, R. Kumar, C. Sridhar, C. J. Wheeler, Y. J. Tsai, R. Border, P. Ramsey, M. Martin and P. L. Felgner, *J. Biol. Chem.*, 1994, **269**, 2550–2561.
64. Z. Zhou, X. Liu, D. Zhu, Y. Wang, Z. Zhang, X. Zhou, N. Qiu, X. Chen and Y. Shen, *Adv. Drug Delivery Rev.*, 2017, **115**, 115–154.
65. P. Tam, M. Monck, D. Lee, O. Ludkovski, E. Leng, K. Clow, H. Stark, P. Scherrer, R. Graham and P. Cullis, *Gene Ther.*, 2000, **7**, 1867–1874.
66. S.-D. Li and L. Huang, *Mol. Pharmaceutics*, 2006, **3**, 579–588.
67. K. Hadinoto, A. Sundaresan and W. S. Cheow, *Eur. J. Pharm. Biopharm.*, 2013, **85**, 427–443.
68. H. Yin, R. L. Kanasty, A. A. Eltoukhy, A. J. Vegas, J. R. Dorkin and D. G. Anderson, *Nat. Rev. Genet.*, 2014, **15**, 541–555.
69. N. Somia and I. Verma, *Nat. Rev. Genet.*, 2000, **1**, 91–99.
70. N. Erathodiyil and J. Y. Ying, *Acc. Chem. Res.*, 2011, **44**, 925–935.
71. W. K. Rhim, J. S. Kim and J. M. Nam, *Small*, 2008, **4**, 1651–1655.
72. X. Zhou, X. Zhang, X. Yu, X. Zha, Q. Fu, B. Liu, X. Wang, Y. Chen, Y. Chen, Y. Shan, Y. Jin, Y. Wu, J. Liu, W. Kong and J. Shen, *Biomaterials*, 2008, **29**, 111–117.
73. C. Hu, Q. Peng, F. Chen, Z. Zhong and R. Zhuo, *Bioconjugate Chem.*, 2010, **21**, 836.
74. F. Degliangeli, P. Kshirsagar, V. Brunetti, P. P. Pompa and R. Fiammengo, *J. Am. Chem. Soc.*, 2014, **136**, 2264–2267.

75. F. Xu, H. Dong, Y. Cao, H. Lu, X. Meng, W. Dai, X. Zhang, K. A. Al-Ghanim and S. Mahboob, *ACS Appl. Mater. Interfaces*, 2016, **8**, 33499–33505.
76. D. He, K.-W. Wong, Z. Dong and H.-W. Li, *J. Mater. Chem. B*, 2018, **6**, 7773–7793.
77. H. Dong, J. Zhang, H. Ju, H. Lu, S. Wang, S. Jin, K. Hao, H. Du and X. Zhang, *Anal. Chem.*, 2012, **84**, 4587–4593.
78. E. Keles, Y. Song, D. Du, W. J. Dong and Y. Lin, *Biomater. Sci.*, 2016, **4**, 1291–1309.
79. S. R. Ryoo, J. Lee, J. Yeo, H. K. Na, Y. K. Kim, H. Jang, J. H. Lee, W. H. Sang, Y. Lee and V. N. Kim, *ACS Nano*, 2013, **7**, 5882–5891.
80. K. Kalantar-zadeh, J. Z. Ou, T. Daeneke, M. S. Strano, M. Pumera and S. L. Gras, *Adv. Funct. Mater.*, 2015, **25**, 5086–5099.
81. Q. H. Wang, K. Kalantar-Zadeh, A. Kis, J. N. Coleman and M. S. Strano, *Nat. Nanotechnol.*, 2012, **7**, 699–712.
82. W. Yin, L. Yan, J. Yu, G. Tian, L. Zhou, X. Zheng, X. Zhang, Y. Yong, J. Li and Z. Gu, *ACS Nano*, 2014, **8**, 6922–6933.
83. T. Liu, C. Wang, X. Gu, H. Gong, L. Cheng, X. Shi, L. Feng, B. Sun and Z. Liu, *Adv. Mater.*, 2014, **26**, 3433–3440.
84. W. Dai, H. Dong, B. Fugetsu, Y. Cao, H. Lu, X. Ma and X. Zhang, *Small*, 2015, **11**, 4158–4164.
85. W. Fan, W. Bu, B. Shen, Q. He, Z. Cui, Y. Liu, X. Zheng, K. Zhao and J. Shi, *Adv. Mater.*, 2015, **27**, 4155–4161.
86. W. Tao, X. Zhu, X. Yu, X. Zeng, Q. Xiao, X. Zhang, X. Ji, X. Wang, J. Shi, H. Zhang and L. Mei, *Adv. Mater.*, 2017, **29**.
87. Q. Xi, D. M. Zhou, Y. Y. Kan, J. Ge, Z. K. Wu, R. Q. Yu and J. H. Jiang, *Anal. Chem.*, 2014, **86**, 1361–1365.
88. Z. Wu, G.-Q. Liu, X.-L. Yang and J.-H. Jiang, *J. Am. Chem. Soc.*, 2015, **137**, 6829–6836.
89. L. He, D. Q. Lu, H. Liang, S. T. Xie, X. B. Zhang, Q. L. Liu, Q. Yuan and W. H. Tan, *J. Am. Chem. Soc.*, 2018, **140**, 258–263.
90. C. Wu, S. Cansiz, L. Zhang, I.-T. Teng, L. Qiu, J. Li, Y. Liu, C. Zhou, R. Hu, T. Zhang, C. Cui, L. Cui and W. Tan, *J. Am. Chem. Soc.*, 2015, **137**, 4900–4903.
91. W. Pan, B. Liu, X. N. Gao, Z. Z. Yu, X. H. Liu, N. Li and B. Tang, *Nanoscale*, 2018, **10**, 14264–14271.
92. M. Ou, J. Huang, X. Yang, K. Quan, Y. Yang, N. Xie and K. Wang, *Chem. Sci.*, 2016, **8**, DOI: 10.1039.C1036SC03162E.
93. B. Wang, Z. Chen, D. Ren and Z. You, *Sens. Actuators, B*, 2019, **279**, 342–350.
94. T. Xie, M. Li and Y. T. Long, *Chem. Commun.*, 2017, **53**, 7768–7771.
95. L. S. Lin, Z. X. Cong, J. B. Cao, K. M. Ke and X. Chen, *ACS Nano*, 2014, **8**, 3876.
96. Y. Cen, W. J. Deng, Y. Yang, R. Q. Yu and X. Chu, *Anal. Chem.*, 2017, **89**, 10321–10328.

97. C. K. Choi, J. Li, K. Wei, Y. J. Xu, L. W. Ho, M. Zhu, K. K. To, C. H. Choi and L. Bian, *J. Am. Chem. Soc.*, 2015, **137**, 7337–7346.
98. J. Hemphill and A. Deiters, *J. Am. Chem. Soc.*, 2013, **135**, 10512–10518.
99. C. P. Liang, P. Q. Ma, H. Liu, X. Guo, B. C. Yin and B. C. Ye, *Angew. Chem., Int. Ed.*, 2017, **56**, 9077–9081.
100. D. Li, Y. Wu, C. Gan, R. Yuan and Y. Xiang, *Nanoscale*, 2018, DOI: 10.1039/c8nr05229h.
101. Y. Wang, P. D. Howes, E. Kim, C. D. Spicer, M. R. Thomas, Y. Lin, S. W. Crowder, I. J. Pence and M. M. Stevens, *ACS Appl. Mater. Interfaces*, 2018, **10**, 28290–28300.
102. F. Yang, Y. Cheng, Y. Cao, H. Dong, H. Lu, K. Zhang, X. Meng, C. Liu and X. Zhang, *Chem. Sci.*, 2019, **10**, 1709–1715.
103. C. Xue, S. X. Zhang, C. H. Ouyang, D. Chang, B. J. Salena, Y. Li and Z. S. Wu, *Angew. Chem., Int. Ed. Engl.*, 2018, **57**, 9739–9743.
104. X. D. Meng, K. Zhang, W. H. Dai, Y. Cao, F. Yang, H. F. Dong and X. J. Zhang, *Chem. Sci.*, 2018, **9**, 7419–7425.

CHAPTER 4

In Situ *Analysis of Intracellular Telomerase Activity*

4.1 Clinical Significance of Telomerase

In most eukaryotes, the terminal DNA sequence at the end of the chromosome consists of a bunch of simple repeats.[1] These repeats, along with the associated proteins, constitute special complexes called the telomere, which provides a protective cap to prevent the degradation of the chromosomes.[2] Hence the telomere is essential for stabilizing the ends of chromosomes.[3] The repeated sequence varies between different species.[4,5] In human cells, the repeated sequence is a hexamer, TTAGGG. The linear chromosomal DNAs need a special mechanism to maintain the replication of their repeated ends.[1] Telomerase, as a ribonucleoprotein reverse transcriptase, is responsible for maintaining the length of the telomeres by adding the repeated sequence to the termini of the telomeres.[4] In ciliated protozoa and yeasts, the average length is maintained and the expression of telomerase is properly regulated. In contrast, in most human somatic cells, the expression of telomerase is extensively inhibited as the activity of telomerase cannot be detected and the telomeres are shortened with successive cell divisions.[5] However, in human cancer cells and in immortalized cell lines, telomerase activity is obviously upregulated, which has attracted considerable attention from researchers regarding the study of the potential of telomerase for cancer theranostics.

Telomerase was first identified in 1985 by Greider and Blackburn.[6] This ribonucleoprotein enzyme contains a single RNA subunit and the associated protein domain. The core components of human telomerase include a 451 bp integral RNA (human telomerase RNA, hTR), dyskerin (DKC1) and human telomerase reverse transcriptase (hTERT).[7] Telomerase uses the RNA subunit as the template to extend the chromosome ends by producing single-strand repeats. The working principle of human telomerase is shown in Figure 4.1. First,

Detection Science Series No. 16
In Situ Analysis of Cellular Functional Molecules
By Huangxian Ju, Bo Tang and Lin Ding

Published by the Royal Society of Chemistry, www.rsc.org

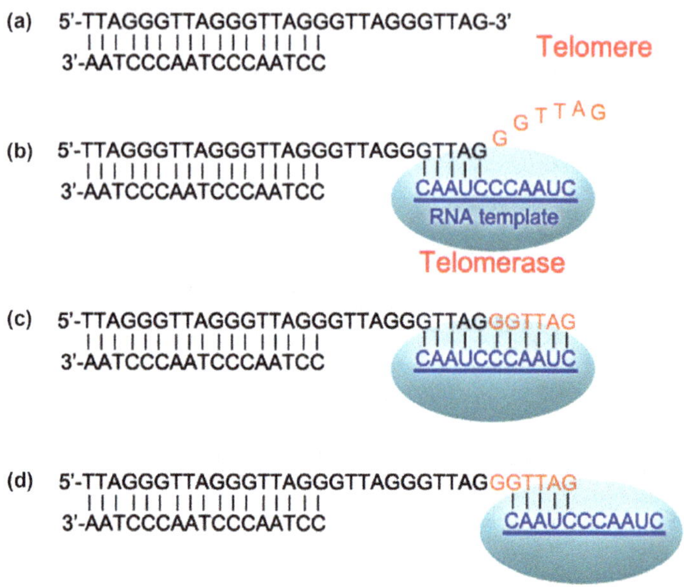

Figure 4.1 Working principle of human telomerase. (a) The structure of telomere; (b) and (c) bases added based on the RNA template; (d) relocation of telomerase.

the RNA template of telomerase catches the 3′ end of the single-strand telomere (the single stand can also be an artificially synthesized DNA primer). Then a hybrid structure is formed at the 3′ end of the telomere and the extension of the 3′ end to the template is achieved. Six nucleotides are then added to the 3′ end of the telomere. Subsequently, the telomerase moves along the telomere strand towards the newly formed 3′ end and this elongation process continues as the single strand of the telomere extends longer and longer. Normally, more than one repeat will be added to the 3′ end of the telomere. By subsequent reverse transcription, the telomere returns to the double-strand form.

Experimental evidence shows that cancer cells have a very long lifespan, with the ability to proliferate indefinitely.[8] In contrast, normal cells can divide for only a limited number of cycles. The reason for cellular ageing is that the 3′ end of a linear DNA strand cannot be copied perfectly without any loss.[9] During the process of cell division, telomeres shorten by more than 50 bp owing to the end replication problem caused by DNA polymerases.[10] After several rounds of cell division, the telomeres become shorter and shorter, until eventually the cells cannot divide any more. In this case, these cells become senescent. In most cases, cell growth stops before the arrival of a cancerous mutation. However, in cancer cells, the length of the telomere can be maintained by activated telomerase. Therefore, the growth arrest caused by the loss of telomere could be used for cancer therapy.[11]

Since there is a very strong connection between telomerase activity and malignancy in almost all types of cancer, telomerase has been considered as a prominent cancer biomarker for cancer diagnosis and a potential target for cancer therapy.[12] With the significant developments in biomedical research on telomerase, more and more bioanalytical strategies for the detection of telomerase activity have been established. In the following part of this chapter, we introduce some of the useful telomerase assays and discuss their advantages and limitations. Finally, we discuss the challenges that remain in developing rapid, sensitive and highly efficient detection methods for testing telomerase activity. In particular, we are especially interested in designing telomerase detection methods for the *in situ* analysis of telomerase activity at the single-cell level, which may be of help in elucidating the heterogeneity of telomerase expression in different cells, and thus improving our understanding of telomerase-related cell biology and pathways.

4.2 Traditional Analytical Methods for Telomerase

After Greider and Blackburn's discovery of telomerase in 1985,[6] in the following 5 years telomerase came to be recognized as an exciting cancer diagnostic and prognostic biomarker and also a promising cancer therapeutic target.[13–15] Subsequently, sensitive assays for telomerase analysis were rapidly developed, and the progress in assays for telomerase detection will be of great help in exploring the complex relationship between telomerase and cancer.

The amount of telomerase in human cells is very low. Even in cancer cells, the content of telomerase is less than 100 molecules per cell.[16] The earliest methods required a large amount of sample or had to use radioactive materials, which limited the development of telomerase detection methods.[17] Subsequently, amplification-based methods for telomerase detection were introduced. The TRAP (telomere repeat amplification protocol) assay, originally developed by Kim *et al.* in 1994,[18] is a typical amplification-based method. Using the TRAP method, the activity of telomerase could be obtained from just a few cells. The working principle of TRAP is shown in Figure 4.2. To improve the sensitivity of detection, the TRAP method is mainly based on the polymerase chain reaction (PCR) principle. In the TRAP assay, telomerase adds telomeric repeats to the 3′ end of the non-telomeric oligonucleotide primer (TS), then the elongated products are amplified by the downstream primer (CX) *via* PCR cycling. Using an immortal cell extract, a positive signal can be detected by the TRAP assay.

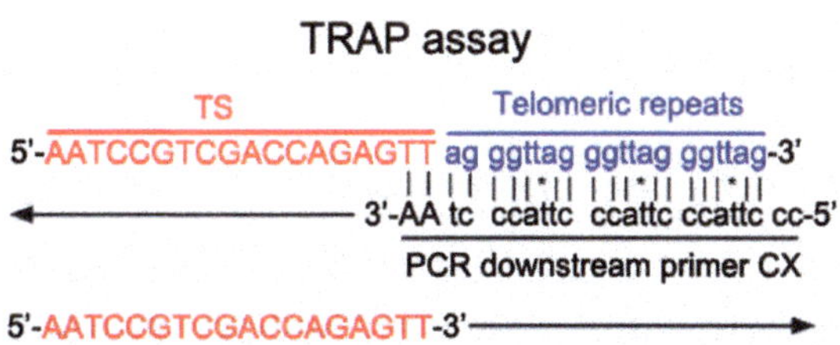

Figure 4.2 Working principle of the TRAP assay.

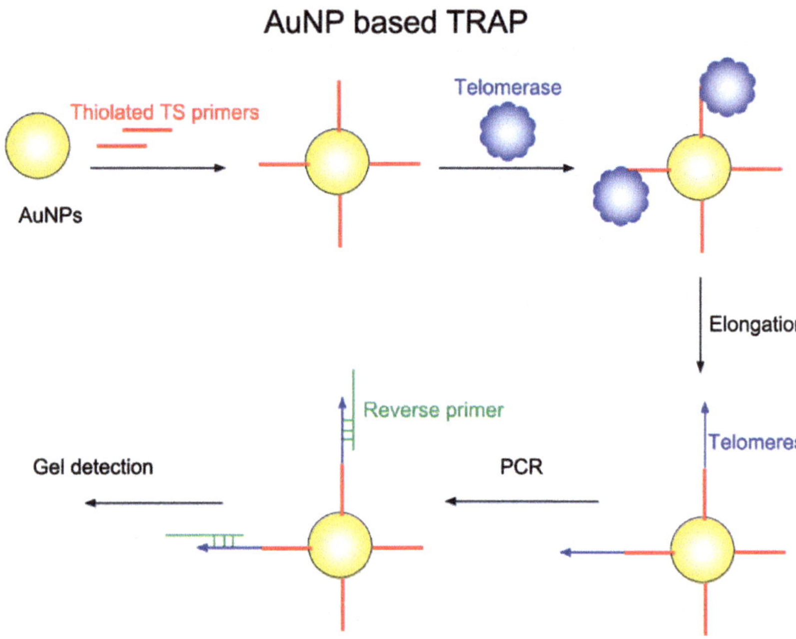

Figure 4.3 Working principle of the AuNP-based TRAP assay.

However, the original TRAP method is time consuming and suffers from PCR-related mistakes. To make things worse, impurities from the tissue extract may inhibit the TRAP process. These issues have been resolved to a certain extent by recent studies that make TRAP-based methods more flexible and less expensive.

The detection limit and specificity of the traditional TRAP assay are not satisfactory. To resolve these problems, nanotechnology has been applied to improve the TRAP method. For example, Xiao *et al.* used gold nanoparticles (AuNPs) as the core and then functionalized the surface of the AuNPs with TS oligonucleotide primer to substitute the traditional single-strand TS primer (Figure 4.3).[19] This AuNP-based TRAP method could greatly improve the specificity owing to the elimination of primer–dimer interference. Using this improved TRAP method, telomerase activity could be conveniently detected from only 50 positive MCF-7 breast cancer cells. A possible explanation of the improvement is that the protein in cell lysate can be absorbed by the AuNP owing to the high ratio of surface area to volume, hence the PCR reaction is greatly promoted.

4.3 *In Situ* Imaging of Intracellular Telomerase Activity Based on Functional Nanoprobes

The TRAP-based methods use cell lysate for the detection of telomerase activity and thus fail to provide *in situ* information on telomerase activity in single cells. To solve this problem, functional nanomaterials were applied to establish an intracellular detection method because of their powerful ability to construct multifunctional platforms for intracellular study.

Nanomaterials have shown a strong ability to construct delivery platforms for intracellular research. Mesoporous silica nanoparticles (MSNs) are especially attractive owing to their unique pore structure, biocompatibility and ease of functionalization. In particular, the large pore volume and surface area of MSNs make it possible to load large numbers of functional molecules.[20]

Ju's group developed a telomerase-responsive MSN to realize *in situ* "off–on" imaging of intracellular telomerase activity.[20] As shown in Figure 4.4, the MSN is used to build a telomerase-responsive probe. A specially designed

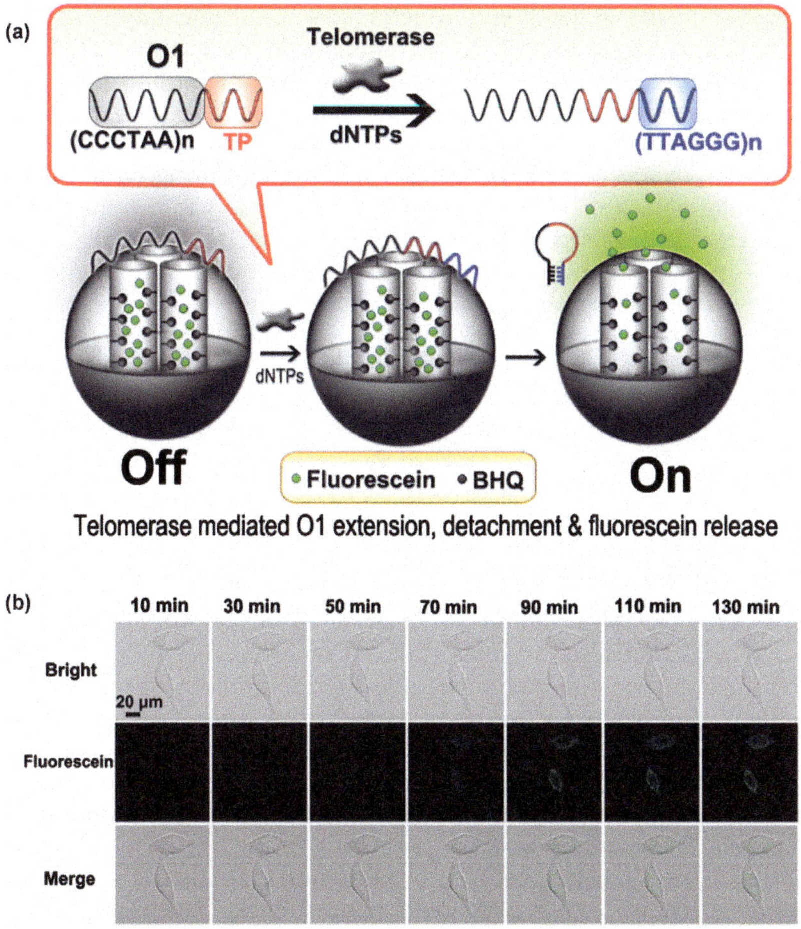

Figure 4.4 (a) Schematic illustration of MSN probe-based detection of telomerase. (b) Confocal image of HeLa cells incubated with 15 μL of MSN probe (1 mg mL^{-1}) after different times.
Reproduced from ref. 20 with permission from American Chemical Society, Copyright 2013.

wrapping DNA (O1) is attached to the surface of the MSN, which works like a bio-gate to block the pores. In the presence of telomerase, the 3′ end of O1 is elongated with telomeric repeats, leading to the detachment of O1 from the MSN probe surface. Fluorescein is filled in the pores of MSN and black hole fluorescence quencher (BHQ) molecules are covalently immobilized on the inner walls. Therefore, the detachment of O1 leads to the release of fluorescein, hence the fluorescence is turned "on". The sequence of O1 is designed as 5′-(CCCTAA)$_n$ AATCCG TCGAGC AGAGTT-3′, where the underlined part is the telomerase primer (TP). In the absence of telomerase, the O1-sealed MSN probe is in its "off" state owing to the quenching effect of the immobilized BHQ on the fluorescence of fluorescein. In the presence of telomerase, the O1 sequence is elongated in the presence of dNTPs, adding a telomeric repeated sequence at the 3′ end, which is just complementary to its 5′ end as designed. The hybridization of the complementary sequences at the 5′ and 3′ ends of O1 will produce a rigid hairpin-like DNA structure and then leads to the detachment of the extended O1 from the MSN surface. Thus the released fluorescein will turn the fluorescence "on". Using human cervical cells (HeLa) as a model, *in situ* imaging of telomerase activity in living HeLa cells was successfully achieved. Microscopic images showed that the fluorescence signals are mainly observed in the cytoplasm around the nucleus. The MSN probes can also be applied to distinguishing cancer cells from normal cells. Greater telomerase activity in BEL-7402 liver cancer cells compared with QSG-7701 normal cells can be observed.

In conclusion, owing to the excellent delivery ability of nanomaterials, we anticipate that nanoparticle-based methods will be of help in understanding the biological roles of telomerase in cancer pathways and provide powerful platforms for *in vivo* cancer diagnostics and therapy.

4.4 *In Situ* Imaging of Intracellular Telomerase Activity Through FRET

Although the previously reported telomerase-responsive MSN probe realizes the non-invasive analysis of intracellular telomerase, its application is limited owing to the complicated preparation and multiple-stage response. The sensitivity and specificity are also affected by the non-specific release of fluorescein. To achieve the highly sensitive and specific *in situ* imaging of intracellular telomerase activity with minimal background noise, the fluorescence resonance energy transfer (FRET) effect is used to reduce the background fluorescence signal.

The FRET phenomenon between two different molecules is an important physical effect, with great potential for developing biosensors. Recently, FRET has been widely used in biomedical analysis and drug discovery. The mechanism of FRET involves a pair consisting of a donor fluorophore and a nearby acceptor chromophore. As the donor fluorophore is in an excited state, it can transfer the excitation energy to the nearby acceptor *via* dipole–dipole interaction. The resonance energy transfer is similar to coupled

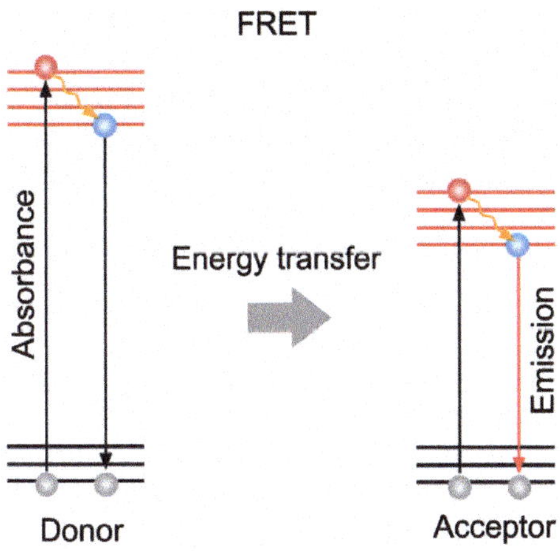

Figure 4.5 Jablonski diagram showing the FRET process.

oscillators vibrating at the same frequency, where an excited donor fluorophore is treated as an oscillating dipole exchanging energy with the acceptor chromophore having a similar frequency. Therefore, a pair of molecules with the FRET effect is called a donor–acceptor pair.[21,22] Usually, under excitation with incident light, a donor fluorophore can absorb energy and transfer the excitation energy to the acceptor as follows:

$$D + h\nu \rightarrow D^*$$

$$D^* + A \rightarrow D + A^*$$

$$A^* \rightarrow A + h\nu'$$

where D is the donor and A is the acceptor.

As shown in Figure 4.5, the coupled energy transfer from the donor to the acceptor in FRET can be illustrated by a Jablonski diagram. By choosing a proper donor and acceptor, the excited-state energy from the donor can be directly transferred to the acceptor without loss. Owing to the particular physical properties of FRET, it can therefore be used to monitor various biomarkers or biopathways in living cells, which is extremely favourable for the *in situ* imaging analysis of intracellular telomerase.

4.4.1 FRET Based on Organic Quenchers

Based on FRET, Ju's group designed an intracellular cascade amplification method for the sensitive imaging of intracellular telomerase activity.[23] Inspired by the DNA nature of telomerase primer, this method for the *in situ* detection of telomerase activity was designed to combine the biological function of telomerase with isothermal nucleic acid-based signal amplification. To overcome the difficulties in forming a linkage of the telomerase extension function with the DNA assembly and to translate specifically the telomerase activity to a DNA signal without interference from complex intracellular components, a cascade approach was developed for the visualization of telomerase activity in living cells by linking telomerase-responsive primer extension with catalysed hairpin assembly (CHA)-based signal amplification in series through a DNA transmitter. As shown in Figure 4.6, a liposome nanoprobe was designed that encapsulated two function modules: a telomerase-targeting responder–transmitter DNA complex (HPT) module and a CHA signal amplification module. For the HPT part, a nicked hairpin DNA structure is composed of a telomerase primer-tagged segment (HP) and a transmitter segment (T) from the 5′ to the 3′ end. The 3′-end of HP along with T is designed to be complementary to the 5′ end of HP, forming the stem part of the hairpin. The sequence of HP is 5′-AACCCT AACCCT AACTCT GCTCGA CGGATT CCCCCCCCCC AATCCG TCGAGC AGAGTT-3′, where the underlined part represents the telomerase primer (TP). The second part contains a pair of metastable DNA hairpins (H1, H2) and a hybridized reporter complex (R-FQ) with strands respectively

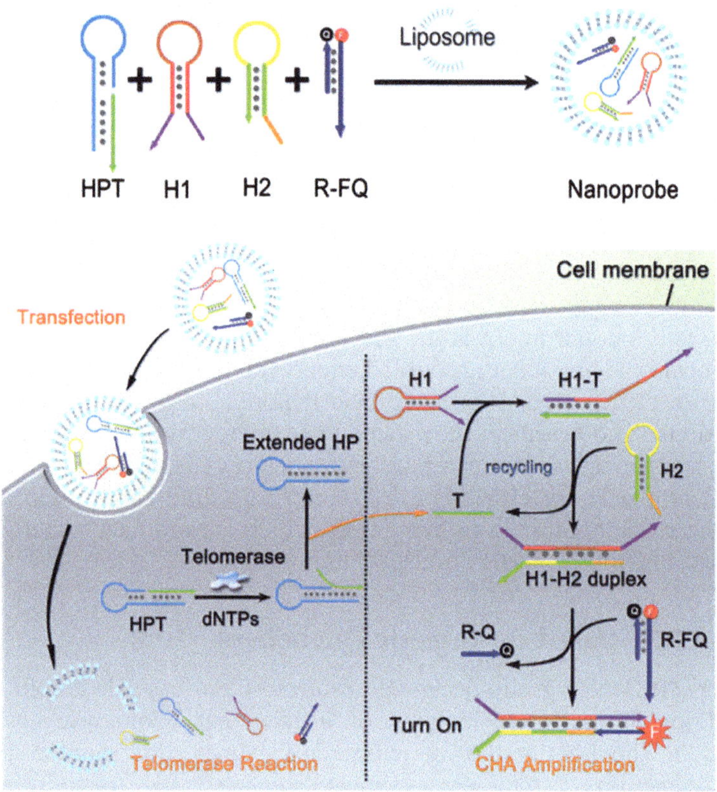

Figure 4.6 Schematic illustration of the nanoprobe fabrication and the cascade strategy for telomerase imaging in living cells.
Reproduced from ref. 23 with permission from Elsevier, Copyright 2016.

labelled with a fluorophore (F, FAM) and quencher (Q, Dabcyl), thus constituting a CHA configuration. Using liposomes as the transport cargo, the functional parts can be conveniently transfected into living cells and then the HPT released can be recognized by telomerase to elongate the 3′ end. Subsequently, the transmitter DNA is released, which acts as a catalyst of CHA amplification. Hence the telomerase activity can be imaged *in situ* with high sensitivity. The proposed strategy can be further employed for monitoring the change in intracellular telomerase activity in response to drugs to determine drug efficacy. This method can be also used to compare telomerase activity among different cell lines and to differentiate normal cells from tumour cells.

4.4.2 FRET Based on AuNPs

In addition to FRET based on organic molecular quenchers, AuNP-based FRET methods have also been developed for intracellular telomerase imaging. Ju's group designed a robust AuNP probe functionalized by a nicked molecular beacon (MB).[24] As shown in Figure 4.7, AuNPs are used as the substrate for immobilization and protection of DNA strands and also the quencher for FRET. The specially designed MB contains a nick on the 5′ end stem, which divides the MB into two parts: a shorter telomerase primer (TSP) and a longer sequence (l-MB) for the construction of a loop with a thiol-functionalized 3′ end for modification to the AuNP. The fluorescence of Cy5 at the 5′ end of l-MB is quenched by the AuNP core. In the presence of

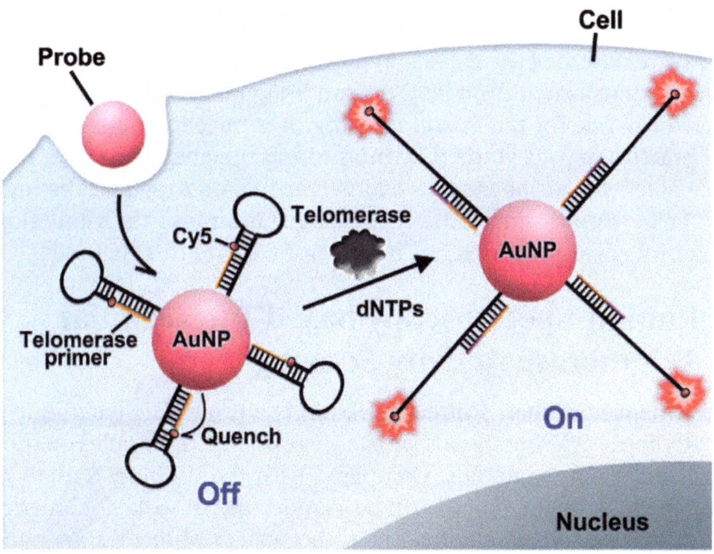

Figure 4.7 Schematic illustration of nicked molecular beacon-functionalized AuNPs for *in situ* analysis of intracellular telomerase.
Reproduced from ref. 24 with permission from American Chemical Society, Copyright 2014.

telomerase, the TSP is elongated, which can open the hairpin structure, leading to the departure of Cy5 from the AuNP. Thus the fluorescence is turned "on". This probe possesses synthetic convenience, intracellular stability, cytocompatibility and good specificity for the recognition of telomerase to trigger telomerase primer elongation on MB, which leads to inner chain substitution hybridization to open the hairpin and light up the telomerase. The practicality of the probe with recognition and enzymatic amplification functions has been demonstrated by dynamically monitoring the dose-dependent change in telomerase activity in response to a telomerase-related drug, *in situ* quantifying the intracellular telomerase activity and distinguishing different tumour cells and tumour cells from normal cells. This strategy uses a nicked MB to construct a novel functionalized probe for the *in situ* detection of intracellular telomerase activity at single-cell level. The designed probe possesses synthetic convenience, intracellular stability, cytocompatibility and good specificity for the recognition of telomerase. The triggered TSP elongation on MB leads to inner chain substitution hybridization to open the hairpin and lights up the telomerase in living cells. The fluorescence switch can be achieved in cytoplasm through one-step incubation of the living cells with the probe, which is extremely convenient and practical. We anticipate that this probe will accelerate the elucidation of the basic role of telomerase in various biological pathways and contribute to the clinical diagnosis and therapeutic monitoring of cancer.

Ju's group also designed a smart vesicle kit for *in situ* telomerase detection based on AuNP.[25] The vesicle kit encapsulates a telomerase primer (TSP) and a Cy5-tagged molecular beacon (MB) functionalized AuNP (probe). In the presence of telomerase and dNTPs, the TSP can be extended to open the loop of the MB by hybridization of the extension product with the loop of MB, leading to a fluorescent state switch from "off" to "on" (Figure 4.8). This approach shows good performance for the *in situ* imaging of intracellular telomerase activity and its practicality was verified by monitoring the change in cytoplasmic telomerase activity in response to a telomerase-related drug and distinguishing tumour cells from normal cells. The understanding of the vital telomerase-involved biological processes has been improved *via* this method.

4.5 Raman Spectroscopy-based Intracellular Telomerase Activity Sensing

Surface-enhanced Raman scattering (SERS) is a promising technique that enhances Raman scattering signals *via* the absorption of molecules on the surface of rough metal nanomaterials. Compared with the ordinary Raman method, the enhancement factor of SERS can be as much as 10^{10}–10^{11}-fold higher owing to the strong electromagnetic field near the surface of metal substrates or the charge transfer between the Raman molecules and the metal substrates, thus providing an extremely sensitive platform for detecting very small amounts of molecules.[26] In addition, SERS can provide spectral fingerprint signatures of the molecules to be measured, which reduces the interference of non-specific

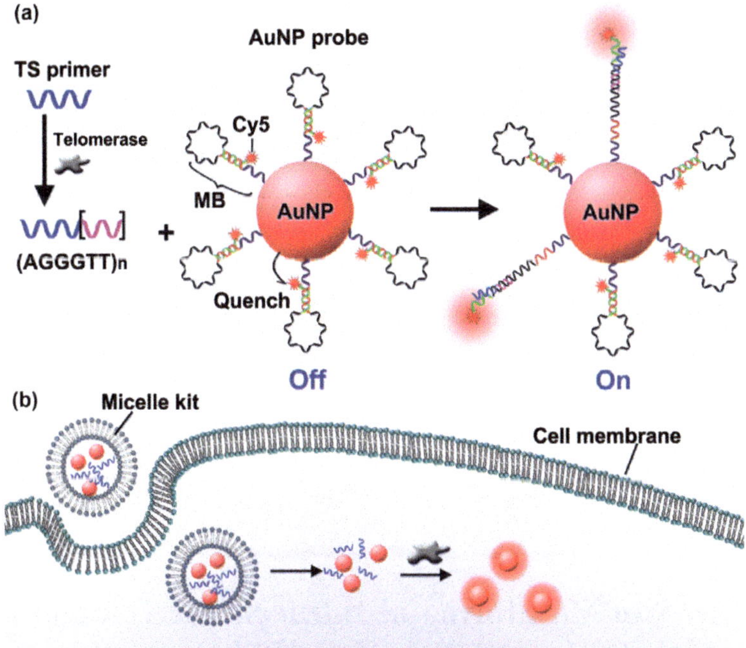

Figure 4.8 Schematic illustration of the intracellular analysis of telomerase with an MB-functionalized AuNP probe and TSP co-encapsulated liposome (vesicle kit). (a) Telomerase-triggered TSP elongation and subsequent fluorescence recovery of the probe. (b) Internalization of vesicle kit for the *in situ* detection of telomerase.
Reproduced from ref. 25 with permission from American Chemical Society, Copyright 2014.

molecules.[27] Based on the advances in SERS, this technique has been applied to the detection of telomerase activity in tumour cells.

Kuang's group used a nanoparticle assembly-based SERS method to detect the intracellular telomerase *in situ*.[28] They designed a pyramid "Py-SERS probe", using a Cy5 molecule as the SERS reporter. As shown in Figure 4.9, the DNA frames embedded with TS primer are applied for the preparation of AuNP pyramids. Four single-strand DNA (ssDNA, S1–4 for AuNP assembly) molecules are used, which are complemented by a Cy5-modified reporter sequence and TS primer. The SERS probes obtained are then used for telomerase detection. In the presence of telomerase and dNTPs, the TS primer can be elongated with TTAGGG repeats. The Cy5-modified reporter sequence is then replaced with the extended sequence, hence the Cy5-modified sequence is released from the pyramid, which reduces the Raman signal while the fluorescence signal increases. In K562 tumour cells, the designed SERS probe can reflect the intracellular telomerase activity. As a result, the detection limit of telomerase is 6.2×10^{-15} IU. Hence SERS realized the highly sensitive and *in situ* detection of telomerase in living cells.

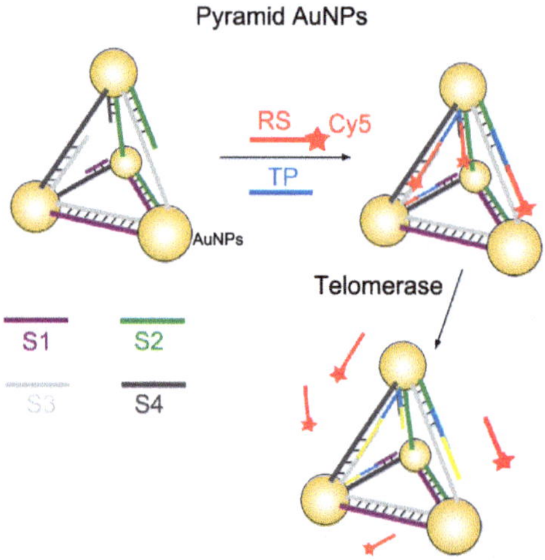

Figure 4.9 Schematic illustration of the pyramid-based intracellular detection of telomerase.

4.6 *In Situ* Monitoring of Intracellular Telomerase Activity Using AIEgen-based Bioprobes

Aggregation-induced emission (AIE) is a phenomenon discovered by Tang and co-workers[29–31] that involves a special class of dye molecules that show high luminescence when their molecules are aggregated. Luminogenic materials with AIE attributes have attracted much interest since their discovery in 2001. The AIE effect is completely different from the traditional aggregation-caused quenching (ACQ) effect. Traditionally, the luminescence of organic luminophores is often weakened or even quenched at high concentrations in solution, a phenomenon widely known as concentration quenching. The main cause of this quenching effect is associated with the formation of aggregates, which has frequently been referred to as ACQ.[29] The ACQ effect is common to most aromatic hydrocarbons and their derivatives because of their planar aromatic rings. In contrast, the AIE luminogens (AIEgens) are non-emissive under dispersed conditions but become highly luminescent when the molecules are aggregated in high concentration. The AIE effect is precisely the opposite of the ACQ effect, which makes it favourable for biochemical detection and cell imaging.

Zhuang *et al.* developed an AIEgen-based strategy for the *in situ* monitoring of intracellular telomerase activity in which a positively charged AIEgen (TPE-H) is used to link with silole-R for telomerase activity detection.[32] Using both the blue fluorescent silole-R and red fluorescent TPE-H, a two fluorescent signal-based telomerase detection strategy can be obtained with improved sensitivity and specificity (Figure 4.10). Silole-R is highly sensitive to the existence of telomerase. However, the emission wavelength of silole-R is 478 nm (blue emission), which could be disturbed

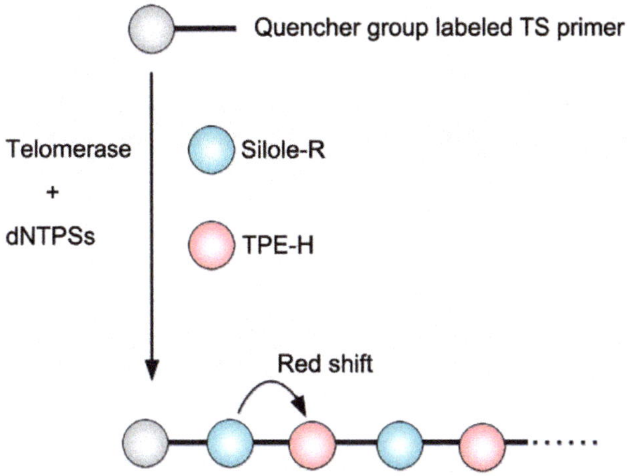

Figure 4.10 Schematic illustration of the two fluorescent signal-based bioprobe for the detection of extracellular and intracellular telomerase activity.

by the autofluorescence of cells or organisms. Hence the use of silole-R alone is not suitable for imaging telomerase activity in living cells. On the other hand, the red emission (605 nm) of TPE-H avoids interference from blue autofluorescence. With the energy transfer between silole-R and TPE-H, the two fluorescent signal-based biosensor shows a wider linear range and lower background for telomerase detection in both bulk solution and living cells.

4.7 Perspective

Because of the importance of telomerase, assays for its detection have shown a rapidly developing trend in recent years, resulting in the accumulation of abundant information on the design and fabrication of telomerase sensors. Tremendous progress has been achieved towards establishing strategies for the *in situ* imaging of intracellular telomerase activity. Despite the important achievements, the use of telomerase as a reliable cancer biomarker is still a challenge, and the existing telomerase detection assays still suffer from instability, low repeatability and low sensitivity. These aspects may become the driving force for telomerase research in future.

As a final remark, telomerase research is still challenging. However, since telomerase activity is closely related to pathologies such as cancer, age-related diseases and premature ageing, the potential application of telomerase-based theranostics is exciting and promising. A greater understanding of telomerase may lead to the realization of these aims and we hope that single-cell analysis methods for telomerase detection can point the way to future developments.

Acknowledgement

Dr Ruocan Qian is thanked for expert assistance with the writing of this chapter.

References

1. E. H. Blackburn, *Nature*, 2000, **408**, 53.
2. P. W. Hammond and T. R. Cech, *Nucleic Acids Res.*, 1997, **25**, 3698.
3. Y. Li, B. W. Liu, X. Li and Q. L. Wei, *Biosens. Bioelectron.*, 2010, **25**, 2543.
4. X. Zhou and D. Xing, *Chem. Soc. Rev.*, 2012, **41**, 4643.
5. T. M. Nakamura, G. B. Morin, K. B. Chapman, S. L. Weinrich, W. H. Andrews, J. Lingner, C. B. Harley and T. R. Cech, *Science*, 1997, **277**, 955.
6. C. W. Greider and E. H. Blackburn, *Cell*, 1985, **43**, 405.
7. C. Autexier and N. F. Lue, *Annu. Rev. Biochem.*, 2006, **75**, 493.
8. J. W. Shay, Y. Zou, E. Hiyama and W. E. Wright, *Hum. Mol. Genet.*, 2001, **10**, 677.
9. H. Vaziri and S. Benchimol, *Curr. Biol.*, 1998, **8**, 279.
10. A. G. Bodnar, M. Ouellette, M. Frolkis, S. E. Holt, C. P. Chiu, G. B. Morin, C. B. Harley, J. W. Shay, S. Lichtsteiner and W. E. Wright, *Science*, 1998, **279**, 349.
11. J. W. Shay, *Mol. Med. Today*, 1995, **1**, 378.
12. J. W. Shay and S. Bacchetti, *Eur. J. Cancer*, 1997, **33**, 787.
13. C. B. Harley, *Nat. Rev. Cancer*, 2008, **8**, 167.
14. N. D. Hastie, *Nature*, 1990, **346**, 866.
15. D. Broccoli, J. W. Young and T. D. Lange, *Proc. Natl. Acad. Sci. U. S. A.*, 1995, **92**, 9082.
16. S. B. Cohen, M. E. Graham, G. O. Lovrecz, N. Bache, P. J. Robinson and R. R. Reddel, *Science*, 2007, **315**, 1850.
17. G. B. Morin, *Cell*, 1989, **59**, 521.
18. N. W. Kim, M. A. Piatyszck, K. R. Prowse, C. B. Harley, M. D. West, P. L. Ho, G. M. Coviello, W. E. Wright, S. L. Weinrich and J. W. Shay, *Science*, 1994, **266**, 2011.
19. Y. Xiao, K. Y. Dane, T. Uzawa, A. Csordas, J. R. Qian, H. T. Soh, P. S. Daugherty, E. T. Lagally, A. J. Heeger and K. W. Plaxco, *J. Am. Chem. Soc.*, 2010, **132**, 15299.
20. R. Qian, L. Ding and H. Ju, *J. Am. Chem. Soc.*, 2013, **135**, 13282.
21. J. R. Lakowicz, *Principles of Fluorescence Spectroscopy*, Kluwer, New York, 2nd edn, 1999.
22. E. A. Jares-Erijman and T. M. Jovin, *Nat. Biotechnol.*, 2003, **21**, 1387.
23. L. Yan, J. Hui, Y. Liu, Y. Guo, L. Liu, L. Ding and H. Ju, *Biosens. Bioelectron.*, 2016, **86**, 1017.
24. R. Qian, L. Ding, L. Yan and H. Ju, *J. Am. Chem. Soc.*, 2014, **136**, 8205.
25. R. Qian, L. Ding, L. Yan and H. Ju, *Anal. Chem.*, 2014, **86**, 8642.
26. C. Wang, H. Yang, S. Wu, Y. Liu, W. Wei, Y. Zhang, M. Wei and S. Liu, *TrAC, Trends Anal. Chem.*, 2018, **105**, 404.
27. N. Guarrotxena, B. Liu, L. Fabris and G. C. Bazan, *Adv. Mater.*, 2010, **22**, 4954.

28. L. G. Xu, S. Zhao, W. Ma, X. L. Wu, S. Li, H. Kuang, L. B. Wang and C. L. Xu, *Adv. Funct. Mater.*, 2016, **26**, 1602.
29. J. Luo, Z. Xie, J. W. Y. Lam, L. Cheng, H. Chen, C. Qiu, H. S. Kwok, X. Zhan, Y. Liu, D. Zhu and B. Z. Tang, *Chem. Commun.*, 2001, 1740.
30. B. Z. Tang, X. Zhan, G. Yu, P. P. S. Lee, Y. Liu and D. Zhu, *J. Mater. Chem.*, 2001, **11**, 2974.
31. Y. Hong, J. W. Y. Lam and B. Z. Tang, *Chem. Soc. Rev.*, 2011, **40**, 5361.
32. Y. Zhuang, C. Shang, X. Lou and F. Xia, *Anal. Chem.*, 2017, **89**, 2073.

CHAPTER 5

In Situ Analysis of the Intracellular Caspase Family

5.1 Introduction to the Intracellular Caspase Family

Apoptosis is a naturally occurring and highly organized process of programmed cell death through which redundant, damaged and/or abnormal cells are eliminated from an organism to maintain cellular homeostasis.[1] Dysregulation of apoptosis contributes to the development and progression of various diseases, including cancer, autoimmune disorders, myocardial ischaemia and reperfusion injury, chronic heart failure and neurodegenerative disease.[2] Apoptosis is also a desired outcome of many cancer treatments, including irradiation,[3] chemotherapy,[4] gene therapy,[5] thermal therapy[6] and photodynamic therapy,[7] which are designed to clear the transformed cells by the induction of apoptosis. Therefore, observation of apoptosis can provide very valuable information on diseases in the living body and also on the therapeutic efficacy of drugs during treatment. In particular, direct visualization and quantification of apoptosis in cancer cells can be utilized for predicting anticancer efficacy and optimizing the selection of anticancer drugs.

Caspases are a family of cysteine–aspartyl-specific proteases that are essential for the execution of apoptosis and can be used as significant biomarkers of cell death. Under normal physiological conditions, caspases are synthesized as inactive procaspases. They can be activated in the presence of various extrinsic and/or intrinsic stimuli to induce apoptosis. Caspase-regulated apoptosis is a cascade process that involves multiple caspases at different stages and pathways.[8] To date, at least 14 mammalian caspases have been identified and caspases involved in apoptosis can be broadly divided into two groups: the initiator caspases (caspase-2, -8, -9 and -10) and the effector caspases (caspase-3, -6 and -7).[9] The initiator caspases are the

Detection Science Series No. 16
In Situ Analysis of Cellular Functional Molecules
By Huangxian Ju, Bo Tang and Lin Ding

Published by the Royal Society of Chemistry, www.rsc.org

first to be activated in a pathway by specific death stimuli from receptors, lysosome or the mitochondria. The fundamental function of the initiator caspases is to activate downstream effector caspases, which can then hydrolyse peptide bonds of aspartic acid residues of numerous regulatory and structural proteins within the cell. These cleavages lead to inactivation of protein substrates and initiate signalling pathways that promote the controlled disassembly of cells and finally cellular destruction and death.[10] Given the central position of caspases in the apoptotic pathway, the development of imaging methods for the *in situ* analysis of the caspase family is highly desirable for understanding the different biological processes and evaluating therapeutic efficacy in a wide range of preclinical applications.

5.2 Imaging of Caspase Activity in Living Cells

To achieve caspase imaging detection, optical imaging probes are the most appropriate choice. Compared with traditional imaging methods such as magnetic resonance imaging (MRI), X-ray radiography, computed tomography (CT), positron emission tomography (PET) and ultrasonography (US), optical imaging probes are inexpensive, non-radioactive and simple. Additionally, to improve the sensitivity and specificity, optical imaging probes can be designed to maximize the target signal while minimizing the background signal,[11] resulting in higher target-to-background ratios than in conventional imaging strategies. Although the optical imaging approach is emerging as a promising *in situ*, real-time and high-resolution modality,[12] there remain several significant challenges. The imaging effect of optical imaging probes depends greatly on the optical properties of fluorophores and also the signal conversion mechanisms. Many optical imaging probes for the *in situ* detection of caspase have been developed to meet the different requirements. In this chapter, we summarize several caspase optical imaging probes with different signal conversion mechanisms.

5.2.1 FRET-based Sensors for Imaging of Intracellular Caspase Activity

Fluorescence-based biosensors have become key players in the monitoring and identification of cellular molecular dynamics, cell physiology and cell-to-cell interactions since the advent of fluorescent proteins in 1992.[13] Fluorescence resonance energy transfer (FRET)-based sensors, involving radiationless transfer from a "donor" fluorophore to an "acceptor" fluorophore or quencher when the distance between donor and acceptor is within 1–10 nm are promising imaging tools for monitoring molecular dynamics in living cells. A variety of fluorescent probes bearing donor and acceptor molecules at the terminus of the caspase protein-cleavable segment could be developed to detect apoptosis in living cells through the FRET effect. Based on the above design strategy, Shaulov-Rotem *et al.* developed a quenched fluorescence activity-based probe (ABP) that is selective for caspase-3 activity and emits a fluorescent signal after covalently modifying its target.[14] The quenched fluorescence ABP covalently binds its target small molecules in an activity-dependent manner. Different from conventional fluorescence ABPs, the quenching probe includes an additional quenching unit. When the target protease binds to the probe, the quenching unit is removed. Therefore, after binding to the activated enzyme, the probe changes from a non-fluorescent free probe to an enzyme-bound fluorescent complex. This specific probe can be used as a tool to study the level of caspase-3 activity and localization changes and it was found that during the apoptotic cell death of cancer cells, caspase-3 is active in the endoplasmic reticulum.[14] Moreover, a dual-FRET-based fluorescent probe was constructed based on a similar design strategy for the sequential detection of matrix metalloproteinase MMP-2 and caspase-3. The novel fluorescent probe FAM-K(GRVGLPG-Dabcyl)SDEVDSK(Dabcyl) consists of a dual fluorescence resonance-quenched fluorophore pair and two enzyme-specific peptide ligands. The fluorescence of FAM (fluorescein amidite) in the probe is initially quenched to Dabcyl by the double FRET method. Meanwhile, Pro–Leu–Gly–Val–Arg (PLGVR) and Asp–Glu–Val–Asp (DEVD) were used as cleavable substrates between fluorophore pairs as specific peptide sequences of MMP-2 and caspase-3, respectively. Extracellular MMP-2 could first identify and slice the PLGVR part of the probe. Then, once cells had been exposed to apoptotic stimulation, intracellular activated caspase-3 became the second trigger for DEVD sequence cutting, further enhancing the FAM fluorescence and indicating intracellular caspase-3 activity. By using this probe, doxorubicin (Dox)- and UV-induced apoptosis signals can be monitored successfully *in situ*.[15]

Yoshifumi *et al.* successfully developed transgenic mice stably expressing a FRET-based fluorescent probe by using insulator sequences that protect the genes from inappropriate signals generated by the surrounding chromatin environment. The mouse not only stably expresses a fluorescent-resonant gene based on genetically encoded FRET for visualizing the caspase activity, but also simultaneously performs time-lapse imaging of apoptosis and morphogenesis of living embryo cells. In conclusion, caspase activity and functional *in vivo* imaging analysis in this transgenic mouse have great potential for elucidating the consequences and significance of tissue homeostasis and caspase activation and apoptosis in disease.[16] Yuan *et al.* developed a simple but unique fluorescent probe for accurate caspase-3 detection through self-validation. The probe uses a coumarin fluorophore as an energy donor, a pyrophoric agent with aggregation-induced emission (AIEgen) as a quencher and caspase-3 specifically for cleaving DEVD as a linker. Since energy transfer and free molecule movement of AIEgen dissipate receptor energy, the probe is non-fluorescent. When caspase-3 is present, based on the separation and release of AIEgen residues from the donor and quenching agent, the probe generates dual signals: coumarin shows strong green fluorescence and AIEgen shows strong red fluorescence. Based on the dual signal, the probe is able to monitor caspase-3 activity in solution and living cells in real time with high efficiency, opening up new avenues for self-validation diagnostic, imaging and drug screening applications.[17] FRET-based caspase biosensors also show promise for application in disease studies. Figueroa *et al.* reported a FRET-based caspase sensor that is preferentially cleaved by caspase-3, -6 or -9, and the FRET-based sensor shows a powerful capability to detect stimuli-dependent differential activation of caspases and to distinguish local from global caspase activation in live neuronal cells.[18] Luo *et al.* designed a caspase-3-activatable fluorescent peptide (Dabcyl-KFFFDEVDK-FAM, FRET-Pep) for self-reporting of the therapeutic efficiency induced by camptothecin (Figure 5.1).[19] The peptide sequence KFFFDEVDK brings fluorescein amidite (FAM) as the donor and Dabcyl as the acceptor in close proximity to achieve the FRET effect, resulting in the fluorescence of FRET-Pep being turned off. The fluorescence of FRET-Pep can be turned on upon reaction with caspase-3 by enzymatic hydrolysis of DEVD. In the process of drug-induced apoptosis, the therapeutic efficacy could be reported by fluorescence imaging *in vivo*, *in situ* and in real time using FRET-Pep.

Further, biomaterials also can be used as quenchers to construct smart nanoprobes for monitoring caspase protease in living cells. Oishi *et al.*

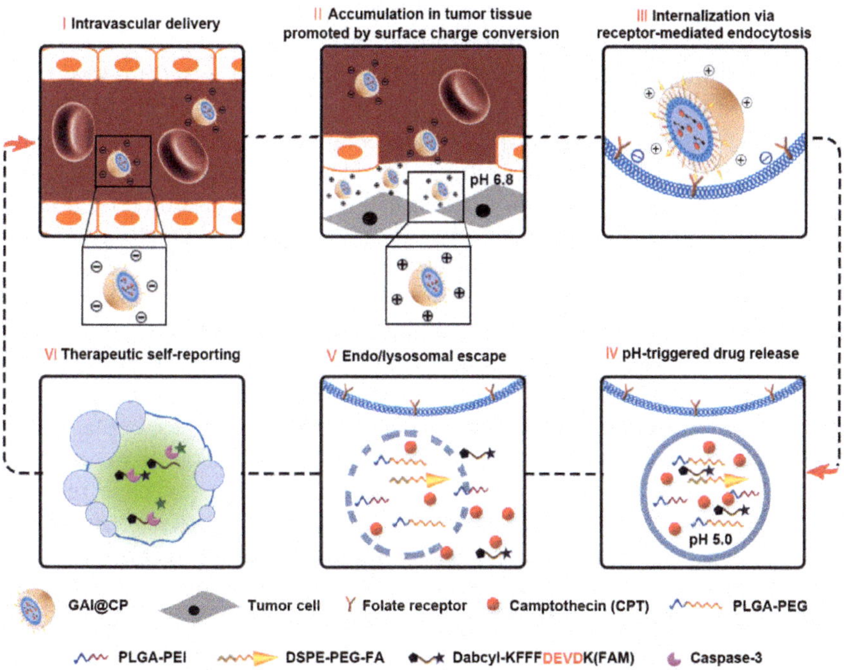

Figure 5.1 Design of programmed nanoparticles for accurate tumour cell eradication and therapeutic self-reporting with a fluorescent probe based on FRET.[19]
Reproduced from ref. 19, http://dx.doi.org/10.7150/thno.18187, with permission from Ivyspring.

reported an innovative smart apoptosis nanoprobe with caspase-3-responsive properties.[20] The designed nanoprobe is based on a PEGylated nanogel containing gold nanoparticles (AuNPs) as the quencher in the cross-linked pH-responsive polyamine gel core and fluorescein isothiocyanate (FITC)-labelled DEVD peptide grafted at the PEG chain ends. The FITC-DEVD-nanogel-AuNP showed remarkable quenching/dequenching of the fluorescence signal according to the caspase-3 activation in living cells. Graphene and graphene oxide (GO) have unique mechanical, electronic and optical properties[21] and great efforts have been devoted to exploring GO-based biosensors. Tian *et al.* developed a drug-loaded nanovehicle as a theranostic system for the simultaneous targeted delivery of anticancer drugs and self-feedback of therapeutic efficacy *in vitro* and *in vivo*. Cellular experiments demonstrated that drug-loaded nanocarriers could specifically transport drugs and release them into cancer cells with high expression of folate receptors to avoid damage to normal cells. Moreover, confocal fluorescence imaging confirmed that the drug-induced cancer cell death could be visualized with the light-up fluorescence of fluorescein activated by caspase-3. The results of animal experiments showed that the targeted delivery of the drug and the self-assessment of the therapeutic efficacy were further successfully achieved.[22] To output the signal of drug-induced cancer apoptosis, GO and peptide substrates with fluorophore labels have been used to form a FRET-based protease sensor.

Furthermore, variants of fluorescent proteins (FPs) were employed in a FRET-based sensor for signal output. A change in FRET efficiency, due to a change in relative distance or orientation between a more blue-shifted donor FP and a more red-shifted acceptor FP, resulted in a ratiometric change in emission that could be used for monitoring biological parameters in single cells.[23] Ding *et al.* developed and optimized a pair of spectrally distinct FRET-based biosensors for imaging of Ca^{2+} and caspase-3 in cells. They also demonstrated that the imaging of Ca^{2+} dynamics or Ca^{2+} and caspase-3 activity in two subcellular compartments is feasible in the same subcellular compartment using the optimized biosensors without substantial spectral cross-talk between emission channels. This strategy has the advantage of ratiometric imaging of Ca^{2+} and caspase-3 activity in the cytoplasm of a single cell, making it easier to quantify.[24] Wagner and Schreiber designed a dubbed ZipGFP-based executioner caspase fluorogenic protease reporter, which increases fluorescence 10-fold after activation by protease. The results indicate that the ZipGFP-based caspase reporter could detect apoptosis in human cells and visualize apoptosis in the living embryos of zebrafish, which is an important model vertebrate for understanding embryonic development. ZipGFP could be further used to design reporters for many other proteases, which overcome the sensitivity limitations of FRET-based reporters and could be suitable for imaging protease activity in physiological environments.[25] Shcherbakova *et al.* designed the first single-body orange FP with a large Stokes shift (LSSFP), which not only filled the spectral gap existing between the green–yellow and red LSSFP, but also had a fivefold higher brightness than the brightest red LSSFP and was successfully applied to four-colour single-laser fluorescence cross-correlation spectroscopy and simultaneous imaging of two FRET pairs.[26]

5.2.2 Fluorescence Lifetime Imaging Microscopy for Imaging of Intracellular Caspase Activity

The fluorescence lifetime (FLT) is an intrinsic property of the fluorophore that can be considered as a state function, because it does not depend on initial perturbation conditions such as excitation wavelength and duration of light exposure and is also not affected by photobleaching.[27] In addition, the FLT is a parameter that is largely independent of fluorescence intensity and fluorophore concentration. Different fluorophores have different decay times and this period of time is referred to the FLT, which is mainly influenced by the physicochemical environment of the fluorophore, including viscosity, temperature, pH and oxygen concentration. It can be used as a contrast agent to monitor different cell microenvironments. Thus, functional cell behaviour is imaged throughout the living organism in order to be able to study the systemic effects of disease and potential therapies, and to understand further the localization and spread of signalling responses. FLT imaging microscopy (FLIM) is a highly advanced spectroscopic technique that can be used in different fields, especially in chemistry and physics, and it is one of the most precise methods for FRET measurements. FLIM is widely considered to provide a more powerful FRET reading because it only needs to measure the donor fluorescence to avoid problems associated with internal filter effects and spectral cross-talk between the donor and acceptor channels, and allows for the use of different instruments. Measured values are compared with those for samples that are not calibrated. FLT measurements are inherently scaled and therefore independent of fluorophore concentration and excitation/detection efficiency, making them relatively robust to measurements in scattering media such as biological tissue. FLIM can use wide-field detection with time-gated or modulated imaging detectors. It can also be implemented in laser scanning confocal microscopy (LSCM) or multiphoton microscopy, typically using time-correlated single-photon counting (TCSPC).

Ju's group synthesized a highly efficient persistent luminescence nanoprobe (PLNP) for biosensing and lifetime imaging of cell apoptosis by setting an appropriate delay time and gate time (Figure 5.2).[28] In this approach, the luminescence of FITC is effectively suppressed owing to its shorter lifetime than the delay time, which represents the interval between pulsed excitation light and signal acquisition. The long-lived luminescence of the PLNP is captured during the gate time, representing the time for signal acquisition. Accordingly, taking caspase-3, microRNA-21 (miRNA-21) and platelet-derived growth factor (PDGF) as different targets, three different recognition nanoprobes are formed by covalently assembling cyclic arginine–glycine–aspartic acid peptide (cRGD) and FITC-labelled substrate peptide, DNA or

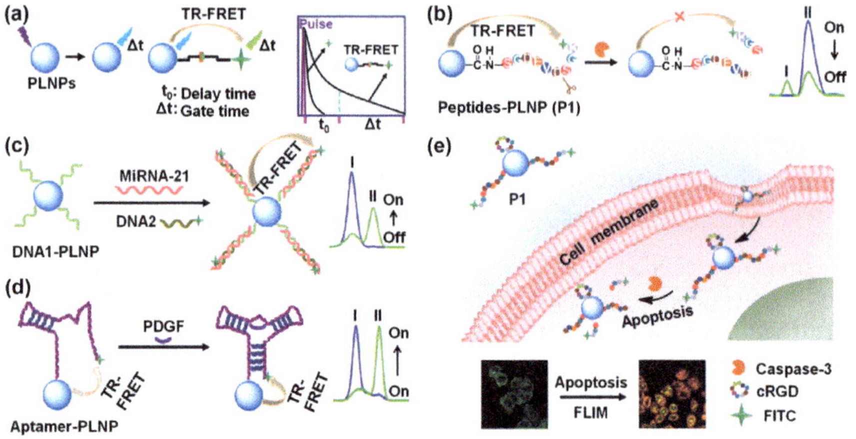

Figure 5.2 Schematic illustration of (a) the PLNP-based TR-FRET principle, (b), (c) and (d) TR-FRET detection strategies for caspase-3, miRNA-21 and PDGF protein, respectively, and (e) lifetime imaging of intracellular caspase-3 activity using P1 during cell apoptosis by FLIM.[28]
Reproduced from ref. 28 with permission from Elsevier, Copyright 2015.

aptamer on PLNPs for different states of fluorescence including "on–off", "off–on" and "on–on" TR-FRET to detect enzyme activity, nucleic acids and proteins, respectively. This method represents a novel protocol for the *in situ* quantification of caspase-3 activity based on the FLT and thus provides a robust approach for the dynamic evaluation of caspase-dependent cell apoptosis. French's group established a method for visualizing apoptosis in live zebrafish using FLIM with optical projection tomography (OPT) to map the FRET biosensor activity in space and time.[29] They used the genetic expression of the caspase-3 FRET biosensor and applied FLIM–OPT to read the biological functions of live transgenic zebrafish larvae. Apoptosis, or programmed cell death, was mapped in 3D by imaging the activity of a FRET biosensor that is cleaved by caspase-3 and the entire process required only 150 s to obtain a 3D FLIM data set. For the first time, FLIM–OPT was used to map the FRET signal changes of the caspase-3 biosensor after gamma irradiation in whole living zebrafish embryos. FRET basically reads two or more times by the intensity of the resonance energy transfer between them. Colocalization of fluorophores (10 nm) can be used to monitor apoptosis in live zebrafish larvae over time and FLIM–OPT can be used to map the activity of caspase-3 by measuring the increase in donor FLT. FLIM–OPT provides a means to visualize the body's response to potential therapies in space and time.

Based on average lifetime measurements, Savitsky's group reported a new technique for detecting intracellular enzyme activity.[30] They used the fusion protein TagRFP-23-KFP to register the caspase-3 activation in live cells, which is based on FLIM technology, allowing one to follow sensor cleavage of the proteolytic enzyme caspase-3. Specifically, FLIM FRET using live cells is performed by confocal fluorescence microscopy. The A549 cell line is transduced with a specially designed lentiviral vector pLVT with a DNA fragment of TagRFP-23-KFP. Computer simulations are performed to estimate the FRET efficiency and to analyse the possible spatial limitations of the reaction between the substrate TagRFP-23-KFP and the caspase-3 dimer. The fusion protein TagRFP-23-KFP was successfully used to record caspase-3 activation based on mean lifespan measurements. This system provides a new approach for high-throughput screening using special readers oriented on measurements of FLT. Luker's group measured changes in fluorescent imaging reporter genes by FLIM to detect caspase-3 activity.[31] Since lifetime measurements are independent of reporter concentration and light scattering in the tissue, FLIM can be stably imaged in a 3D environment. Apoptosis of cancer cells is quantified by using FLIM based on single-cell resolution of caspase-3 activity. In addition, the same group studied the method by which this technique measures caspase-3 activation at single-cell resolution in a variety of complex environments.[32] Real-time activation of the reporter gene for caspase-3 is imaged in response to apoptotic cell death. Caspase-3 reporter gene activity provides valuable insights into the responsiveness of cancer cells to treatment and the overall viability of a single-cell scale. Cleavage of the DEVD linkage sequence alters FRET within the reporter and affects its longevity. Apoptosis is quantified in a variety of environments ranging from 2D planar and 3D spheroid cell culture systems to *in vivo* mouse mammary tumour xenografts to assess cell-by-cell apoptosis in a variety of pharmacological and genetic approaches associated with cancer cell death reactions.

Ruedas-Rama's team demonstrated the unique properties of FLIM in combination with quantum dot (QD) nanoparticles for intracellular sensing.[33] Using pH-sensitive MPAQD and time-resolved fluorescence methods to monitor intracellular pH, FLIM technology allows for the monitoring of pH changes within the cytoplasm, providing sensitive readings for intracellular pH quantification. The photoluminescence (PL) lifetime of mercaptopropionic acid-terminated QDs (MPA-QDs) collected from FLIM images is used to determine intracellular pH. The MP average lifespan of MPA-QDs varies between 8.7 ns ($pH < 5$) and 15.4 ns ($pH > 8$) in media simulating the intracellular environment. These long decay times of QD nanoparticles make them easy to distinguish from intrinsic cellular autofluorescence, thereby increasing the selectivity of sensing applications. For the first time, the authors successfully detected intracellular pH changes of different cell types by detecting the PL decay time of QDs. In particular, the combination of the FLIM method and QD nanoparticles exhibits greatly improved sensitivity compared with other fluorescent dyes used for pH imaging. Uchiyama's group established a method for imaging the temperature distribution in living cells by performing a time-correlated single-photon counting (TCSPC)–FLIM of a novel fluorescent polymer thermometer (FPT).[34] By utilizing the high spatial resolution of FPT, intracellular temperature imaging revealed the thermal distribution of living cells and the higher temperatures in the vicinity of the mitochondria. Fluorescent polymer-based thermometers and fluorescence lifetime imaging microscopy contribute to intracellular temperature mapping. The spatial and temperature resolution of the thermometers is at the diffraction limit level (200 nm) and 0.18–0.58 °C. The intracellular temperature distribution that was observed indicated that the nucleus and centrosome of COS7 cells show significantly higher temperatures than the cytoplasm, and the temperature difference between the nucleus and the cytoplasm varies according to the cell cycle. The mitochondrial heat production is also observed to increase the proximal local temperature. In contrast, the authors' intracellular temperature imaging method has significant advantages with respect to sensitivity, spatial resolution, functional independence and ability to function in living cells. This work provides insights into the regulation mechanisms of intracellular signalling, thereby contributing to intracellular thermodynamic studies and the discovery of important life mechanisms.

5.2.3 AIE Probe for Imaging of Intracellular Caspase Activity

Sensitive and selective fluorescent biosensors are powerful analytical tools for studying biological events in living cells. As studies have progressed, a wide variety of fluorophores have been developed and utilized for analysing the intracellular caspase family. However, conventional fluorescent chromophores have the serious limitation that aggregation will lead to fluorescence quenching. This phenomenon greatly limits the use of fluorescent chromophores. To solve this problem, Tang's group developed a category of luminescent materials that have weak or no fluorescence in a good solvent, but the fluorescence will be significantly enhanced in the state of aggregation.[35] This unusual photophysical phenomenon is denoted aggregation-induced emission (AIE). In contrast to aggregation-caused quenching (ACQ), the novel AIE phenomenon permits the use of dye solutions of any concentration for bioassays. The AIE property is ascribed to the restriction of intramolecular rotations (RIR), restriction of intramolecular vibrations (RIV) and restriction of intramolecular motions (RIM). Luminescent materials with AIE have good biocompatibility, good resistance to photobleaching and high quantum yields. Owing to these excellent optical properties, AIE materials are considered to have great potential for *in vivo* imaging.

In 2012, Liu's group created an AIEgen-peptide bioconjugate named Ac-DEVDK-TPE for the specific light-up detection of caspase and monitoring of cell apoptosis in real time.[36] The bioconjugate consists of two parts: one part is a hydrophilic caspase-specific DEVD peptide and the other is a typical fluorogen with AIE characteristics, a hydrophobic tetraphenylethene (TPE) group. The probe is almost non-fluorescent in aqueous solution but shows a dramatic enhancement in FL in response to caspase-3/-7, which are activated in the apoptotic process and could cut off the DVED-Ac sequence from Ac-DEVDK-TPE. This fluorescence "switch-on" response is attributed to aggregation of cleaved the part of hydrophobic TPE residues, which restricts the intramolecular rotations of TPE moieties and populates the radiative decay channels. The Ac-DEVDK-TPE probe allows real-time monitoring of caspase-3/-7 activities both in solution and in living cells with a high signal-to-noise ratio. Moreover, the probe offers a new approach to real-time monitoring of cell apoptosis and screening drugs.

As studies developed, some researchers found that variations in conformation would greatly affect enzyme–substrate interactions, leading to different reactivities and products. Tang and Liu's group designed and synthesized a dual DVED-labelled AIE probe for monitoring caspase activity to investigate stereoisomeric effects on the probe performance.[37] This probe consists of a dual caspase-specific DEVD peptide and a TPE unit with AIE characteristics. Two stereoisomers, (*Z*)-TPE-2DEVD and (*E*)-TPE-2DEVD, were prepared and purified. The experimental results showed that both stereoisomers can act as light-up probes for the *in situ* analysis of caspase-3. In particular, (*Z*)-TPE-2DEVD has a greater light-up response with a higher signal-to-noise ratio than (*E*)-TPE-2DEVD. The reason is that the cleaved product of (*Z*)-TPE-2DEVD, *i.e.* (*Z*)-TPE-2A, has a higher brightness. This study not only represents a convenient strategy for monitoring caspase-3, but also opens up new thoughts for fluorescent probe design with optimized performance. The aforesaid caspase fluorescent probes are all based on TPE units, but the emission maximum of TPE is situated only at around 470 nm, which limits the further expansion of its range of use. To circumvent this limitation, Tang and Liu's group developed azide-functionalized tetraphenylethenepyridinium (N_3-PyTPE) by the incorporation of pyridinium units into TPE through vinyl functionalization. They then conjugated N_3-PyTPE with a hydrophilic N-terminal acetyl-protected and C-terminal alkyne-functionalized Asp–Glu–Val–Asp (Ac-DEVD) peptide to form a new caspase-3/-7 targeting probe.[38] This new bioconjugate shows a long-wavelength absorption at 405 nm and emission at 636 nm, and can be used for *in vivo* real-time monitoring of caspase activation and cell apoptosis.

Imaging of specific cells is of great significance for disease surveillance and treatment. Monitoring and imaging the apoptosis of targeted cancer cells in real time are critical to the treatment and prognosis of malignancies. In this context, Liu's group conjugated cyclic Arg–Gly–Asp(cRGD) and Ac-DEVD to two sides of a tetraphenylsilole (TPS) unit, which afforded a new asymmetric bioprobe denoted Ac-DEVD-TPS-cRGD.[39] In this probe, cRGD targeted cancer cells and Ac-DEVD could be specifically cleaved by caspase-3/-7. To verify the design, integrin $\alpha_v\beta_3$ receptor-overexpressed U87MG cancer cells were used as an example, and it was demonstrated that the probe can target U87MG cancer cells and has good sensitivity and optical properties. Direct and real-time monitoring of different caspases in a single process is important for understanding disease development. Tang and Liu's group also designed a general platform based on AIE, allowing two caspase activities to be detected by a single fluorescent probe in living cells.[40] The AIE bioconjugate consists of three parts, namely a hydrophilic peptide (DVED-IETD) and two AIEgens with green and red emission excitable at a single wavelength of 405 nm. DVED is the substrate of apoptosis effector caspase-3 and IETD is the substrate of apoptosis initiator caspase-8. The probe has almost no fluorescence in aqueous media, and when the peptide substrate was cleaved by the cascade activation of caspase-3 or -8, the green and red fluorescence could be sequentially turned on. Hence the detection of caspase cascade activation can be realized. Most fluorescent probes monitor only a single process of drug activation or therapeutic response, but simultaneous monitoring of the whole processes is able to provide a clearer picture for the treatment and prognosis of diseases. Tang and Liu's team reported a multifunctional fluorescent probe, denoted TPETP-SS-DEVD-TPS-cRGD, consisting of two AIEgens with distinguishable emissions and two different targeted peptides.[41] In this system, the probe emits red fluorescence when ingested by target cells, and subsequently green fluorescence will be generated by apoptosis-induced caspase-3/-7 activation. Using this probe, the whole process of treatment can be observed.

5.2.4 Two-photon (TP) Imaging of Intracellular Caspase Activity

Fluorescence imaging provides an indispensable means of locating and monitoring biological targets within complex and dynamic intracellular environments.[42] Existing fluorescent probes are usually excited by one photon with a short excitation wavelength, which limits their application in subcellular organelles and deep body tissues owing to the penetration depth, photodamage and spontaneous cellular autofluorescence. Compared with one-photon excitation, two-photon fluorescence microscopy (with excitation at 700–1000 nm) has become more popular in bioimaging owing to the advantages of two-photon technology, including long-wavelength excitation, high 3D resolution, minimal photodamage, less photobleaching and deep tissue penetration.[43–45]

In 2014, Yan *et al.* reported the first example of a two-photon absorption (TPA) nanomicelle-based two-photon fluorescent nanoprobe using for the assay of enzymatic activities in living cells and tissues.[46] The sensing system is composed of a *trans*-4-[*p*-(*N*,*N*-diethylamino)styryl]-*N*-methylpyridinium iodide (DEASPI)/β-cyclodextrin polymer (βCDP) nanomicelle as the TPA fluorophore and the carrier vehicle for delivery of a specific peptide sequence to live cells through fast endocytosis and an adamantine (Ad)-GRRRDEVDK-BHQ2 (black hole quencher 2) peptide (denoted Ad-DEVD-BHQ2) anchored on the DEASPI/βCDP nanomicelle's surface to form a TPA DEASPI/βCDP@Ad-DEVD-BHQ2 nanoconjugate by the βCD/Ad host–guest inclusion strategy. *In vitro* assays revealed that the DEASPI/βCDP nanomicelle-based TPA fluorescent nanoconjugate provides a robust, sensitive and selective sensor for the quantitative detection of caspase-3 even under complex biological conditions. Two-photon confocal fluorescence microscopy experiments with HeLa cells and tissues suggested that the βCDP nanomicelle-based conjugate is efficiently delivered into live cells and acts as a "signal-on" fluorescent sensor for the specific, high-contrast imaging of target biomolecules. The introduction of two-photon excitation could effectively eliminate the self-absorption and autofluorescence of the biological molecules in biological matrices, reduce the photodamage resulting from the high-energy excitation and enhance the penetration depth and spatial resolution, compared with the currently prevailing strategy of one-photon excitation-based fluorescent biosensor construction; it substantially improves the performance of fluorescent sensors in complex biological conditions. Taking into consideration the unique properties of βCDP-based nanomaterials and the TPM technique, this new TPA DEASPI/βCDP nanomicelle-based strategy for developing robust biomolecule sensors is expected to have great potential for *in vitro* and *in vivo* applications in medical research and clinical diagnostics. Encouraged by this result, Ge *et al.* used an AuNP/peptide–TPdye two-photon nanosensor for the assay of caspase-3 in live cells and liver tissues treated by ischaemic reperfusion surgery.[47] The nanosensor is formed by an assembly of TPdye-labelled peptide sequences on the AuNP's surface. The AuNP/peptide–TPdye nanoconjugate exhibits good cell permeability, high stability and good biocompatibility. Compared with the currently prevailing strategy of one-photon excitation-based fluorescent biosensor construction, the introduction of a two-photon dye as the signal reporter effectively eliminates problems of self-absorption and autofluorescence in biological matrices while using a long excitation wavelength and offers larger depth penetration and less photodamage. *In vitro* and *in vivo* assays indicated that the AuNP/peptide–TPdye nanoconjugate provides a robust, sensitive and selective sensor for the measurement of caspase-3 even under complex biological conditions. This new AuNP/peptide–TPdye two-photon nanosensor is expected to have great potential for applications in anti-HIRI drug screening, medical research and clinical diagnostics.

To obtain a better understanding of the mechanisms of programmed cell death, Sergeeva *et al.* exploited multiparameter fluorescence microscopy to analyse energy metabolism, intracellular pH and caspase-3 activation in living cancer cells *in vitro*.[48] The FRET-based sensor mKate2-DEVD-iRFP was used for caspase-3 activity detection and the dynamic changes were visualized by two-photon fluorescence imaging microscopy. The correlations between energy metabolism, intracellular pH and caspase-3 activation and their dynamic changes were studied in CT26 cancer cells during apoptosis. Induction of apoptosis was accompanied by a switch to oxidative phosphorylation, cytosol acidification and caspase-3 activation. The results showed that during the early stage of apoptosis, cell shrinkage was visible within 1 h by light-transmitted microscopy. After 2 h, extensive plasma membrane blebbing occurred, followed by the formation of apoptotic bodies. Before this blebbing and the formation of apoptotic bodies, activation of caspase-3 in the cells could be detected, as shown by two-photon fluorescence imaging. It was also demonstrated that protein kinase inhibitor staurosporine-induced apoptosis was accompanied by a switch to a more oxidative energy metabolism and cytosolic acidification preceding caspase-3 activation.

So far, the visualization of caspase activation has mostly remained at the cellular level, partially owing to a lack of cell-permeable imaging probes that are capable of direct and real-time investigation of endogenous caspase activities in deep tissues. In this respect, two-photon enzymatic probes provide the visualization of subcellular/deep-brain caspase activities. For example, Qian *et al.* developed a suite of two-photon, small-molecule/peptide probes that allow the sensitive and dynamic imaging of individual caspase activities in neurodegenerative models under physiological conditions.[49] This novel location-based strategy with C3RM/C3RE could be further explored to study the spatial dynamics of endogenous caspase-3/-7 under different stimuli. With the features of a low background and high sensitivity endowed by the two-photon caspase-1-detecting probes, this work successfully imaged the enzyme in primary neurons of glucose-deprived mice. Finally, with the ability of directly imaging caspase activities in deep tissues, caspase-3 activation was apparently observed in ethanol-intoxicated mouse brains. The results indicate that these probes are useful tools for studying caspase activation in neurological disease models. Future work will be directed at the development of more selective caspase probes (*e.g.* caspase-6/-9) and the development of the same design strategy for other classes of biologically important proteases.

5.2.5 Nanomaterial-based Probe for Imaging of Intracellular Caspase Activity

Since nanomaterials possess various advantages, including targeted delivery, degradability, good biocompatibility and visualization when modified with imaging agents, the application of nanomaterials in the imaging of intracellular caspase activity has been widely investigated. Recently, metal–organic frameworks (MOFs), an emerging class of porous materials, have attracted considerable research interest with respect to drug delivery, biosensing and cancer therapy. Zhang *et al.* developed a multifunctional nanoprobe based on assembling a porphyrin, folate and a peptide into an MOF cage, which induced photosensitive cell apoptosis, and monitored the therapeutic efficacy *in situ* by activation of caspase-3 (Figure 5.3).[50] This nanoprobe integrates the theranostic functions in a single nanocarrier *via* coupling the apoptosis-inducing capability of photosensitizers with the subsequent caspase activation. After folate receptor (FR)-mediated uptake, the nanoprobe could induce cell apoptosis with caspase-3 activation by generating $^1O_2$ in mitochondria, which cleaves the peptide and lights up the fluorescence of Cy3 labelled on the peptide. According to the diversity of structure, capacity and tuneability for modification of MOFs to protect fluorescence and photosensitivity, the multifunctional nanoprobe is designed for

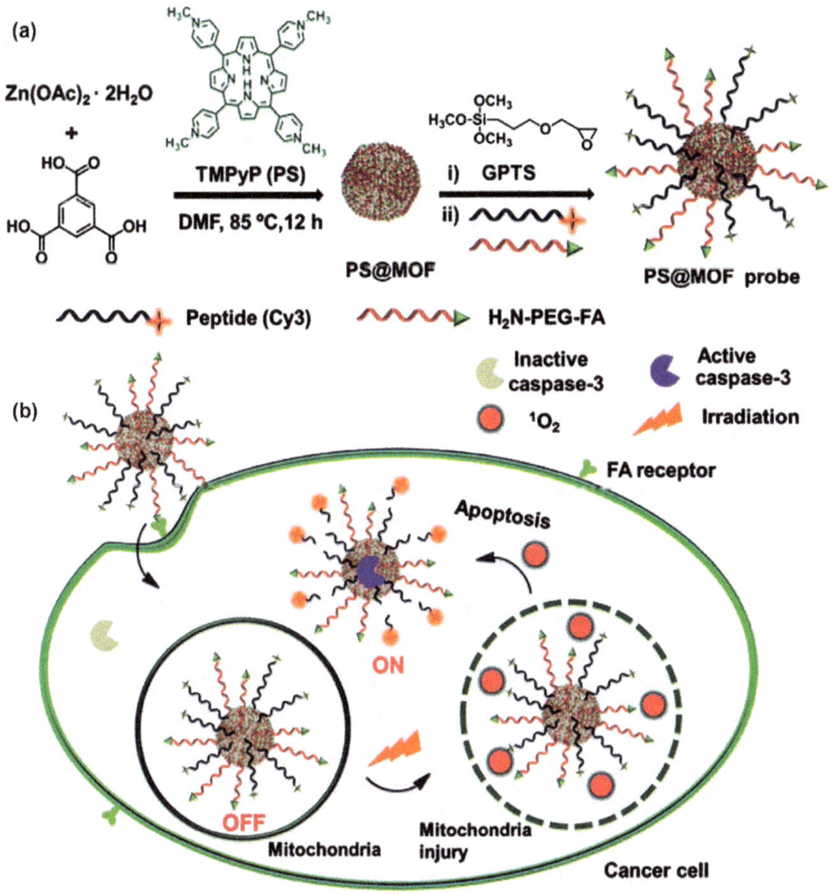

Figure 5.3 Schematic representation of (a) the probe preparation and (b) FA receptor-targeted delivery of the probe for tumour cell apoptosis and imaging of caspase-3 activation for therapeutic monitoring.[50] Reproduced from ref. 50 with permission from the Royal Society of Chemistry.

efficient cell-specific cancer treatment and monitoring of therapeutic effectiveness *via* an "off–on" response dependent on caspase apoptosis imaging.

Wang *et al.* design a novel, sensitive, simple and robust intracellular protease sensor based on the peptide substrates and nanoconjugate of GO for high contrast imaging of apoptotic signals in live cells.[51] GO is a nanocarrier that can deliver the peptide cargos in live cells and a fluorescence quencher. The GO–peptide conjugate, after being transported into cells, may provide greatly enhanced fluorescence imaging. They chose apoptosis-related caspase-3 activation to test this design. The strategy involves a spacer peptide at the N-terminus and a fluorescein amidite (FAM)-labelled lysine at the C-terminus to construct a peptide probe. Meanwhile, this probe conjugates with GO to form GO–peptide, which shows excellent stability and solubility in water and cell growth medium and a capability for *in vitro* assay of caspase-3 activity. Compared with the existing prevailing method of non-covalent assembly for biosensor construction based on graphene, it substantially improves the stability of protease sensors in protein-abundant media, providing a susceptible and selective sensor for the quantitative detection of caspase-3. Liu *et al.* detected activated caspase-3 in apoptotic neurons *in vivo* with a brain-targeted dual-function nanodevice using FRET.[52] This nanodevice is based on dendrigraft poly-L-lysines (DGLs) and links with a brain-targeted peptide, RVG29, and also a caspase-3-cleavable peptide linker (DEVD). The brain-targeted DGLs–RVG29–FRET nanodevice could bind its specific receptors in brain capillary endothelial cells (BCECs) and mediate transcytosis across the blood–brain barrier (BBB); it showed almost no fluorescence in the normal physiological state, but showed fluorescence in the presence of activated caspase-3, which could be used to detect the level of caspase-3 activation in conformity with the degree of apoptosis in rat brain. Li *et al.* adopted the appealing NSET-dependent quenching effect of peptide-coated ultra-small gold nanoclusters (AuNCs) to develop a novel fluorescence lighting-up probe for the detection and bioimaging of caspase 3 activity.[53] In this approach, they first synthesized peptide-coated AuNCs by a one-step biomineralization method; the resulting AuNCs present efficient broad-spectrum quenching properties and can effectively quench the fluorescence of FITC- or TAMRA (5-carboxytetramethylrhodamine)-tagged peptides through the nanosurface energy transfer (NSET) process. To demonstrate the validity of their protease-responsive AuNC nanoprobes for caspase detection, they used this probe to detect the caspase-3 activity in living cells. This approach for detecting the level of caspase-3 activation has the advantages of high cell-penetrating ability and less interference with cell functions, and its synthesis is very easy, green and inexpensive and it also provides an extra therapeutic function *via* a well-defined apoptosis-targeted pathway.

Sensors based on bioluminescence resonance energy transfer (BRET) have recently attracted increasing attention as a viable alternative to FRET-based biosensing. Merkx and co-workers utilized a caspase-cleavable amino acid flexible linker to connect the bright and stable luciferase NanoLuc and the fluorescent acceptor protein mNeonGreen (mNG) to ensure efficient energy transfer in the absence of caspase activity.[54] Three sensors were developed for targeting initiator caspase-8 (Ile–Glu–Thr–Asp) involved in the extrinsic pathway, initiator caspase-9 (Leu–Glu–His–Asp) from the intrinsic pathway and the downstream executioner caspase-3 (Asp–Glu–Val–Asp). The ratiometric and stable bioluminescent signal offered by these sensors is advantageous for high-throughput caspase activity assays and allows long-term imaging of caspase activity in single cells. In addition, these sensors can be adopted as attractive screening high-throughput cell-based drugs to monitor caspase-mediated apoptosis. MoS_2 nanosheets (NSs) have shown great potential for biosensing with unique properties. Li *et al.* developed a covalent biofunctionalization strategy with MoS_2 NSs with high conjugation efficiency of peptides by utilizing polydopamine (PDA) as a nano-bio interface and constructed an intracellular fluorescent biosensor for caspase-3 activity detection with high sensitivity and selectivity.[55] This MoS_2@PDA-PEG-Peptide (MPPP) biosensor is based on FRET with FAM as donor and MoS_2 NSs as acceptor. Caspase-3 is activated to cleave the substrate peptide and the FAM is released from the MoS_2 NSs during cell apoptosis, which results in the recovery of the fluorescence of the MPPP biosensor and allows the visualization of the intracellular apoptosis process. This covalent biofunctionalization of MoS_2 NSs can significantly improve the conjugation efficiency of biomolecules and enhance their stability in complicated environments, which can be used for high-contrast fluorescent imaging of cell apoptosis. Furthermore, it will facilitate the construction of 2D multifunctional nanoplatforms for various biomedical applications.

5.2.6 Bioorthogonal Cyclization-mediated *In Situ* Self-assembly of Small-molecule Probes for Imaging of Caspase Activity

The detection of tumour apoptosis is essential for predicting therapeutic efficacy. Caspase-3 and -7, as the effector caspases, are ideal apoptosis-imaging targets, which play an important role in intrinsic and extrinsic pathways of programmed cell death. Self-assembly of small molecules in living systems has been widely used in biology and medicine, and is also popular in the synthesis of supramolecules and nano/microstructures in living organisms. Rao and co-workers employed an improved cyanobenzothiazole (CBT) first-order condensation bioorthogonal cyclization reaction to monitor the process of the self-assembly of a fluorescent small molecule in apoptotic cells and also indicated that imaging the activity of caspase-3 and -7 in a model of tumour-bearing mice confirmed its applicability *in vivo*.[56] It was confirmed that the bioorthogonal cyclization reaction between free cysteine and D-cysteine and 2-cyano-6-hydroxyquinoline (CHQ) was of good biocompatibility with fast intramolecular condensation kinetics (6×10^{-3} s^{-1}). Also, it is important to mention that this first-order reaction is independent of local probe concentrations and it allows efficient intramolecular macrocyclization in physiological environments, which is a major advantage compared with intermolecular reactions that control self-assembly *in vivo*. Moreover, the small-molecule fluorescent probe [caspase-sensitive nanoaggregation fluorescent probe (C-SNAF)] was able to monitor continuously the tumour response *in vivo* during the whole course of chemotherapy, while the tumour fluorescence increased from the first to the third round of treatment. In addition, the maximum tumour fluorescence level from C-SNAF was closely related to the change in maximum tumour size after chemotherapy. The extent of tumour response was assessed longitudinally in individual mice and valuable information on the efficacy of the administration of the treatment to the individual was evaluated relative to other mice receiving treatment. The results revealed that C-SNAF is a feasible approach for imaging tumour cell apoptosis induced by drugs *in vivo*, with rapid, real-time and non-invasive acquisition of the information. In brief, a series of *in vitro* and *in vivo* experiments illustrated that the active caspase-3/-7 as target could trigger C-SNAF into nanoaggregates to distinguish negative from positive responses. By using three-dimensional structured illumination microscopy (3D-SIM), the fluorescent nanoaggregates are assembled *in situ* and distributed in the whole cytosol and the whole process is visualized directly in apoptotic cells and tumour tissues. The caspase-3/-7-triggered bioorthogonal macrocyclization and nano-aggregation of C-SNAF can be used in tumour therapy, neurodegenerative diseases and organ-transplant rejection. For the first time, a synthetic small molecule probe that could be capable of self-assembly in live animals demonstrates the potential of this novel bioorthogonal cyclization chemistry for molecular imaging. More importantly, a similar self-assembly strategy may be applied to other molecular targets, such as cathepsin and furin, and should find widespread use in the non-invasive imaging of enzyme activity *in vivo*.

The introduction of focused bioorthogonal reactions broadened the applications for imaging caspase activity. A probe that depended on the activity of the enzyme by making a "split luciferin" ligation reaction one of the orthogonal reactions was developed for imaging caspase-3/-7 activity.[57] This bioorthogonal reaction, a condensation between 6-hydroxy-2-cyanobenzothiazole (CBT) and D-cysteine, results in the formation of D-luciferin, which has important significance for imaging *in vivo* and in a biocompatible marking strategy. The luciferin substrate produced by a live mouse is visualized by bioluminescence imaging and can also be accessed in the target tissue using the "caged" luciferin method. Hence this reaction can be applied to caspase-3/-7 with real-time, non-invasive imaging. Free D-cysteine is released from the caspase-3/-7 peptide substrate, Asp–Glu–Val–Asp–D-Cys [DEVD-(D-Cys)], in a caspase-dependent manner. The D-cysteine then reacted with 6-amino-2-cyanobenzothiazole (NH_2-CBT) *in vivo* to produce 6-amino-D-luciferin along with subsequent light emission from luciferase. This probe was used to detect caspase-3/-7 activation in luciferase-expressing mice in an model of inducible liver failure. This *in vivo* model includes lipopolysaccharide (LPS) administered intraperitoneally into the bacterial cell wall and D-galactosamine (D-GalN). Based on split luciferin, the probe DEVDc was found to be more sensitive than the DEVD–aminoluciferin probe, because the former involves caging of the complete luciferin molecule. A similar method was used for the detection of thrombin, in which a GGRc peptide was used in combination with CBT; the resulting signal depends on the concentration of enzyme.[58] In brief, one advantage of this method over the full luciferin scaffolds is better cell-penetrating properties of the CBT derivatives. Further, this strategy is superior to the use of the commercially available DEVD–aminofluorescein substrate for the imaging of caspase-3/-7 activity. Moreover, the split luciferin approach allows the modular construction of biosensors, in which both reaction partners could be caged, to report on multiple biological events. Lastly, the luciferin ligation reaction is three orders of magnitude faster than another bioorthogonal reaction, Staudinger ligation, which suggests further applications for specific molecular targeting *in vivo*.

5.2.7 Bioluminescence Imaging of Intracellular Caspase Activity

Bioluminescence imaging (BLI) is a useful methodology developed over the past decade for investigating ongoing biological processes at the molecular level in living organisms. It is a versatile and sensitive tool based on the detection of optical light generated during enzyme (luciferase)-catalysed oxidation of a substrate (luciferin) when the enzyme is engineered to express *in vivo* as a genetic reporter. This optical imaging approach allows non-invasive, inexpensive and real-time *in situ* analysis with the advantages of no requirement for external excitation light, no photobleaching and no influence of the background signals. Common applications of BLI include studies of protein–protein interactions, tumour cell tracking, bacterial/viral pathogenesis and treatment response. Recently, this powerful technology has been expanded to monitor apoptosis by exploiting the substrate-binding cleft of caspases.

Scabini *et al.* utilized D-luciferin and a new formulation of (*Z*)-DEVD-aminoluciferin, a modified firefly luciferase prosubstrate containing caspase-cleavage peptide DEVD (aspartic acid–glutamic acid–valine–aspartic acid, which is optimal for caspase-3 and -7) with water solubility and low toxicity, to monitor *in vivo* apoptosis in transgenic tumour cells expressing luciferase.[59] Upon activation of caspase-3 or -7, (*Z*)-DEVD-aminoluciferin is recognized and cleaved, hence the active aminoluciferin is released to react with luciferase, producing measurable light. Hence the tumour burden could be assessed by quantitation of the bioluminescent signal, which was normalized by the D-luciferin signal. This approach succeeded in monitoring caspase-3 activation and subsequent apoptosis induction after camptothecin and temozolomide medication in subcutaneous tumour models of colon cancer and glioblastoma, respectively. Treated mice displayed more than two fold induction of the (*Z*)-DEVD-aminoluciferin luminescent intensity when compared with the untreated mice, which was due to the tumor cell death and caspase activation caused by chemotherapeutic drugs. Moreover, early apoptosis induction by caspase-3 activation was demonstrated to be a robust biomarker predicting tumour growth inhibition by anti-cancer drugs.

Merkx's group reported three BRET-based protein sensors for monitoring the activity of caspase-3, -8 or -9.[54] These sensors are composed of the small and stable luciferase NanoLuc and the fluorescent acceptor protein mNeonGreen (mNG), fused together *via* a caspase-cleavable flexible peptide chain. In the non-cleaved state, the energy is transferred from the donor furimamide, the catalytic product of NanoLuc, to the acceptor mNG.

Upon linker cleavage, each bright bioluminescent caspase sensor represented a robust 10-fold decrease in the acceptor/donor emission ratio. These sensors offer a ratiometric and stable bioluminescent signal that is advantageous for high-throughput screening of caspase activity and is suitable for long-term imaging of a single cell. Furthermore, this strategy provides an attractive alternative to the FRET-based imaging system in light-sensitive cells.

Dubikovskaya's group discovered that the selective reaction between D-cysteine and 2-cyanobenzothiazoles could efficiently generate a luciferin substrate for firefly luciferase in living cells and in living animals.[57] This split luciferin ligation reaction is a powerful tool to explain complex cellular processes *in vivo*. They then successfully demonstrated that this reaction could be further applied to real-time imaging of caspase-3 or -7 activity in hepatic apoptosis animal models. Caspase-mediated release of free D-cysteine from the caged group DEVD allows ligation with 6-amino-2-cyanobenzothiazole to produce 6-amino-D-luciferin, which is catalysed by luciferase. The split luciferin reagents produce several-fold higher signals than equimolar amounts of D-luciferin and D-aminoluciferin. For imaging caspase-3/-7 activity, it is superior to the commercially available DEVD–aminoluciferin substrate with a higher signal-to-background ratio, demonstrating great potential for the sensitive imaging of apoptosis.

Kindermann *et al.* presented a bioluminescent activity-based probe (ABP) for caspase-1 monitoring.[60] Based on the reverse design concept, they converted a chemically optimized protease inhibitor, pralnacasan, into a selective substrate ABP, referred to as CM-269, by substituting a cleavable peptide bond for the substrate-mimicking warhead and subsequently attaching aminoluciferin used as a reporter group. Compared with available fluorogenic peptidic caspase-1 substrates, CM-269 exhibits superior selectivity, good kinetic properties and a 1000-fold increase in sensitivity. They applied CM-269 to monitor caspase-1 activity in complex proteomic samples, such as whole-cell lysates of stimulated and non-stimulated monocytic cells. Their results proved that considerable caspase-1 activity is detectable even in non-stimulated immune cells, which helps in the understanding of caspase-1 activation and its role in inflammation.

Chen's group utilized an apoptosis-specific bioluminescence reporter gene, pcFluc-DEVD, to map the kinetics of drug-induced caspase activation and apoptosis through temporal bioluminescence imaging.[61] They genetically modified cell lines, including human head and neck squamous carcinoma UM-SCC-22B cells and murine breast cancer 4T1 cells, with pcFluc-DEVD. The bioluminescence signal, correlated with caspase-3 activity in the transfected cells, was evaluated after the treatment with doxorubicin. Their results showed that 12 days after administration, the BLI signal induced by one dose of Doxil (doxorubicin·HCl liposome injection) reach the maximum whereas the average size of the tumour almost reached the minimum. This protocol featuring non-invasive temporal imaging could dramatically benefit the preclinical evaluation of chemotherapeutics targeting the apoptotic pathway.

A dual-analyte luciferin imaging strategy for selectively monitoring both hydrogen peroxide and caspase-8 activity in a murine model of acute inflammation was reported by Chang and co-workers.[62] They synthesized two complementary caged precursors [peroxy-caged luciferin-2 (PCL-2) and (*Z*)-Ile–Glu–Thr–Asp–D-Cys (IETDC)] with the ability to react with H_2O_2 or active caspase-8 individually. Elevations in two different biological analytes led to the release the uncaged luciferin precursors 6-hydroxy-2-cyanobenzothiazole (HCBT) and D-cysteine that subsequently formed firefly luciferin *in situ,* giving rise to a bioluminescent signal. This system constitutes an AND-type molecular logic gate that responds to the simultaneous presence of H_2O_2 and caspase-8 activity. They used PCL-2 and IETDC to image simultaneously H_2O_2 fluxes and caspase-8 activity, respectively, in living animals with acute inflammation. This method offers a potentially powerful new chemical tool suitable for monitoring simultaneous oxidative stress and inflammation processes based on luciferin-based bioluminescence imaging. In conclusion, the non-invasive and sensitivity offered by BLI open the door to the detection of caspase activation and cell apoptosis in a preclinical setting. Further, novel discoveries and developments in luciferase reporters, substrates, gene-editing techniques and near-infrared emitting systems promise a new era of deeper and more sensitive molecular imaging.

5.3 Simultaneous Imaging of Caspases and Other Proteins

Apoptosis is an important process in multicellular organisms for processes such as cellular clearance, pathologies and immune response and also is an essential programmed cell death pathway for excluding damaged or noxious cells and maintaining homeostasis in biotic systems that involve caspases and other proteins. The process of apoptosis involves many biochemical diversifications, including the translocation of phosphatidylserine (PS) from the cytoplasmic to the extracellular leaflet of the plasma membrane, chromosomal DNA fragmentation and activation of caspases. Irregular apoptosis can lead to cancers, Alzheimer's disease, myocardial infarction, atherosclerosis and many other diseases. Therefore, there is an urgent need for the manufacture of a non-invasive imaging intermedium that permits direct visualization and monitoring of apoptosis *in vivo* and that has great potential value for the early diagnosis of diseases, evaluation of therapy efficacy and screening of apoptosis-related drugs. Further, the simultaneous imaging of caspases and other proteins has great potential for the early diagnosis of diseases and evaluation of therapy efficacy.[63–67] Yang and co-workers developed a biosensor for the *in situ* detection of cytochrome *c* (Cyt *c*) and caspase-9 by immobilizing anti-Cyt *c* or anti-caspase-9 at the interface of gold nanoparticle-doped polydopamine (AuNP/PDA) composites.[68] Polydopamine (PDA) could be used as an enhanced sensing probe for the development of electrochemical sensors and the inherent nitrogen atoms of PDA could efficiently be led into chemically active sites that are continuously used to catch metal nanoparticles for the efficient fabrication of nanocomposites. Owing to the excellent characteristics of AuNPs, AuNP/PDA composites could provide a more homogeneous surface for the conjugation antibodies and subsequently antigen packing. Therefore, AuNP/PDA was used as an immobilization platform to prepare antibody-functionalized AuNP/PDA-based immunosensors. Yang and co-workers[68] utilized curcumin (CCM) as a model drug, which mainly activated the mitochondria-mediated caspase pathway of induction of apoptosis in cancer cells. CCM could be exploited to induce apoptosis in HeLa cells, in which the expression level of Cyt *c* and caspase-9 during the process of apoptosis could change. As a result, the changes in the expression level of Cyt *c* and caspase-9 in the cytosol after curcumin-induced apoptosis were detected by using electrochemical measurements including electrochemical impedance spectroscopic (EIS) analysis, cyclic voltammetry (CV) and differential pulse voltammetry (DPV). The biosensor achieved a detection limit of 0.03 ± 0.01 μM for standard Cyt *c* and 0.08 ± 0.02 μM for standard caspase-9 with a linear range from 0.1 to 100 μM. Most importantly, it could detect as few as 100 cells that expressed Cyt *c* and caspase-9 and also the results are in good accord with standard flow cytometric analysis. This smart Cyt *c* or caspase-9 electrochemical immunosensor offers a rapid and sensitive approach for the monitoring of apoptosis markers with considerable specificity and reproducibility. This novel strategy could be of great importance in clinical research and disease diagnosis.

Tian's group created a multifunctional nanocapsule (MNC) with the aim of targeting lysosomal tumour cell death and real-time detection of drug effects through encapsulating artemisinin (ART) and a dual protease light-up nanoprobe in a folate-functionalized liposome (Figure 5.4).[69] For the evaluation of ART-induced apoptosis, a carboxylic graphene oxide (GO) nanosheet was combined with a cathepsin B (CaB)-specific substrate peptide modified with rhodamine B (GFLGK-RhB) and a caspase-3-specific substrate peptide modified with fluorescein amidite (FAM) to act as a nanoprobe (GO-RhB-FAM). The ART and GO-RhB-FAM were integrated into folic acid (FA)-modified liposome to give the MNC, which could be selectively taken up by FA receptor highly expressed tumour cells. The fluorescence of RhB and FAM was inhibited by GO owing to the energy transfer effect. When the nanoprobe met with CaB and caspase-3 in the apoptosis, these peptides were cleaved. The RhB and FAM were then released from the GO surface and generated dual-signal fluorescence for the simultaneous sensing of CaB and caspase-3. The modifications of GFLGK-RhB and CALDEVD-FAM on the GO surface were characterized by atomic force microscopy (AFM), which confirmed the successful conjugation of the peptide on the GO surface. The fluorescence intensity of the nanosheet increased linearly with increase in CaB concentration in the range 0.25–5 U mL^{-1}. Further, the GO-RhB-FAM was found to provide fluorescence signals that were linearly related to the concentration of caspse-3 in the range 0.25–4 U mL^{-1}. The MNC was stable

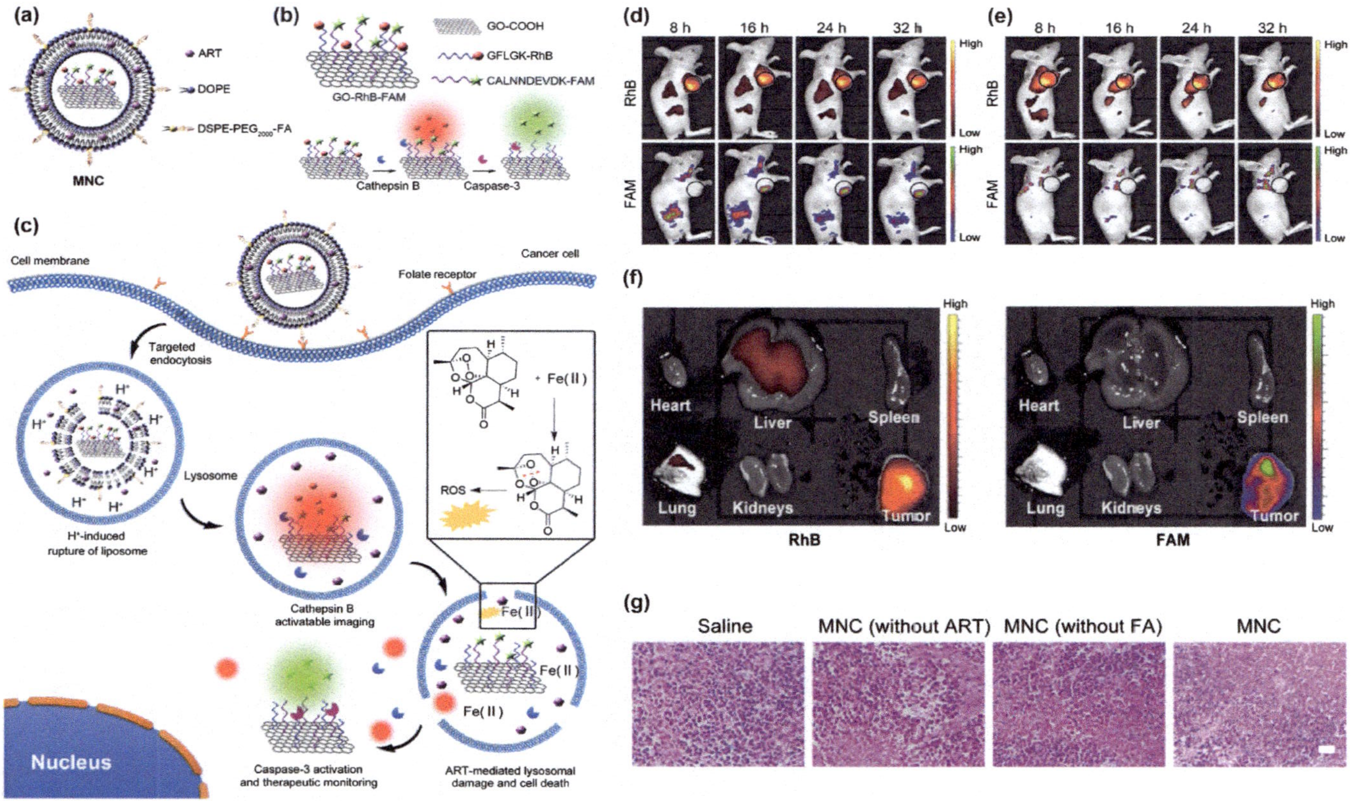

Figure 5.4 Schematic illustration of (a) MNC and (b) GO-RhB-FAM for simultaneous cathepsin B and caspase-3 activatable imaging and (c) dual protease light-up for imaging of lysosomal tumour cell death. *In vivo* fluorescence imaging of the HeLa tumour-bearing mice at different times after intravenous injection of (d) MNC or (e) MNC (without ART). (f) *Ex vivo* apoptosis imaging of tumour and normal tissues harvested from euthanized HeLa tumour-bearing nude mice at 32 h post-injection of MNC. (g) Histological observation of the tumour tissues from different treatment groups stained with H&E for apoptosis assay. Scale bar: 100 µm.[69] Reproduced from ref. 69 with permission from Elsevier, Copyright 2016.

and no significant release of ART was observed, while the rate of release of ART apparently increased in an acidic environment of pH 5.0, which ensured an effective ART delivery and release into the acidic lysosomes of tumour cells. Additionally, the presence of FA on the MNC specifically bound to FA receptor (FR) that was generally overexpressed on the membrane of tumour cells, which resulted in receptor-mediated endocytosis for particular imaging and drug delivery. Furthermore, At 32 h post-injection, the MNC-treated mice were killed and significant RhB and FAM fluorescence activation was found to occur in the tumour whereas the MNC (without ART)-treated tumour-bearing mice showed no FAM fluorescence signal in the tumour region, indicating that the MNC-induced anti-tumour effect was due to ART and the therapeutic efficacy would be self-monitored in real time. The MNC was successfully applied in tumour theranostics and implemented in living imaging in tumour-bearing mice, which was useful for *in vivo* applications, and this study substantially expands the perspective for the further fabrication of a near-infrared (700–1000 nm) nanoplatform to achieve precise tumour theranostics.

5.4 Real-time Monitoring of Caspase Cascade Activation

The evolution of the intracellular caspase family is crucial in cell apoptosis. To evaluate this process, numerous innovative detection probes have been designed. Compared with "always on" probes, activatable imaging probes have the advantage of monitoring cell apoptosis, which could provide valuable insights into the early detection of therapy efficiency and evaluation of disease progression. Activatable probes are designed typically based on a fluorogenic substrate for amplifying a fluorescent signal in response to specific biomolecular recognition in real time. Fluorogenic substrates include peptide substrates specific to the target enzyme, dye and quencher pairs and vectors to optimize stability or enhance the cellular uptake of the probe. Most of the activatable probes used in previous studies are suitable for targeting extracellular enzymes; however, only a few are capable of targeting intracellular enzymes to achieve real-time imaging. Ju and co-workers designed a universal platform for the *in situ* monitoring and evolution of intracellular caspase activation (Figure 5.5).[70] The platform employed a gold nanostructure as the nanocarrier to induce cell apoptosis by photothermal therapy, two kinds of dye-labelled peptides specific to upstream caspase-9 and downstream caspase-3 as the signal switch and FA as a targeting moiety. The peptides were cleaved by intracellular caspase activation and the fluorescence of the dyes was lit up sequentially, which could be used not only to quantify intracellular caspase-9 and -3 activities, but also for real-time monitoring of the therapeutic effect of a targeted cancer cell as a novel therapeutic response tool. Further, therapeutic efficiency could also be assessed by the recovered fluorescence in real time by monitoring caspase activity in a living mouse.

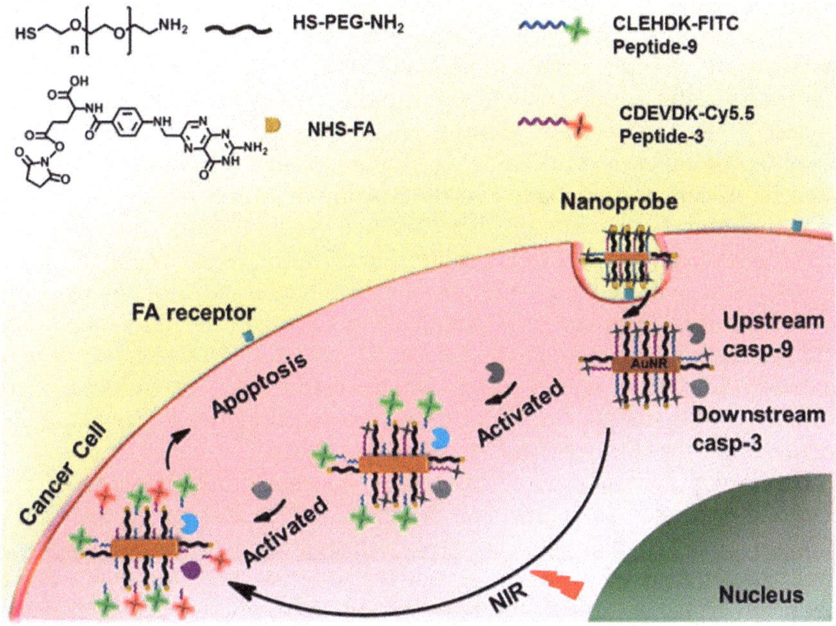

Figure 5.5 Schematic illustration of the integrated platform for *in situ* and real-time monitoring of the evolution of the caspase family activated *via* near-infrared photothermal therapy.[70]
Reproduced from ref. 70 with permission from the Royal Society of Chemistry.

Taking advantage of the high specificity of peptide substrates for caspases, this method is applicable to other protease assays by changing the substrate peptide sequence and provides a powerful tool to evaluate the intracellular cascade caspase activation. Zhu *et al.* designed a simple, versatile and robust one-step platform that could efficiently deliver highly quenched fluorogenic probes with distinctive pairs of dyes and quenchers to the initiator caspase-8 and the effector caspase-3 in MDA-MB-435 cells, and permits real-time imaging of multiple intracellular caspase activities during the apoptotic process induced by a tumour necrosis factor (TNF)-related apoptosis-induced ligand (TRAIL) in living cells. By applying the platform to study intracellular apoptotic processes, cell-based high-throughput screening systems for drug discovery might be realized in the future.[71] Although the above approach is only applicable with two caspases, by designing a series of different fluorogenic probes possessing different dyes, more complicated multiplexed imaging, targeting three or more caspases, can be performed.

Further, combined nanoparticles have also been used in the real-time monitoring of caspase cascade activation. Tang and co-workers reported an Au–Fe_2O_3 nanoparticle (NP) conjugated with integrin-targeting peptide (RGD) and fluorescein isothiocyanate (FITC)-labelled capsase-3 peptide (DEVD) for initiated catalytic formation of $^{\cdot}OH$ in HepG2 cells.[72]

This remarkable NP was capable of clearly visualizing the ·OH-induced apoptosis process within cells. Moreover, the catalytic efficiency of an Au–Fe_2O_3 NP is higher than that of individual γ-Fe_2O_3 NPs because the polarization effect at the Au–Fe_2O_3 interface is better. Such bifunctional Au–Fe_2O_3 NPs can serve as a multifunctional platform for simultaneous targeting, therapeutic and imaging functions and therefore have great potential in cancer diagnostic and therapeutic applications. In addition to the forms already mentioned, there are many other forms of probes and platforms that permit real-time monitoring of caspase. The probe is almost non-fluorescent in aqueous solution but display a significant fluorescence enhancement in response to caspase-3/-7, which are activated in the apoptotic process and able to cleave the DEVD moieties.[36] The aggregation of cleaved hydrophobic TPE residues, which restricts the intramolecular rotations of TPE's phenyl rings and populates the radiative decay channels, could explain this fluorescence "turn-on" response.

The development of specific bioprobes for the *in vivo* imaging of cell apoptosis and drug screening will be facilitated by further tuning of the emission spectrum of an AIE fluorogen to the red and near-infrared regions. Sun *et al.* constructed a protein sensor that consisted of a tandemly fused enhanced cyan fluorescent protein (ECFP), monomeric red fluorescent protein (mCherry) and monomeric yellow fluorescent protein (Venus) to evaluate simultaneously the activities of caspase-3/-9 in single cells by fluorescence cross-correlation spectroscopy (FCCS). Moreover, time-dependent monitoring of single cells by laser scanning fluorescence microscopy (LSM) revealed decreases in cross-correlation amplitudes between ECFP and mCherry and between mCherry and Venus, which predicted the fate of the cell in the early apoptotic stage.[73] When, where and how enzymes are activated at the tissue and organism levels are not clearly understood. One important reason is the lack of tools that are capable of directly imaging them in real time in deep tissues and small animals. Traditional methods such as immunoassays and colorimetric/fluorimetric techniques are not able to provide accurate information about their dynamics. Andrews *et al.* reported the first application of fluorescence lifetime imaging combined with optical projection tomography (FLIM OPT) to read out biological functions in live transgenic zebrafish larvae using a genetically expressed FRET biosensor. FLIM OPT visualizes apoptosis in live zebrafish larvae *via* changes in caspase-3 activation following gamma irradiation, monitored at 24 h post-fertilization. Significant apoptosis was observed at 3.5 h post-irradiation and only 150 s were required to acquire the 3D FLIM dataset, compared with the 300 s required to image a single optical section using LSCM with TCSPC, potentially imposing significantly less stress on the organism.[29] Tang and co-workers also developed a multicolour fluorescent nanoprobe denoted GNP-Se-Casp based on Au–Se bonding for real-time *in situ* monitoring sequential activation of caspase-3, -8 and -9 during cell apoptosis (Figure 5.6).[74] The nanoprobe exhibits high sensitivity and specificity, which effectively prevent the interference from biothiol in living cells.

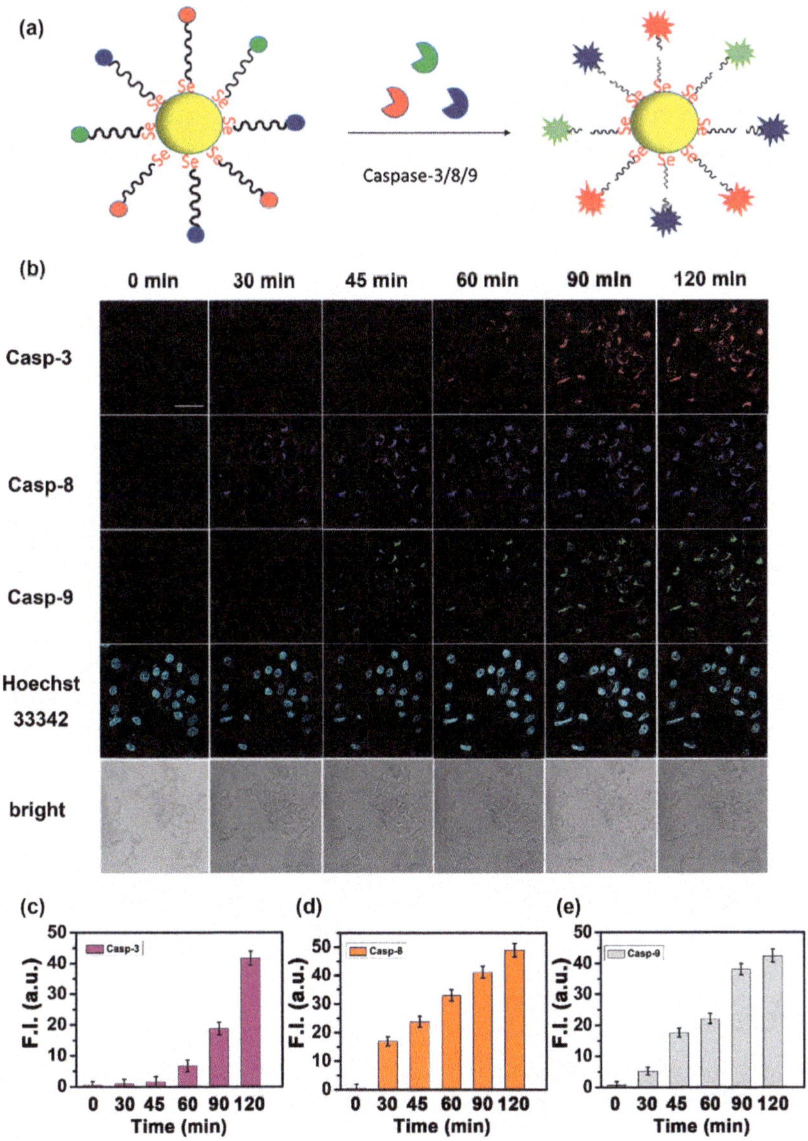

Figure 5.6 Schematic illustration of (a) the multicolour gold–selenium bonding fluorescent nanoprobe for (b) real-time fluorescence imaging of the cell apoptosis in HeLa cells. Scale bar: 50 μm. (c), (d) Quantitative fluorescence intensities of the nanoprobe induced by caspase-3, -8 and -9 activation in HeLa cells.[74]
Reproduced from ref. 74 with permission from American Chemical Society, Copyright 2019.

The nanoprobe was successfully utilized for imaging of the caspase cascade in cancer cell death, providing a novel method for the study of the complex relationships of signal molecules in signalling pathways. Briefly, the development of the activatable probes and platforms has great potential in real-time monitoring of the caspase cascade in living cells, which helps in understanding the complex apoptosis signal mechanisms and exploiting the potential role in both preclinical and clinical applications.

5.5 Perspective

In this chapter, the different types of optical imaging strategies proposed for evaluating intracellular caspases activity and cascade activation and also some duplex imaging methods simultaneously detecting caspases and other proteins have been discussed in detail. Multiplex fluorescence bioimaging methods and simultaneous determinations can effectively and sensitively detect the activity of caspases, which are available for apoptosis evaluation and inhibition monitoring. Based on cascade amplification strategies, the sensitivity and specificity can be improved to detect caspase activity. The observation of temporal and spatial variations of caspase activity in the cell and regulation of apoptosis in a proper way are very important for the discovery of new therapeutic strategies for the treatment of cancer, neurodegenerative diseases, *etc.* Significant tactics involve efficient screening of potential caspase inhibitors and the investigation of new anticancer methods and drugs, bringing new opportunities to tumour research. We hope that the described biosensors and imaging probes can be introduced as commercial tools in clinical laboratories together with other diagnostic biological devices.

Acknowledgement

Dr Jiangwei Tian is thanked for expert assistance with the writing of this chapter.

References

1. M. Poreba, K. Groborz, M. Navarro, S. J. Snipas, M. Drag and G. S. Salvesen, *Cell Death Differ.*, 2019, **26**(2), 229.
2. J. Hickson, S. Ackler, D. Klaubert, J. Bouska, P. Ellis, K. Foster, A. Oleksijew, L. Rodriguez, S. Schlessinger, B. Wang and D. Frost, *Cell Death Differ.*, 2010, **17**(6), 1003.
3. C. E. Shi, C. Q. You and L. Pan, *Nanotechnology*, 2019, **3**, 2.
4. M. K. Barton, *Ca-Cancer J. Clin.*, 2017, **67**(1), 3.
5. Z. Y. He, Y. G. Zhang, Y. H. Yang, C. C. Ma, P. Wang, W. Du, L. Li, R. Xiang, X. R. Song, X. Zhao, S. H. Yao and Y. Q. Wei, *Hum. Gene Ther.*, 2018, **29**(2), 223.
6. A. S. Thakor and S. S. Gambhir, *Ca-Cancer J. Clin.*, 2013, **63**(6), 395.
7. J. Tian, L. Ding, H. J. Xu, Z. Shen, H. Ju, L. Jia, L. Bao and J. S. Yu, *J. Am. Chem. Soc.*, 2013, **135**(50), 18850.

8. A. M. Hunter, E. C. Lacasse and R. G. Korneluk, *Apoptosis*, 2007, **12**(9), 1543.
9. X. Huang, S. Lee and X. Chen, *Am. J. Nucl. Med. Mol. Imaging*, 2011, **1**(1), 3.
10. A. U. Lüthi and S. J. Martin, *Cell Death Differ.*, 2007, **14**(4), 641.
11. H. Kobayashi, M. Ogawa, R. Alford, P. L. Choyke and Y. Urano, *Chem. Rev.*, 2010, **110**(5), 2620.
12. M. Gao, F. Yu, C. Lv, J. Choo and L. Chen, *Chem. Soc. Rev.*, 2017, **46**(8), 2237.
13. S. Zadran, S. Standley, K. Wong, E. Otiniano, A. Amighi and M. Baudry, *Appl. Microbiol. Biotechnol.*, 2012, **96**(4), 895.
14. Y. Shaulov-Rotem, E. Merquiol, T. Weiss-Sadan, O. Moshel, S. Salpeter, D. Shabat, F. Kaschani, M. Kaiser and G. Blum, *Chem. Sci.*, 2016, **7**(2), 1322.
15. S. Y. Li, L. H. Liu, H. Cheng, B. Li, W. X. Qiu and X. Z. Zhang, *Chem. Commun.*, 2015, **51**(77), 14520.
16. Y. Yuan, R. Zhang, X. Cheng, S. Xu and B. Liu, *Chem. Sci.*, 2016, **7**(7), 4245.
17. Y. Yamaguchi, N. Shinotsuka, K. Nonomura, K. Takemoto, K. Kuida, H. Yosida and M. Miura, *J. Cell Biol.*, 2011, **195**(6), 1047.
18. H. Zheng and E. H. Koo, *Mol. Neurodegener.*, 2011, **6**(1), 27.
19. Y. Luo, L. Huang, Y. Yang, X. Zhuang, S. Hu, H. Ju, B. Y. Yu and J. Tian, *Theranostics*, 2017, **7**(5), 1245.
20. M. Oishi, A. Tamura, T. Nakamura and Y. Nagasaki, *Adv. Funct. Mater.*, 2009, **19**(6), 827.
21. L. Feng, L. Wu and X. Qu, *Adv. Mater.*, 2013, **25**(2), 168.
22. J. Tian, Y. Luo, L. Huang, Y. Feng, H. Ju and B. Y. Yu, *Biosens. Bioelectron.*, 2016, **80**, 519.
23. R. E. Campbell, *Anal. Chem.*, 2009, **81**(15), 5972.
24. Y. Ding, H. W. Ai, H. Hoi and R. E. Campbell, *Anal. Chem.*, 2011, **83**(24), 9687.
25. B. K. Wagner and S. L. Schreiber, *Cell Chem. Biol.*, 2016, **23**(1), 3.
26. D. M. Shcherbakova, M. A. Hink, L. Joosen, T. W. J. Gadella and V. V. Verkhusha, *J. Am. Chem. Soc.*, 2012, **134**(18), 7913.
27. M. Y. Berezin and S. Achilefu, *Chem. Rev.*, 2010, **110**(5), 2641.
28. L. Zhang, J. Lei, J. Liu, F. Ma and H. Ju, *Biomaterials*, 2015, **67**, 323.
29. N. Andrews, M. C. Ramel, S. Kumar, Y. Alexandrov, D. J. Kelly, S. C. Warren, L. Kerry, N. Lockwood, A. Frolov, P. Frankel, L. Bugeon, J. McGinty, M. J. Dallman and P. M. French, *J. Biophotonics*, 2016, **9**(4), 414.
30. A. P. Savitsky, A. L. Rusanov, V. V. Zherdeva, T. V. Gorodnicheva, M. G. Khrenova and A. V. Nemukhin, *Theranostics*, 2012, **2**(2), 215.
31. J. M. Buschhaus, A. E. Gibbons, K. E. Luker and G. D. Luker, *Curr. Protoc. Cell Biol.*, 2017, **77**, 21.12.1.
32. J. M. Buschhaus, B. Humphries, K. E. Luker and G. D. Luker, *Cells*, 2018, **7**(6), E57.

33. A. Orte, J. M. Alvarez-Pez and M. J. Ruedas-Rama, *ACS Nano*, 2013, 7(7), 6387.
34. K. Okabe, N. Inada, C. Gota, Y. Harada, T. Funatsu and S. Uchiyama, *Nat. Commun.*, 2012, **3**, 705.
35. J. Mei, N. L. Leung, R. T. Kwok, J. W. Lam and B. Z. Tang, *Chem. Rev.*, 2015, **115**(21), 11718.
36. H. Shi, R. T. Kwok, J. Liu, B. Xing, B. Z. Tang and B. Liu, *J. Am. Chem. Soc.*, 2012, **134**(43), 17972.
37. J. Liang, H. Shi, R. T. Kwok, M. Gao, Y. Yuan, W. Zhang, B. Z. Tang and B. Liu, *J. Mater. Chem. B*, 2014, **2**(27), 4363.
38. H. Shi, N. Zhao, D. Ding, J. Liang, B. Z. Tang and B. Liu, *Org. Biomol. Chem.*, 2013, **11**(42), 7289.
39. D. Ding, J. Liang, H. Shi, R. T. K. Kwok, M. Gao, G. Feng and B. Liu, *J. Mater. Chem. B*, 2014, **2**(2), 231.
40. Y. Yuan, C. J. Zhang, R. T. K. Kwok, D. Mao, B. Z. Tang and B. Liu, *Chem. Sci.*, 2017, **8**(4), 2723.
41. Y. Yuan, C. J. Zhang, R. T. K. Kwok, S. Xu, R. Zhang, J. Wu, B. Z. Tang and B. Liu, *Adv. Funct. Mater.*, 2015, **25**(42), 6586.
42. M. Hu, L. Li, H. Wu, Y. Su and P. Yang, *J. Am. Chem. Soc.*, 2011, **133**(31), 12009.
43. M. Jiang, X. Gu, J. W. Y. Lam, Y. Zhang, R. T. K. Kwok, K. S. Wong and B. Z. Tang, *Chem. Sci.*, 2017, **8**(8), 5440.
44. S. Wang, W. Wu, P. Manghnani, S. Xu, Y. Wang, C. C. Goh, L. G. Ng and B. Liu, *ACS Nano*, 2019, **13**(3), 3095.
45. H. Li, Y. Li, Q. Yao, J. Fan, W. Sun, S. Long, K. Shao, J. Du, J. Wang and X. Peng, *Chem. Sci.*, 2018, **10**(6), 1619.
46. H. Yan, L. He, W. Zhao, J. Li, Y. Xiao, R. Yang and W. Tan, *Anal. Chem.*, 2014, **86**(22), 11440.
47. Q. Ge, N. Wang, J. Li and R. Yang, *Med. Chem. Commun.*, 2017, **8**, 1435.
48. T. F. Sergeeva, M. V. Shirmanova, O. A. Zlobovskaya, A. I. Gavrina, V. V. Dudenkova, M. M. Lukina, K. A. Lukyanov and E. V. Zagaynova, *Biochim. Biophys. Acta, Mol. Cell Res.*, 2017, **1864**(3), 604.
49. L. Qian, C. W. Zhang, Y. Mao, L. Li, N. Gao, K. L. Lim, Q. H. Xu and S. Q. Yao, *Sci. Rep.*, 2016, **6**(1), 26385.
50. L. Zhang, J. Lei, F. Ma, P. Ling, J. Liu and H. Ju, *Chem. Commun.*, 2015, **51**(54), 10831.
51. H. Wang, Q. Zhang, X. Chu, T. Chen, J. Ge and R. Yu, *Angew. Chem., Int. Ed.*, 2011, **50**(31), 7065.
52. Y. Liu, Y. Hu, Y. Guo, H. Ma, J. Li and C. Jiang, *J. Controlled Release*, 2012, **163**(2), 203.
53. Y. Li, P. Li, R. Zhu, C. Luo, H. Li, S. Hu, Z. Nie, Y. Huang and S. Yao, *Anal. Chem.*, 2016, **88**(22), 11184.
54. A. den Hamer, P. Dierickx, R. Arts, J. S. P. M. de Vries, L. Brunsveld and M. Merkx, *ACS Sens.*, 2017, **2**(6), 729.
55. X. Li, Y. Li, Q. Qiu, Q. Wen, Q. Zhang, W. Yang, L. Yuwen, L. Weng and L. Wang, *J. Colloid Interface Sci.*, 2019, **543**, 96.

56. D. Ye, A. J. Shuhendler, L. Cui, L. Tong, S. S. Tee, G. Tikhomirov, D. W. Felsher and J. Rao, *Nat. Chem.*, 2014, **6**(6), 519.
57. A. Godinat, H. M. Park, S. C. Miller, K. Cheng, D. Hanahan, L. E. Sanman, M. Bogyo, A. Yu, G. F. Nikitin, A. Stahl and E. A. Dubikovskaya, *ACS Chem. Biol.*, 2013, **8**(5), 987.
58. A. Godinat, A. A. Bazhin and E. A. Goun, *Drug Discovery Today*, 2018, **23**(9), 1584.
59. M. Scabini, F. Stellari, P. Cappella, S. Rizzitano, G. Texido and E. Pesenti, *Apoptosis*, 2011, **16**(2), 198.
60. M. Kindermann, H. R. Voser, D. Cagli, U. Repnik, C. Miniejew, P. R. E. Mittl, G. Kosec, M. G. Grutter, B. Turk and K. U. Wend, *Chem. Biol.*, 2010, **17**(9), 999.
61. G. Niu, L. Zhu, D. N. Ho, F. Zhang, H. K. Gao, Q. M. Quan, N. Hida, T. Ozawa, G. Liu and X. Y. Chen, *Theranostics*, 2013, **3**(3), 190.
62. G. C. Van de Bittner, C. R. Bertozzi and C. J. Chang, *J. Am. Chem. Soc.*, 2013, **135**(5), 1783.
63. J. M. Brown and L. D. Attardi, *Nat. Rev. Cancer*, 2005, **5**(3), 231.
64. G. Niu and X. Chen, *J. Nucl. Med.*, 2010, **51**(11), 1659.
65. S. J. Riedl and Y. Shi, *Nat. Rev. Mol. Cell Biol.*, 2004, **5**(11), 897.
66. H. Okada and T. Mak, *Nat. Rev. Cancer*, 2004, **4**(8), 592.
67. J. J. Chiang and K. Truong, *Biotechnol. Lett.*, 2005, **27**(16), 1219.
68. Q. Wen, X. Zhang, J. Cai and P. H. Yang, *Analyst*, 2014, **139**(10), 2499.
69. L. Huang, Y. Luo, X. Sun, H. Ju, J. Tian and B. Y. Yu, *Biosens. Bioelectron.*, 2017, **92**, 724.
70. L. Zhang, J. Lei, J. Liu, F. Ma and H. Ju, *Chem. Sci.*, 2015, **6**(6), 3365.
71. L. Zhu, X. L. Huang, K. Y. Choi, Y. Ma, F. Zhang, G. Liu, S. Lee and X. Y. Chen, *J. Controlled Release*, 2012, **163**(1), 55.
72. W. Gao, L. F. Ji, L. Li, G. W. Cui, K. H. Xu, P. Li and B. Tang, *Biomaterials*, 2012, **33**(14), 3710.
73. F. Sun, S. Mikuni and M. Kinjo, *Biochem. Biophys. Res. Commun.*, 2011, **404**(2), 706.
74. X. Liu, X. Song, D. Luan, B. Hu, K. Xu and B. Tang, *Anal. Chem.*, 2019, **91**, 5994.

CHAPTER 6

Visualization of Intracellular Glycosylation

6.1 Introduction to Intracellular Glycosylation

Most eukaryotic proteins are glycosylated with various oligosaccharides,[1] which play important roles in a wide variety of biological processes. The oligosaccharides are mainly composed of 10 kinds of monosaccharides,[2] which are connected through glycosyltransferase (GT)-catalysed substitution from nucleotide pyrophosphate. The oligosaccharides on proteins can be further pruned by glycosidases (GAs) to maintain specific glycoforms, which are essential for the regulation of the structure, function and localization of proteins. The glycosylation of proteins can be classified into two major forms, the *N*-linked glycans and the *O*-linked glycans. In intracellular environments, *O*-GlcNAcylation is the only form of the *O*-linked glycans and most *N*-linked glycans contain the sialylation process. Thus, *O*-GlcNAcylation and sialylation are the two main events of intracellular glycosylation.

6.1.1 *O*-GlcNAcylation

Many intracellular proteins contain the monosaccharide *O*-linked *N*-acetylglucosamine (*O*-GlcNAc) attached to serine or threonine residues *via* *O*-GlcNAcylation.[3] *O*-GlcNAcylation is essential for cell survival and plays important roles in many biological processes, such as transcription, translation and cell division.[4] Generally, the *O*-GlcNAcylation is dynamically regulated by two key enzymes, *O*-GlcNAc transferase (OGT)[5] and *O*-GlcNAcase (OGA),[6] which lead to the addition or removal of GlcNAc residues to or from proteins, respectively. The *O*-GlcNAc modification is often subject to a complex competition process with protein phosphorylation,

Detection Science Series No. 16
In Situ Analysis of Cellular Functional Molecules
By Huangxian Ju, Bo Tang and Lin Ding

Published by the Royal Society of Chemistry, www.rsc.org

which normally occurs at the same or nearby sites and affects important signalling pathways. Further, *O*-GlcNAc may influence the activity of kinases and phosphatases,[7] and also the glucose level available to the cell due to the hexosamine pathway.[8] All these features implicate *O*-GlcNAc in diverse cellular signalling and disease states.

Although interest in *O*-GlcNAc continues to grow, the study of this modification is still a great challenge. The main reason is that *O*-GlcNAcylation is a small uncharged and sub-stoichiometric modification.[9] *O*-GlcNAc is chemically and enzymatically labile,[10] hence is difficult to label it by chemical or biological methods. Methods based on mass spectrometry are also difficult for the study of *O*-GlcNAcylated proteins owing to the loss of *O*-GlcNAc caused by collision-induced dissociation,[11] and it cannot be used for *in situ* visualization of *O*-GlcNAcylation inside cells. In addition, the great challenge of generating *O*-GlcNAc antibodies and the difficulty in predicting *O*-GlcNAc modification sites further impede the study of *O*-GlcNAcylation.[12–14] However, some recent developments in chemical and biological methods have achieved the monitoring and manipulation of *O*-GlcNAcylation inside cells. In these methods, *O*-GlcNAc was functionalized with unnatural chemical reporter groups through metabolism. Following subsequent ligation of an exogenously delivered probe through a bioorthogonal reaction, the *O*-GlcNAcylated proteins can be either imaged or isolated. Some other methods based on *O*-GlcNAc-recognizable probes have also been used for the *in situ* monitoring of *O*-GlcNAcylation inside living cells.

6.1.2 Sialylation

As another important glycosylation mode, sialylation refers to the addition of sialic acid (SA) to the terminal position of the glycan chain on glycosylated proteins. The addition of SA to the glycan chain often occurs in the endoplasmic reticulum or the Golgi apparatus, and is mediated by different sialyltransferases (STs).[15] STs can be classified into different families according to the different carbohydrate linkages.[16] The activity of STs can lead to variations of tumour-associated carbohydrates on the cell surface.[17] SA at the terminus of the glycan chain can be removed by neuraminidases (Neus), also called sialidases, which are located in different cellular compartments, such as lysosomes, the cytosol, the plasma membrane and mitochondria.[18,19] Neus play important roles in regulating SA expression and further reflect the physiological and pathological states of cells.[20] Hence the assessment of STs and Neus inside living cells is of great importance for the study of the sialylation of proteins, which is a promising approach for the development of cancer-related diagnostic and therapeutic tools.

Common *in vitro* methods for the study of sialylation need to separate STs or Neus from cellular lysate and transfer radio- or fluorescently labelled sugars to the corresponding sugar nucleotides.[21–26] These methods are limited by the complicated separation protocol and they cannot provide the dynamic activity of the corresponding enzymes inside living cells. Recently, some fluorescence resonance energy transfer (FRET)-based methods have been developed to achieve the *in situ* visualization of ST or Neus activities on the cell surface[27] or inside living cells. In these methods, some enzyme-responsive probes were designed. By using different cellular delivery strategies, STs and Neus located in different cellular compartments were visualized and their activities were dynamically monitored.

6.2 Visualization of Intracellular *O*-GlcNAcylation

6.2.1 Protein-specific *O*-GlcNAcylation

An "on–off" signal switch is essential for intracellular visualization, because the excess exogenous fluorescent probes can bring a strong fluorescent background. FRET, which is based on a pair of fluorescent donor and receptor, is an ideal signal switch for intracellular imaging. Only when the donor and receptor are close enough can FRET occur. Hence FRET is a convenient tool for the visualization of intracellular protein-specific *O*-GlcNAcylation.

Lin *et al.* developed a FRET-based method to image *O*-GlcNAcylation inside a single cell (Figure 6.1).[28] They tagged the target protein with the genetically encoded enhanced green fluorescence protein (EGFP) as the fluorescence donor and labelled the *O*-GlcNAc moieties on proteins with azide through metabolic labelling.[29,30] By subsequent conjugation of the azide with tetramethylrhodamine (TAMRA) alkyne as the fluorescence

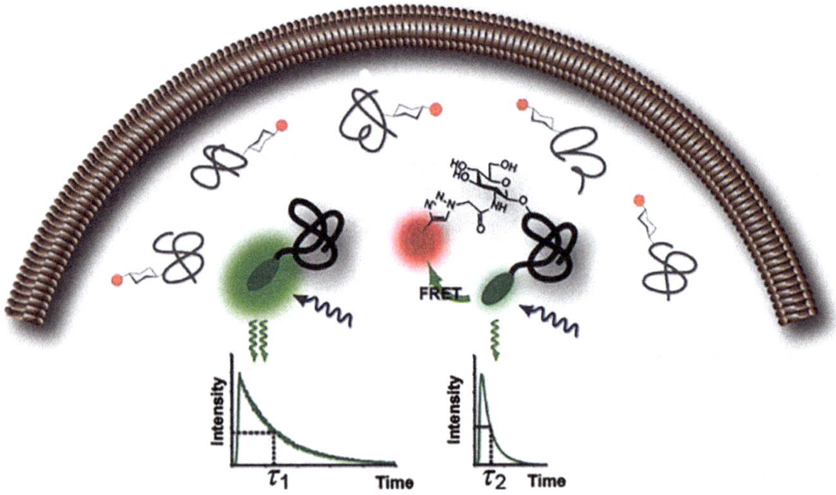

Figure 6.1 Schematic illustration of protein-specific imaging of *O*-GlcNAcylation by FLIM–FRET microscopy in single cells.[28]
Reproduced from ref. 28 with permission from John Wiley and Sons, © 2015 WILEY-VCH Verlag GmbH & Co. KGaA, Weinheim.

acceptor through BTTAA-assisted click chemistry,[31] the *O*-GlcNAc on the target protein can be imaged by fluorescence lifetime imaging microscopy (FLIM).[32]

Lin *et al.*[28] first detected the microtubule-associated protein tau, which can lead to the formation of intracellular neurofibrillary tangles.[33] They[28] genetically fused EGFP to the C-terminus of the tau in CHO cells, which were then treated with peracetylated *N*-azidoacetylgalactosamine ($Ac_4GalNAz$).[34] TAMRA alkyne was then conjugated to the azide moieties to form the EGFP–TAMRA FRET donor–acceptor pair. *In vitro* western blotting indicated the successful dual labelling of the tau protein (Figure 6.2A). In the FLIM images of a single cell, an obvious decrease of the donor fluorescence lifetime was observed (Figure 6.2B–D) compared with the cell treated with a non-azide metabolic reagent.

After confirming the feasibility of the designed method, Lin *et al.*[28] used it to image the changes in tau *O*-GlcNAcylation by inhibiting the OGT or OGA. $Ac_45SGlcNAc$ and thiamet-G are inhibitors of *O*-GlcNAc transferase (OGT) and *O*-GlcNAcase (OGA),[35–37] respectively. After the cells had been treated with $Ac_45SGlcNAc$ or thiamet-G, the donor fluorescence lifetime exhibited an

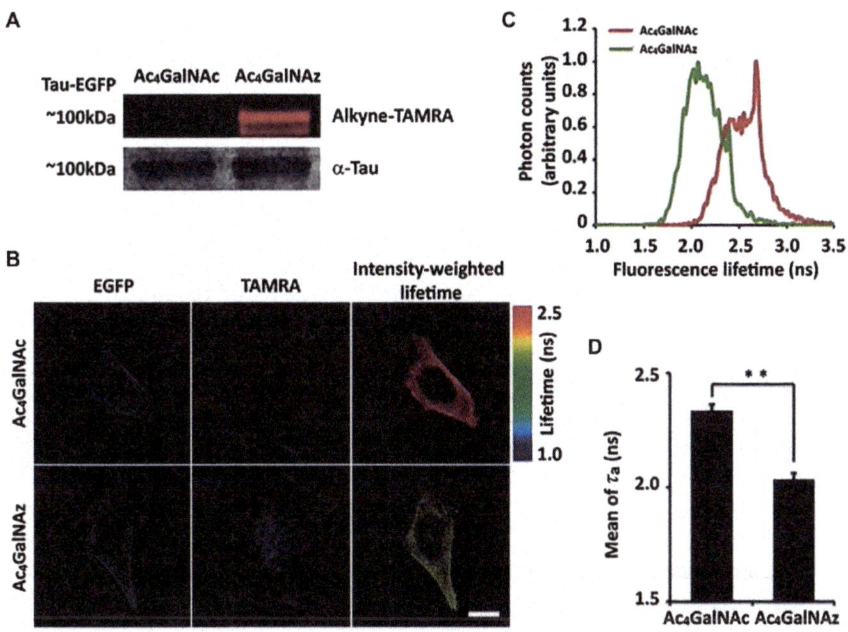

Figure 6.2 (A) *In vitro* characterization of dual-labelled tau. (B) FLIM–FRET images of $Ac_4GalNAc$- or $Ac_4GalNAz$-treated tau–EGFP-expressing CHO cells. Scale bar: 10 mm. (C) Histograms of the distribution of τ values in a single cell treated with Ac4GalNAc or Ac4GalNAz. (D) Statistical analysis of the average lifetime of a single cell (τ_a).[28]
Reproduced from ref. 28 with permission from John Wiley and Sons, © 2015 WILEY-VCH Verlag GmbH & Co. KGaA, Weinheim.

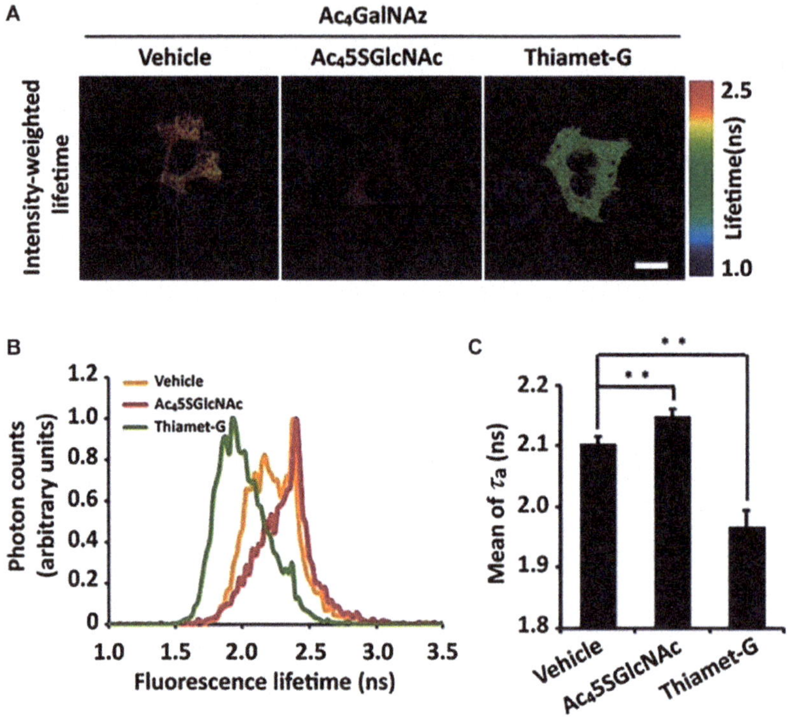

Figure 6.3 (A) Intensity-weighted EGFP lifetime images of tau *O*-GlcNAc inside the cells induced by the OGT and OGA inhibitors. Scale bar: 10 mm. (B) Histograms of the distribution of EGFP lifetime values in cells treated with vehicle, Ac45SGlcNAc and thiamet-G. (C) Statistical analysis of the average lifetime of a single cell (τ_a).[28]
Reproduced from ref. 28 with permission from John Wiley and Sons, © 2015 WILEY-VCH Verlag GmbH & Co. KGaA, Weinheim.

obvious increase or decrease (Figure 6.3), which indicated that FLIM–FRET imaging could be used for screening small molecules that can specifically alter the *O*-GlcNAcylation of the target protein. The authors further used the method to visualize *O*-GlcNAcylation of other proteins, such as β-catenin[38] in MDCK II cells, which indicated the broad applicability of the designed method.

However, the method developed by Lin *et al.* suffered from cytotoxicity of the BTTAA-assisted click chemistry, and thus can not be applied to living cells. Doll *et al.* developed an improved method by using a peracetylated methylcyclopropene-tagged GlcNAc (Ac_4GlcNCyoc) to label intracellular *O*-GlcNAc metabolically with a methylcyclopropene group (Figure 6.4).[39] They also used EGFP to genetically label the target protein as the FRET donor. The acceptor fluorophore conjugated with tetrazine (Tz) was attached to the metabolically incorporated sugar by an inverse electron demand Diels–Alder (DAinv) reaction.[40–42] Only *O*-GlcNAcylation of the target protein

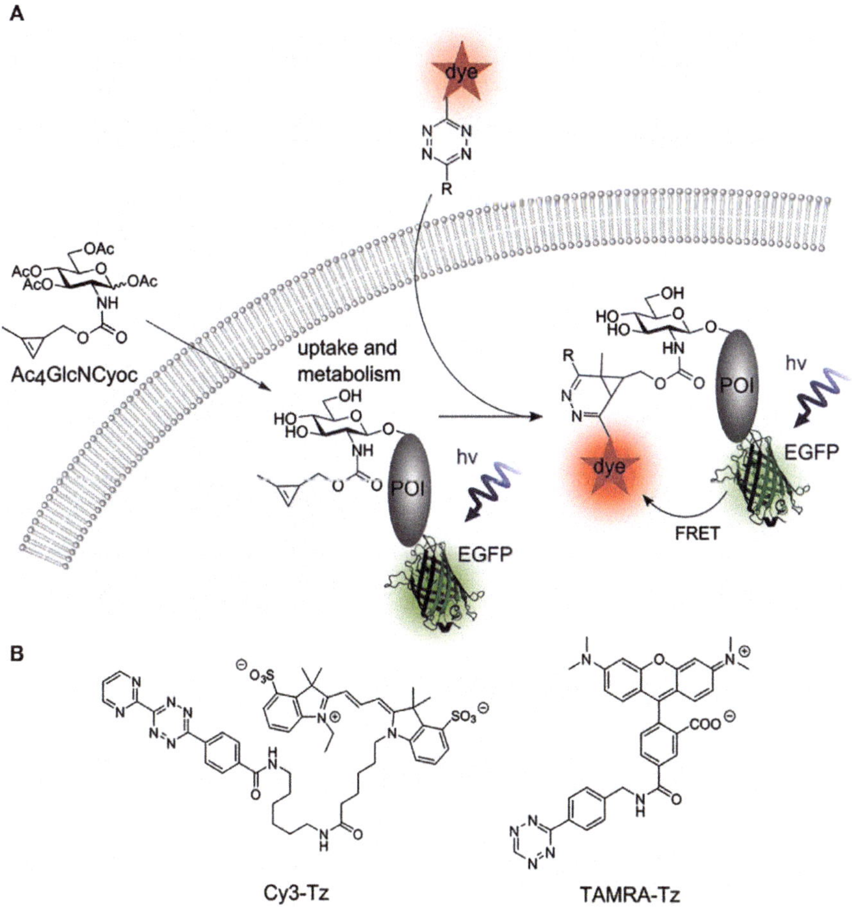

Figure 6.4 (A) Schematic illustration of protein-specific imaging of *O*-GlcNAcylation in living cells. (B) Structures of Cy3–tetrazine (Cy3–Tz) and TAMRA–tetrazine (TAMRA–Tz).[39]
Reproduced from ref. 39 with permission from John Wiley and Sons, © 2015 WILEY-VCH Verlag GmbH & Co. KGaA, Weinheim.

can lead to FRET between the EGFP donor and the acceptor fluorophore, which could be monitored by FLIM.

Doll *et al.* first used OGT as the model protein to establish their method.[39] OGT is an *O*-GlcNAcylated protein without any other type of glycosylation.[43] They transfected HEK293T cells with an EGFP–OGT expression vector or an expression vector for EGFP as control. The cells were then treated with Ac_4GlcNCyoc or Ac_4GlcNAc as control. The cells obtained were lysed and subjected to western blotting using anti-GFP antibody and labelled with Cy3–Tz. The glycosylated proteins were detected by measuring the fluorescence of Cy3 (Figure 6.5A), which indicated that EGFP was not glycosylated and the OGT was successfully modified with Ac_4GlcNCyoc. To detect

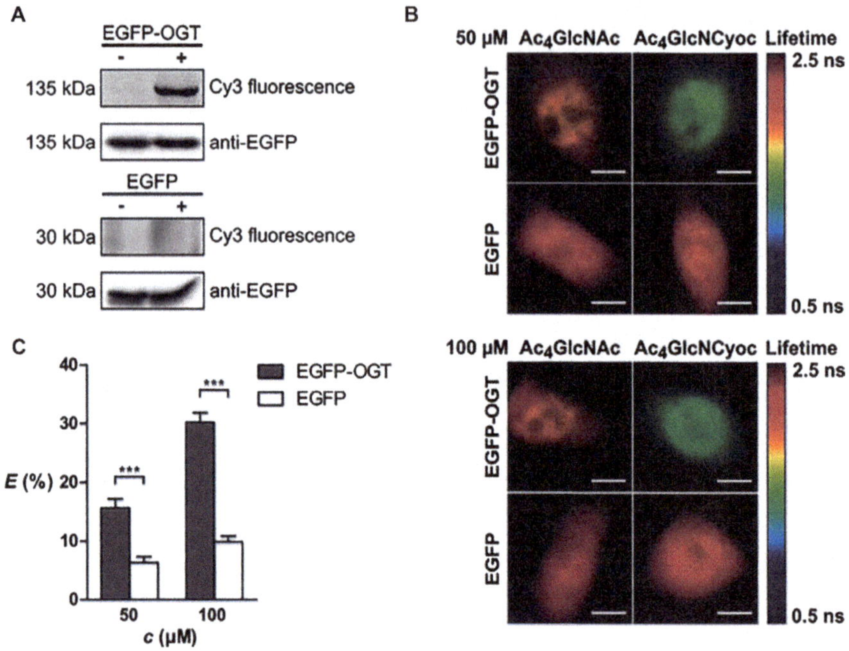

Figure 6.5 (A) *In vitro* confirmation of the incorporation of $Ac_4GlcNCyoc$. (B) Pseudocoloured and intensity-weighted fluorescence modulation lifetime images of EGFP–OGT- and EGFP-transfected cells. Scale bars: 10 μm. (C) FRET efficiencies (E) for cells using sugar concentrations (c) of 50 and 100 μM.[39]
Reproduced from ref. 39 with permission from John Wiley and Sons,

the *O*-GlcNAcylation of OGT in living cells, they used cell-permeable TAMRA–Tz dye instead of Cy3–Tz, which did not influence cell viability. After treatment with TAMRA–Tz, the fluorescence lifetime of the EGFP–OGT-transferred and $Ac_4GlcNCyoc$-treated cell exhibited a strong decrease compared with to the EGFP-transferred or $Ac_4GlcNAc$-treated cell (Figure 6.5B). To determine the FRET efficiencies, they compared the fluorescence lifetimes of $Ac_4GlcNCyoc$- and $Ac_4GlcNAc$-treated cells for each protein (Figure 6.5C) and the results indicated that intermolecular FRET is significantly weaker than protein-specific intramolecular FRET.

Since this method could be successfully used for the visualization of *O*-GlcNAcylated OGT, they applied the method to a further five types of intracellular proteins (the forkhead transcription factor Foxo1, the tumour suppressor p53, the serine/threonine kinase 1 Akt1, the cytoplasmic actin-binding protein vinculin and the calcium/calmodulin-dependent protein kinase CAMK4).[39] The western blot results indicated that Foxo1–EGFP, p53–EGFP, Akt1–EGFP and CAMK4–EGFP were modified with $Ac_4GlcNCyoc$, whereas EGFP–vinculin was not (Figure 6.6A). The FLIM images exhibited notable changes in the fluorescence lifetimes of Foxo1–EGFP, p53–EGFP and

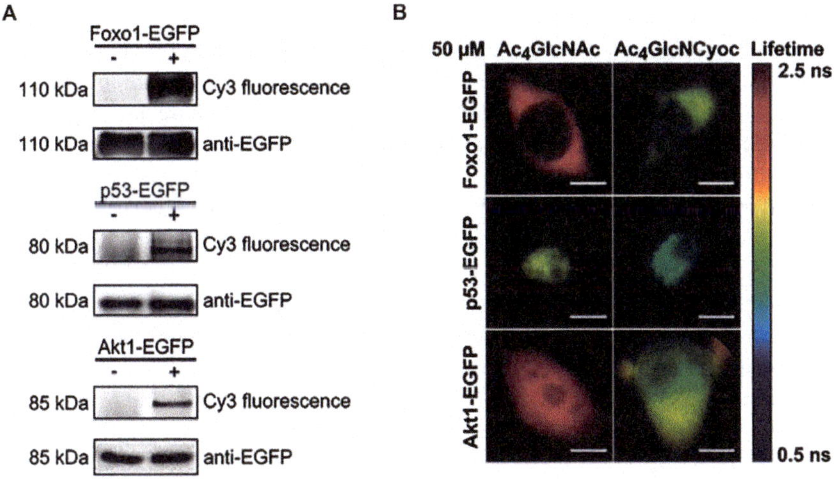

Figure 6.6 Glycosylation of Foxo1, p53 and Akt1 visualized in living cells. (A) *In vitro* confirmation of the incorporation of Ac_4GlcNCyoc into Foxo1–EGFP, p53–EGFP and Akt1–EGFP. (B) Representative fluorescence modulation lifetime images of Foxo1–EGFP-, p53–EGFP- and Akt1–EGFP-transfected cells. Scale bars: 10 μm. (C) Comparison of the FRET efficiencies (E) of all of the studied proteins using sugar concentrations (c) of 50 or 100 μM.[39] Reproduced from ref. 39 with permission from John Wiley and Sons, © 2015 WILEY-VCH Verlag GmbH & Co. KGaA, Weinheim.

Akt1–EGFP upon treatment with 50 μM Ac_4GlcNCyoc (Figure 6.6B). Further, the localization of Foxo1–EGFP was mainly in cytoplasmic, p53–EGFP was exclusively found in the nucleus and Akt1–EGFP was present in both the cytoplasm and nucleus (Figure 6.6B). Whereas the fluorescence lifetimes of Foxo1–EGFP and p53–EGFP were spatially invariant throughout the cell, that of Akt1–EGFP decreased more strongly in the nucleus than in the cytoplasm (Figure 6.6B). Further, the FRET efficiencies for EGFP–OGT, Foxo1–EGFP, p53–EGFP and Akt1–EGFP were well above the background FRET obtained from EGFP at both concentrations. Finally, to ensure that the incorporation of Ac_4GlcNCyoc is dependent on OGT activity, they overexpressed EGFP–OGT, which led to increased incorporation of the modified sugar. Moreover, treatment of the cell lysates with the OGT inhibitor ($Ac_4$5SGlcNAc) resulted in reduced incorporation of Ac_4GlcNCyoc. In addition, they incubated the lysates of Ac_4GlcNCyoc-treated cells with biotin–Tz and adjusted the pH to 12, which resulted in the disappearance of both the *O*-GlcNAc and biotin signals on western blots. These results indicated that Ac_4GlcNCyoc is enzymatically incorporated into *O*-GlcNAcylated proteins.

6.2.2 General *O*-GlcNAcylation

In mammalian cells, more than 1000 kinds of proteins have been identified as *O*-GlcNAcylated proteins.[44] Although FRET-based methods can be used for the visualization of protein-specific *O*-GlcNAcylation, they need to genetically label the proteins with EGFP, and thus cannot give an overview of the *O*-GlcNAcylation state within the whole cell. Generally, the *O*-GlcNAcylation state inside the living cell is mainly composed of *O*-GlcNAcylated proteins and dissociated GlcNAc residues, whereas other components such as GlcNAc phosphates are very limited.[45] In view of this, Chen *et al.* designed a functional dual-colour indicator, which could simultaneously visualize the level of *O*-GlcNAcylated proteins and the total amount of GlcNAc groups inside living cells.[46]

This indicator was constructed on a poly-GlcNAc (PGlcNAc)-coated gold nanoparticle (AuNP) (PGAu) by assembling fluorescent dye-labelled succinylated wheat germ agglutinin (FSWGA) and then another fluorescent dye-labelled GlcNAc (FGlcNAc) (Figure 6.7A). These fluorescent dyes possessed negligible overlapping emission and were quenched by the AuNP.

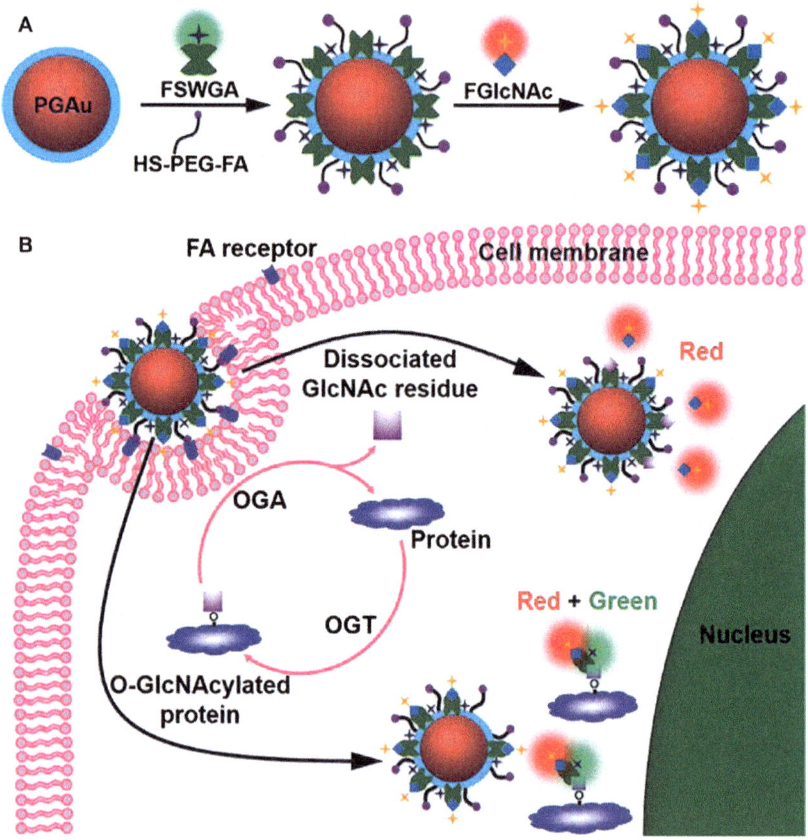

Figure 6.7 Schematic illustration of (A) preparation of functional dual-colour indicator and (B) *in situ* visualization of intracellular OGlcNAcylation.[46] Reproduced from ref. 46 with permission from American Chemical Society, Copyright 2018.

Poly(ethylene glycol) terminated with sulfhydryl and folic acid (HS-PEG-FA) was also doped on its surface to improve its stability and endocytosis efficiency. FSWGA possessed a dimeric subunit structure to connect PGlcNAc and FGlcNAc.[47–50] Different from the native wheat germ agglutinin,[51] FSWGA contains partially obstructed binding sites on each subunit by succinylation to maintain its specificity to GlcNAc.[52] The competitive binding of free GlcNAc residue to the outside subunit led to the release of FGlcNAc and thus lit up the red fluorescence of Alexa Fluor 647 on FGlcNAc. Meanwhile, *O*-GlcNAcylated proteins could competitively bind the inside subunit of FSWGA, which released the conjugate of FSWGA with FGlcNAc to light up both the green fluorescence of fluorescein isothiocyanate (FITC) on FSWGA and the red fluorescence of FGlcNAc (Figure 6.7B).

Chen *et al.* first tested the competitive release ability of the indicator *in vitro* (Figure 6.8).[46] The indicator was incubated with cysteine (the main biogenic thiol inside the cell), glucose, GlcNAc, human p53 protein (a model *O*-GlcNAcylated protein)[39] and MCF-7 cell lysates. The indicator did not exhibit any self-fluorescence and remained non-fluorescent when incubated with cysteine or glucose, indicating its stability towards biogenic thiol and specificity to GlcNAc. The mixture of the indicator with cell lysates showed obvious green and red fluorescence of both FITC and Alexa Fluor 647, which

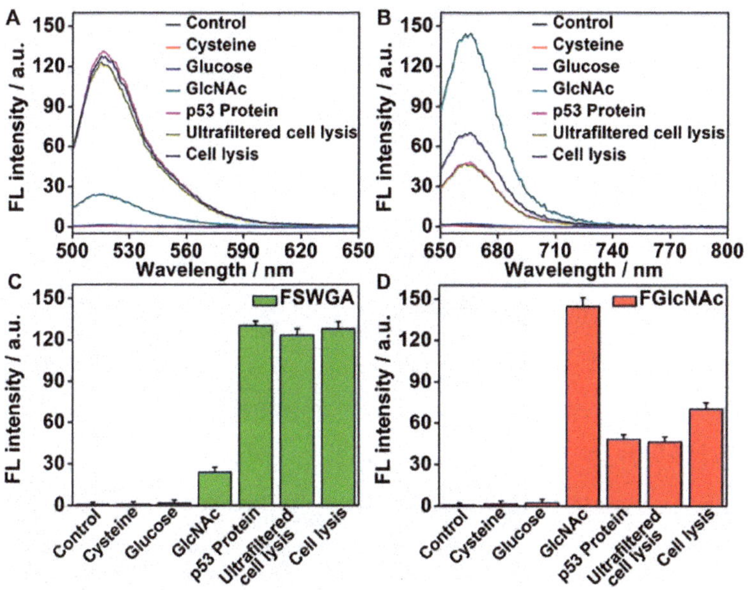

Figure 6.8 (A, B) Fluorescence spectra of probe (control) and probes treated with cysteine, glucose, GlcNAc, p53 protein, ultrafiltered cell lysates and cell lysates in the emission range of (A) FITC and (B) Alexa Fluor 647. (C, D) Peak intensities at (C) 516 and (D) 666 nm from (A) and (B), respectively.[46] Reproduced from ref. 46 with permission from American Chemical Society, Copyright 2018.

indicated the release of FSWGA and FGlcNAc due to the competitive conjugation of intracellular *O*-GlcNAcylated proteins and dissociated GlcNAc residue to FSWGA in the indicator. Furthermore, the green fluorescence from the released FSWGA was similar to that incubated with p53 protein and ultrafiltered cell lysates and the latter two mixtures showed weaker red fluorescence from released FGlcNAc (Figure 6.8C and D). Interestingly, the incubation of the probe with GlcNAc exhibited only the strong red fluorescence from the released FGlcNAc, while the green fluorescence of FSWGA was very weak. Hence they concluded that the competitive conjugation of *O*-GlcNAcylated proteins to FSWGA resulted in the release of the conjugate of FSWGA with FGlcNAc from the probe and the competitive binding of GlcNAc and dissociated GlcNAc residue to FSWGA released only FGlcNAc.

Since the competitive release ability of the indicator was confirmed, Chen *et al.* investigated the location of the indicator in cells by confocal nuclear and confocal lysosome colocalization imaging (Figure 6.9).[46] The nuclei and

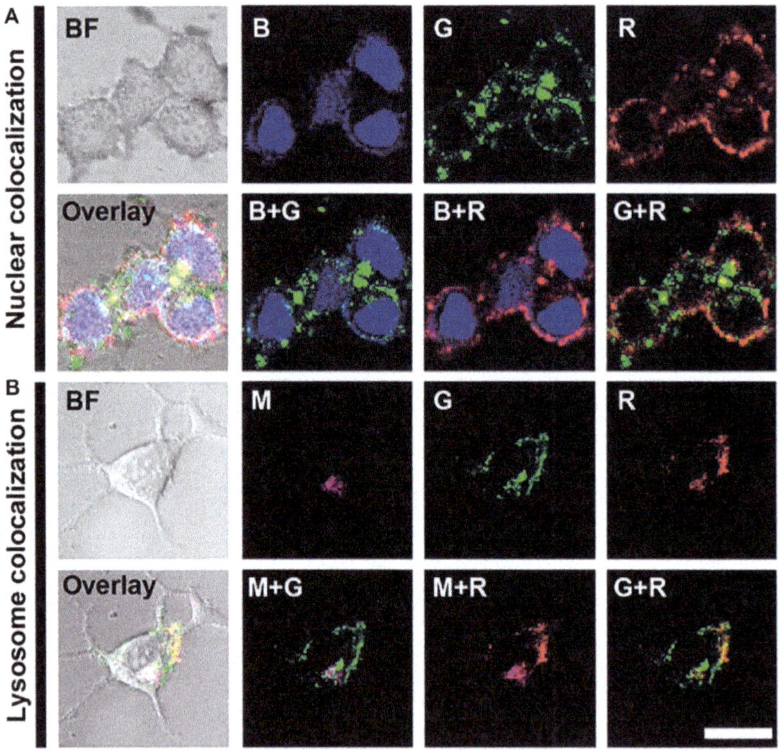

Figure 6.9 (A) Confocal nuclear and (B) confocal lysosome colocalization imaging. BF, bright field; B, blue channel of DAPI; M, mauve channel of RFP; G, green channel of FITC; R, red channel of Alexa Fluor 647. Scale bar: 20 μm.[46] Reproduced from ref. 46 with permission from American Chemical Society, Copyright 2018.

lysosomes in MCF-7 cells were dyed with 4′,6-diamidino-2-phenylindole (DAPI) and lysosomes-RFP, which did not influence the excitation of FITC and Alexa Fluor 647, respectively. The dyed cells were then incubated with 8 nM indicator for 4 h and subjected to confocal fluorescence imaging. The fluorescence from the probe observed with the green (G) and red (R) channels could overlap in both colocalization images (Figure 6.9); however, they did not overlap with the fluorescence from DAPI-dyed nuclei observed with the blue (B) channel (Figure 6.9A), indicating that no probe entered the nuclei. Meanwhile, a large area of G- and R-channel fluorescence from the probe was beyond the fluorescence area from lysosomes-RFP dyed lysosomes observed with the M channel and only partial overlapping was observed (Figure 6.9B), which indicated that most of the probe was dispersed in the cytoplasm and only a negligible amount of probe was caged in the lysosome. The latter did not affect the qualitative monitoring results. Hence the designed functional dual-colour indicator was located in the cytoplasm and could be used for *in situ* monitoring of the targets in cytoplasm.

Chen *et al.* then used the indicator to monitor the variation of *O*-GlcNAcylation in living cells, which can be regulated by OGT shRNAs, OGT or OGA inhibitors and exogenous GlcNAc.[46] After the cells had been treated with OGT shRNA[53] or OGT inhibitor (alloxan),[54] and then incubated with indicator, an obvious decrease in the green fluorescence was observed in the confocal fluorescence images, whereas the change of the red fluorescence was negligible compared with scrambled shRNA-treated and untreated cells (Figure 6.10), indicating that the total amount of intracellular GlcNAc groups was stable and the equilibrium of OGT/OGA-mediated *O*-GlcNAcylation tilted towards less *O*-GlcNAcylated proteins. In contrast, the treatment of OGA inhibitor (PUGNAc)[55–57] led to an increase in green fluorescence due to the inhibited dissociation of *O*-GlcNAcylated proteins, which maintained a higher level of *O*-GlcNAcylated proteins at the same total amount of intracellular GlcNAc groups. However, the treatment of exogenous GlcNAc increased both the red and green fluorescence owing to the increasing amount of GlcNAc in the cytoplasm, which produced more *O*-GlcNAcylated proteins due to the *O*-GlcNAcylation equilibrium. All these data indicated the feasibility of the indicator for the simultaneous monitoring of the level of *O*-GlcNAcylated proteins and the total amount of GlcNAc groups in a cell.

Finally, Chen *et al.*[46] used the designed method to study the biological relativity of intracellular *O*-GlcNAcylation with cell-surface E-cadherin and EpCAM (Figure 6.11A), which can respectively suppress invasion[58] or enhance the proliferation of cancer cells.[59] MCF-7 cells were subjected to treatment for 0, 6, 18 and 28 h with 5 mM sodium butyrate (NaBu), an antitumour drug used to induce better differentiation of tumour cells.[60,61] The treated cells were incubated with 8 nM probe for 4 h and then 20% phycoerythrin (PE)-conjugated E-cadherin or EpCAM antibody for 30 min. Both the green FITC fluorescence positively related to intracellular

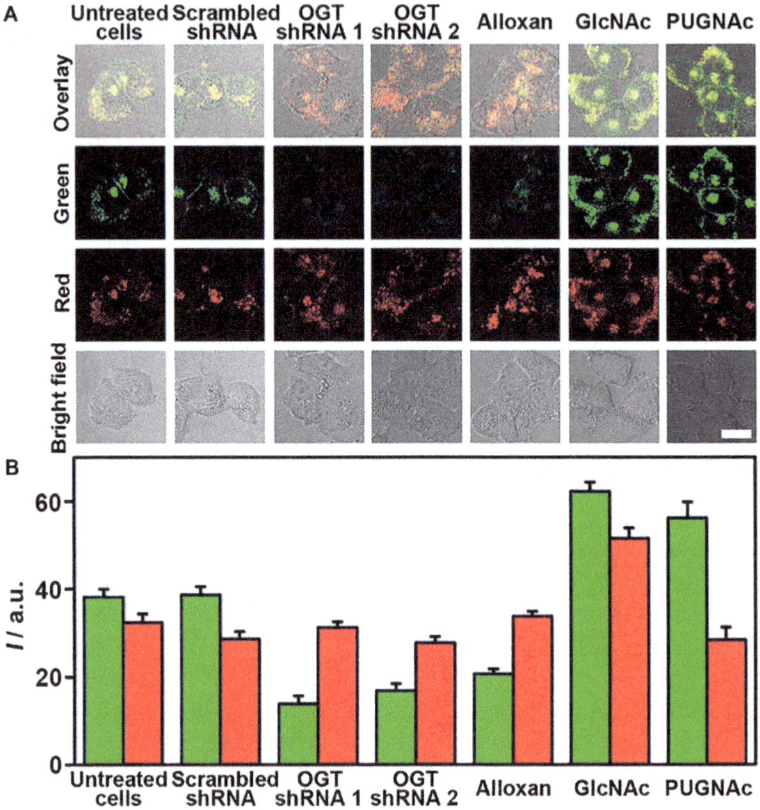

Figure 6.10 (A) Confocal fluorescence images of untreated, scrambled shRNA, OGT shRNA 1, OGT shRNA 2, alloxan, GlcNAc and PUGNAc treated MCF-7 cells after incubation with 8 nM probe for 4 h. Scale bar: 20 μm. (B) Mean intensity of green and red channels obtained from (A).[46]
Reproduced from ref. 46 with permission from American Chemical Society, Copyright 2018.

O-GlcNAcylated proteins and the PE fluorescence from the labelled EpCAM (Figure 6.11C) decreased, whereas the red Alexa Fluor 647 fluorescence positively related to intracellular total GlcNAc and the PE fluorescence from the labelled E-cadherin (Figure 6.11B) increased. The results indicated that the level of *O*-GlcNAcylated proteins decreased whereas the total GlcNAc groups increased. This was consistent with the fact that better differentiated cells have significantly lower OGT expression and higher OGA expression.[62] Further, the increase in total GlcNAc groups along with better differentiation of MCF-7 cells implied the relativity of the intracellular GlcNAc synthesis with cell surface E-cadherin and EpCAM expression.

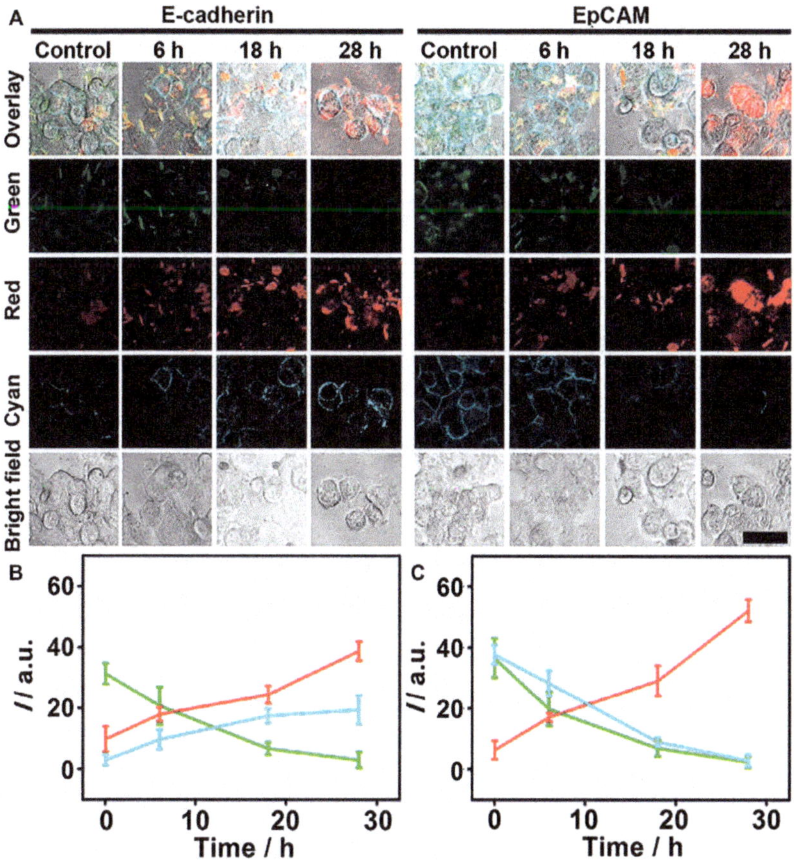

Figure 6.11 (A) Confocal fluorescence images of MCF-7 cells treated with 5 mM NaBu for 0, 6, 18 and 28 h after incubation with 8 nM probe for 4 h and then 20% PE-conjugated E-cadherin or EpCAM antibody in PBS for 30 min. Scale bar: 40 μm. (B, C) Mean intensity of green, red and cyan channels obtained from (A).[46]
Reproduced from ref. 46 with permission from American Chemical Society, Copyright 2018.

6.3 Visualization of Intracellular Sialylation-related Enzymes

6.3.1 Intracellular STs

Visualization of the activities of STs in living cells is important for the study of sialylation. FRET is still the first choice for the design of *in situ* visual methods. Bao *et al.* encapsulated tetramethylrhodamine isothiocyanate-labelled asialofetuin (TRITC-AF) and fluorescein isothiocyanate-labelled APBA (FITC-APBA) in a liposome-based delivery vesicle for the *in situ*

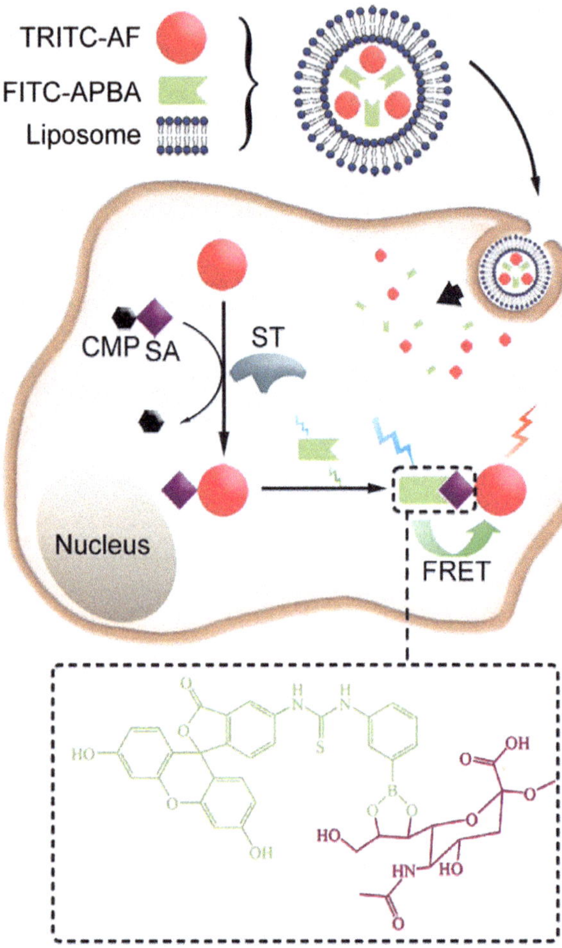

Figure 6.12 Schematic illustration of sensing vesicle, FRET sensing principle and structure of the FITC–APBA/SA complex.[63]

analysis of ST activity (Figure 6.12).[63] AF was chosen as the substrate for ST.[64] FITC and TRITC acted as the donor and acceptor of a FRET pair,[65] respectively. After the vesicle had been delivered into cells, the intracellular ST could transfer the SA from cytidine-5′-monophosphosialic acid (CMP-SA), which exists in the Golgi apparatus, to the terminal position of glycan chains of the TRITC-AF. Upon sialylation of the AF, the sialylation product could be recognized by the released FITC-APBA[66] to bring FITC and TRITC close enough to achieve FRET. The signal intensity depended on the formation of the sialylated TRITC-AF and hence could be used for the assessment of ST activity.

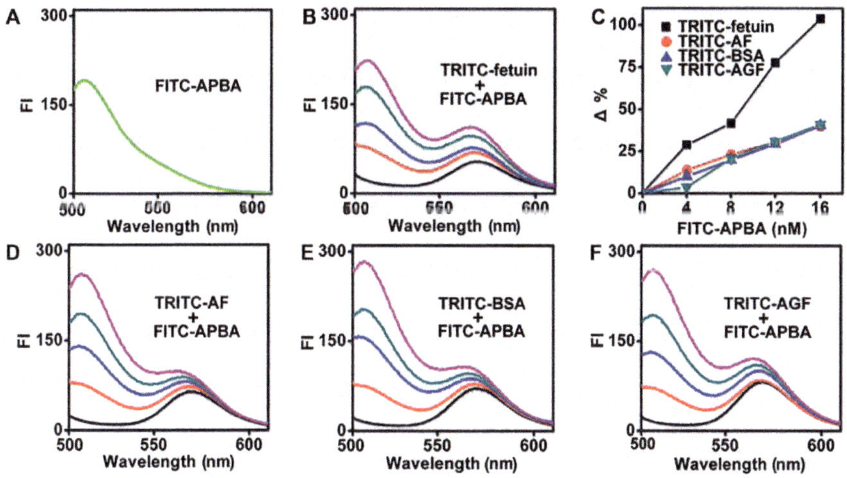

Figure 6.13 FRET responses of TRITC-proteins after incubation with FITC-APBA. (A) 12 nM FITC-APBA and (B) TRITC-fetuin at a TRITC amount of 160 nM after incubation with 0, 4, 8, 12 and 16 nM FITC-APBA (from bottom to top) for 30 min under donor excitation. (C) Plots of FI change (Δ%) at 572 nm *versus* FITC-APBA concentration. (D) TRITC-AF, (E) TRITC-BSA and (F) TRITC-AGF at a TRITC amount of 160 nM after incubation with 0, 4, 8, 12 and 16 nM FITC-APBA (from bottom to top) for 30 min under donor excitation.[63]

Bao *et al.* first tested the FRET between FITC-APBA and TRITC-proteins *in vitro*.[63] FITC-APBA and TRITC-fetuin showed maximum fluorescence emission at 519 and 572 nm, respectively (Figure 6.13). After incubating the mixture of TRITC-fetuin and FITC-APBA for 30 min, the fluorescence intensity (FI) of the mixture at 572 nm became 1.3 times larger than the sum of the FIs of TRITC-fetuin and FITC-APBA tested alone. The FI increased with increase in concentration of FITC-APBA (Figure 6.13B). As controls, a mixture of FITC-APBA with TRITC-AF, TRITC-BSA (bovine serum albumin) or TRITC-AGF (asialo-agalacto-fetuin, the degalactosylated form of AF) displayed a higher FI of FITC at 519 nm and a smaller FI change of TRITC at 572 nm at the same FITC-APBA concentration (Figure 6.13D–F), indicating the quenching of FITC fluorescence and enhancement of TRITC fluorescence by efficient energy transfer from FITC to TRITC due to the binding of SA on fetuin with APBA. The fluorescence enhancement of TRITC-AF, TRITC-BSA or TRITC-AGF upon addition of FITC-APBA suggested the presence of a relatively weaker FRET response (Figure 6.13C), which was attributed to the adsorption of FITC-APBA on TRITC-proteins. Considering the effect of FITC fluorescence on the FI measurement at 572 nm and the FRET response, an acceptor-to-donor concentration ratio of 13 was selected, at which the FRET response was sufficient for cellular imaging experiments.

Bao *et al.* then demonstrated the enzymatic activity of ST to transfer the SA from CMP-SA to TRITC-AF by gel electrophoresis analysis and FITC-APBA-mediated FRET assay.[63] Only the coexistence of the three components could generate a new band with higher molecular weight than TRITC-AF, which corresponded to the sialylated product (Figure 6.14A, lanes 7 and 8). After TRITC-AF had been treated with α-2,3-ST from *Pasteurella multocida* or α-2,6-ST from *Photobacterium damsela* in the presence of CMP-SA and then reacted with FITC-APBA, the FI increased by 38.4 or 20.6% (Figure 6.14B), indicating the formation of the sialylated product of TRITC-AF, which led to the FRET response.

After confirming the FRET and the ST activity of the designed probes, Bao *et al.* delivered the ST substrate and recognition probe into cells by encapsulating FITC-APBA and TRITC-AF in Lipofectamine 2000 to form a sensing vesicle.[63] After incubation for 120 min, vesicle-incubated human cervical carcinoma (HeLa) cells showed bright fluorescent spots of FITC and TRITC (Figure 6.15A). The fluorescence of FITC spread over the whole cell area owing to the release of FITC-APBA from the vesicles, while the fluorescence of TRITC-AF in the perinuclear area was much stronger than that in the nuclear area owing to the much larger size of AF than APBA. In contrast, in the absence of liposome, although the TRITC fluorescence could also be observed in the perinuclear area, the incubated cells showed much brighter FITC fluorescence on the cell surface and very low FITC fluorescence in the cytoplasm and nuclear area, which resulted from the strong binding of APBA to SA exposed on the cell surface and SA-mediated weak uptake of FITC-APBA[67,68] (Figure 6.15B). To verify the function of SA in cellular

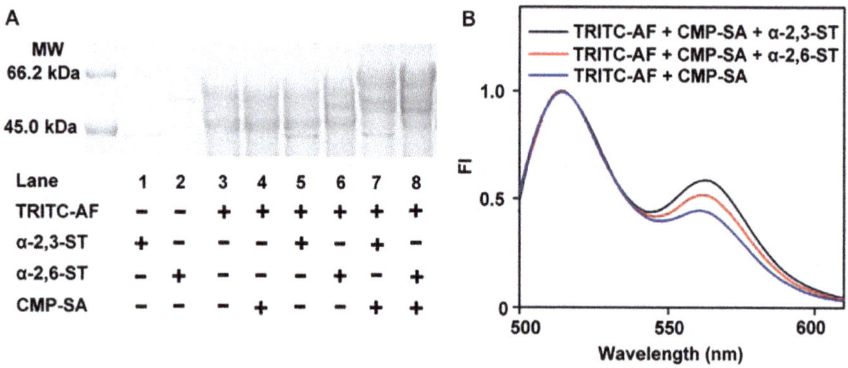

Figure 6.14 (A) Sodium dodecyl sulfate polyacrylamide gel electrophoresis (SDS-PAGE) of Tris–HCl buffer containing marked species after incubation at 37 °C for 4 h. (B) Normalized fluorescence spectra of three mixtures after incubation at 37 °C for 4 h and then reaction with FITC-APBA for 30 min, $\lambda_{ex} = 488$ nm.[63]

delivery, sialidase was used to cleave the SA from the cell surface. After cleaving SA, the vesicle-incubated HeLa cells showed the same fluorescence signals as those without SA cleavage (Figure 6.15A), whereas sialidase and then FITC-APBA/TRITC-AF-treated cells showed an obviously different fluorescence distribution from those without sialidase treatment (Figure 6.15B). The cleavage of SA from the cell surface greatly decreased the fluorescence signals both on the cell surface and in the cell area. Hence in the presence of liposome the cellular delivery was unrelated to cell-surface SA, which excluded the effect of cell-surface SA on the detection of ST activity. Although the fluorescence of TRITC-AF in the absence of liposome could also be observed in the cytoplasm of sialidase-treated cells owing to the protein receptor-mediated endocytosis, the liposome-assisted TRITC-AF delivery ensured the escape of the vesicle from endosome/lysosome to cytosol[69] and its entry into the Golgi complex.[70] The former could be demonstrated by staining the lysosomes and nucleus of TRITC-AF/liposome-incubated HeLa cells with Lyso-Tracker Green DND-26 and Hoechst 33342, respectively. The confocal images indicated the successful escape of TRITC-AF from endosome/lysosomes,[71] hence it distributed throughout the cytoplasm (Figure 6.15C). The staining experiment with Golgi Green verified the entry of TRITC-AF into the Golgi complex (Figure 6.15D), where the presence of ST and CMP-SA ensured the sialylation of TRITC-AF.

Bao *et al.* also addressed the intracellular FRET by acceptor photobleaching.[63] After acceptor photobleaching, enhanced emission from FITC was observed in the dequenching area of HeLa cells incubated with liposome containing FITC-APBA and TRITC-fetuin (Figure 6.16), indicating the recovery of FITC due to the photobleaching of TRITC and the FRET from FITC to TRITC before the photobleaching.

Since the intracellular delivery and the FRET signal of the probes had been verified, Bao *et al.*[63] assessed the ST activity in living cells by the designed method. ST activity is related to cell proliferation and cell cycle, especially the G1 phase.[72,73] After the G1 phase, HeLa cells were transfected with sensing vesicles for 2.0 h and then incubated in growth medium for 1.0 h, and the confocal image showed an obvious FRET signal (Figure 6.17A). The FRET efficiency extracted across the cytoplasm region increased from 14.5 to 64.7% with a statistically significant change ($p < 0.05$) during the incubation period from 1.0 to 2.5 h (Figure 6.17B). This result demonstrated the feasibility of the proposed strategy for non-invasive tracing of the ST-mediated sialylation process. The specificity of TRITC-AF sialylation in the Golgi complex of living cells was demonstrated with TRITC-fetuin, TRITC-BSA and TRITC-AGF. After transfecting the G1 phase HeLa cells with the vesicle containing these proteins and FITC-APBA, the FRET efficiency decreased with increase in incubation time in the growth medium (Figure 6.17C and F), which was the opposite of that in the TRITC-AF sialylation process (Figure 6.17B). The FRET efficiency for TRITC-fetuin remained at a high level in the observed period (76.4–69.4%) owing to the conjugation of FITC-APBA and SA exposed on TRITC-fetuin. However, TRITC-BSA or TRITC-AGF showed a very low FRET efficiency, which resulted from the adsorption of FITC-APBA on these proteins. The decrease in FRET efficiency could be attributed to the desorption of non-covalently linked FITC-APBA from TRITC-proteins. After 2.5 h of incubation, the FRET efficiency of 64.7% for TRITC-AF was close to that for TRITC-fetuin (Figure 6.17B and C), indicating

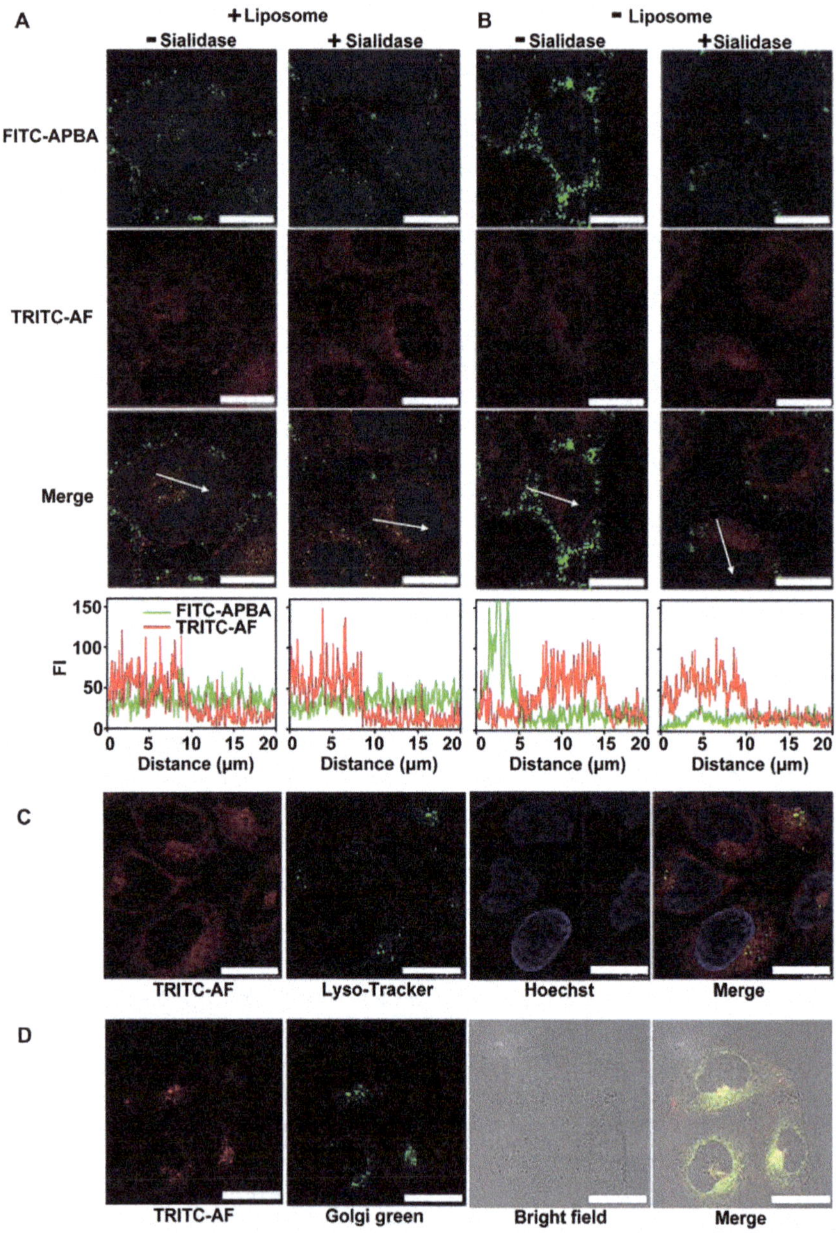

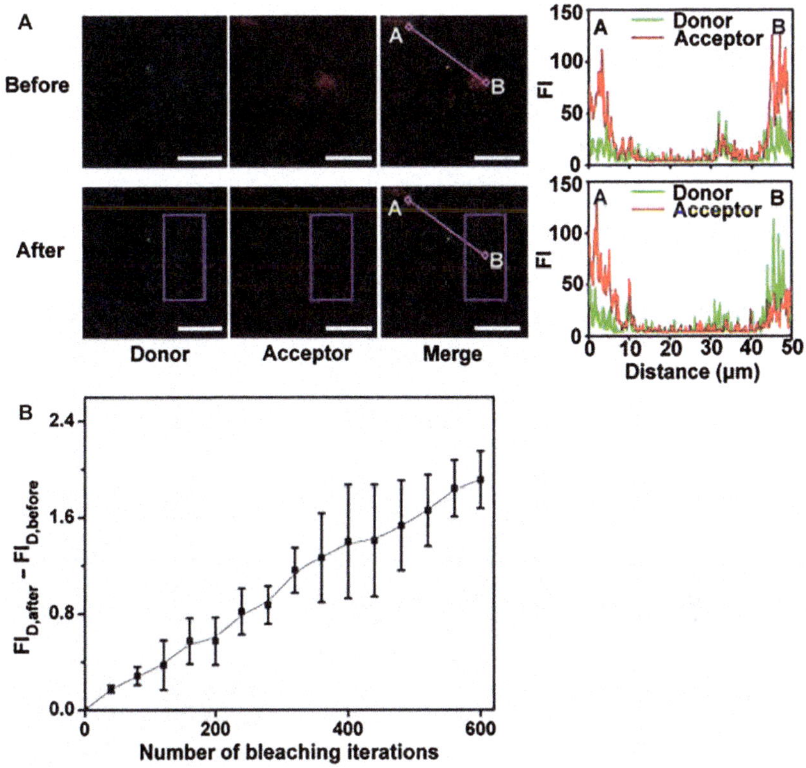

Figure 6.16 Confocal images of HeLa cells incubated with FITC-APBA- and TRITC-fetuin-encapsulated liposome before and after acceptor photobleaching in the marked rectangular area 600 times (A, left) and linear FI scans of donor and acceptor from point A to point B (A, right). Scale bars: 20 μm. (B) FI change of donor during acceptor photobleaching. Data and error bars show the average and standard deviation of three replicates.[63]

Figure 6.15 Confocal images of sialidase-untreated and -treated HeLa cells after delivery of 320 nM TRITC-AF and 12 nM FITC-APBA into cells in (A) the presence and (B) the absence of liposome. FIs of FITC-APBA and TRITC-AF were plotted as a function of distance along with white arrows. (C) Confocal images of HeLa cells incubated successively with liposome and 320 nM TRITC-AF for 120 min, Lyso-Tracker Green DND-26 for 30 min and Hoechst 33342 for 5 min. (D) Confocal and bright-field images of HeLa cells incubated with liposome and 320 nM TRITC-AF for 120 min and then Golgi Green (NBD-C6 ceramide) for 30 min. Scale bars: 20 μm.[63]

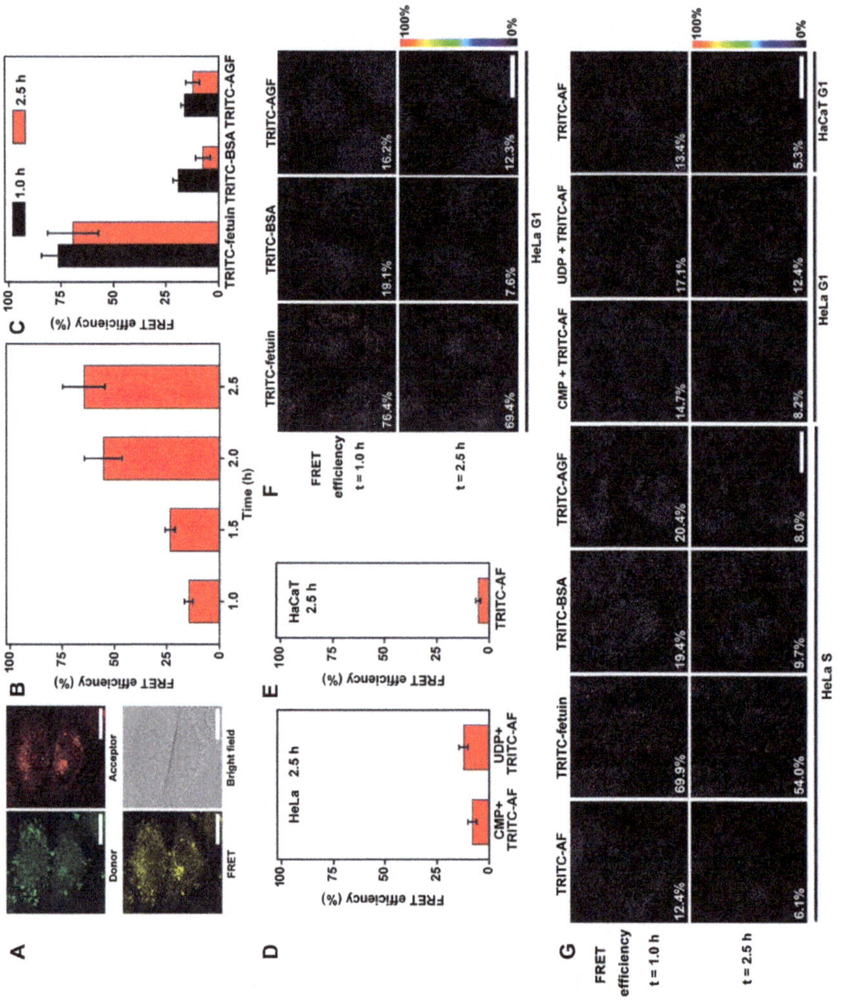
A
Donor
Acceptor
FRET
Bright field
B
FRET efficiency (%)
Time (h)
C
1.0 h
2.5 h
TRITC-fetuin TRITC-BSA TRITC-AGF
D
HeLa 2.5 h
CMP+ TRITC-AF
UDP+ TRITC-AF
E
HaCaT 2.5 h
TRITC-AF
F
FRET efficiency
t = 1.0 h
t = 2.5 h
TRITC-fetuin
TRITC-BSA
TRITC-AGF
76.4%
19.1%
16.2%
69.4%
7.6%
12.3%
HeLa G1
100%
0%
G
FRET efficiency
t = 1.0 h
t = 2.5 h
TRITC-AF
TRITC-fetuin
TRITC-BSA
TRITC-AGF
CMP + TRITC-AF
UDP + TRITC-AF
TRITC-AF
12.4%
69.9%
19.4%
20.4%
14.7%
17.1%
13.4%
6.1%
54.0%
9.7%
8.0%
8.2%
12.4%
5.3%
HeLa S
HeLa G1
HaCaT G1
100%
0%

the formation of sialylated TRITC-AF in living cells in the presence of intracellular ST.

Bao *et al.* further employed the proposed strategy for monitoring the ST activity of G1-phase HeLa cells during drug treatment.[63] CMP and uridine 5′-diphosphate (UDP), inhibitors of ST activity[74] and SA biosynthesis,[75] respectively, were chosen as model drugs. After drug treatment, these cells displayed a much lower FRET efficiency (8.2 and 12.4%, respectively) (Figure 6.17D and G) than those without treatment after 2.5 h of incubation in the growth medium (Figure 6.17B), just like the negative controls (Figure 6.17C), indicating the effective inhibition of CMP of the ST activity and the influence of UDP on SA biosynthesis. To discriminate tumour and normal cells by monitoring the intracellular ST activity, they transfected the G1-phase normal human skin keratinocytes (HaCaT cells) with the sensing vesicles, which showed a FRET efficiency of 5.3% (Figure 6.17E and G), indicating low expression of the ST level in normal cells.

6.3.2 Intracellular Neus

Neus activities have a great influence on the level of cellular glycosylation. Neus on cell membrane could be detected with some simple fluorescent probes,[76,77] but the activities of intracellular Neus, which are mainly located in lysosomes (Lyso-Neu), are much more difficult to visualize. The real-time visualization of Lyso-Neu activity faces three main challenges: (1) design of a Neu activity-responsive substrate, (2) efficient delivery of the Neu substrate to lysosomes and (3) production of Neu-triggered "off–on" signals in living cells. Aiming to address these challenges, Bao *et al.* designed a cell-permeable and lysosome-accessing nanoprobe to deliver a Lyso-Neu-specific substrate efficiently into lysosomes and release the cleavage product into the cytosol, which could be used for neutral pH-enhanced fluorescence detection of Lyso-Neu activity in living cells (Figure 6.18).[78] Efficient delivery was achieved using

Figure 6.17 (A) Confocal and bright-field images of G1-phase HeLa cells after being transfected with sensing vesicle for 2.0 h and then cultured in growth medium for 1.0 h. Scale bars: 20 μm. (B) Time course of FRET efficiency averaged across the cytoplasm region during sialylation of TRITC-AF in G1-phase HeLa cells. (C) FRET efficiency in G1-phase HeLa cells after being incubated with liposome vesicles containing FITC-APBA and TRITC-proteins for 2.0 h and then cultured in growth medium for 1.0 and 2.5 h. (D) and (E) FRET efficiency in G1-phase HeLa cells treated with CMP and UDP as inhibitors and normal G1-phase HaCaT cells after incubation with sensing vesicles. Data and error bars show the average and standard deviation of three replicates. (F) FRET efficiency images of G1-phase HeLa cells. (G) FRET efficiency images and the corresponding quantitative data of S-phase and G1-phase HeLa cells treated with inhibitors and G1-phase HaCaT cells. Scale bars: 20 μm.[63] Reproduced from ref. 63, https://doi.org/10.1038/srep10947, under the terms of the CC BY 4.0 licence, http://creativecommons.org/licenses/by/4.0/.

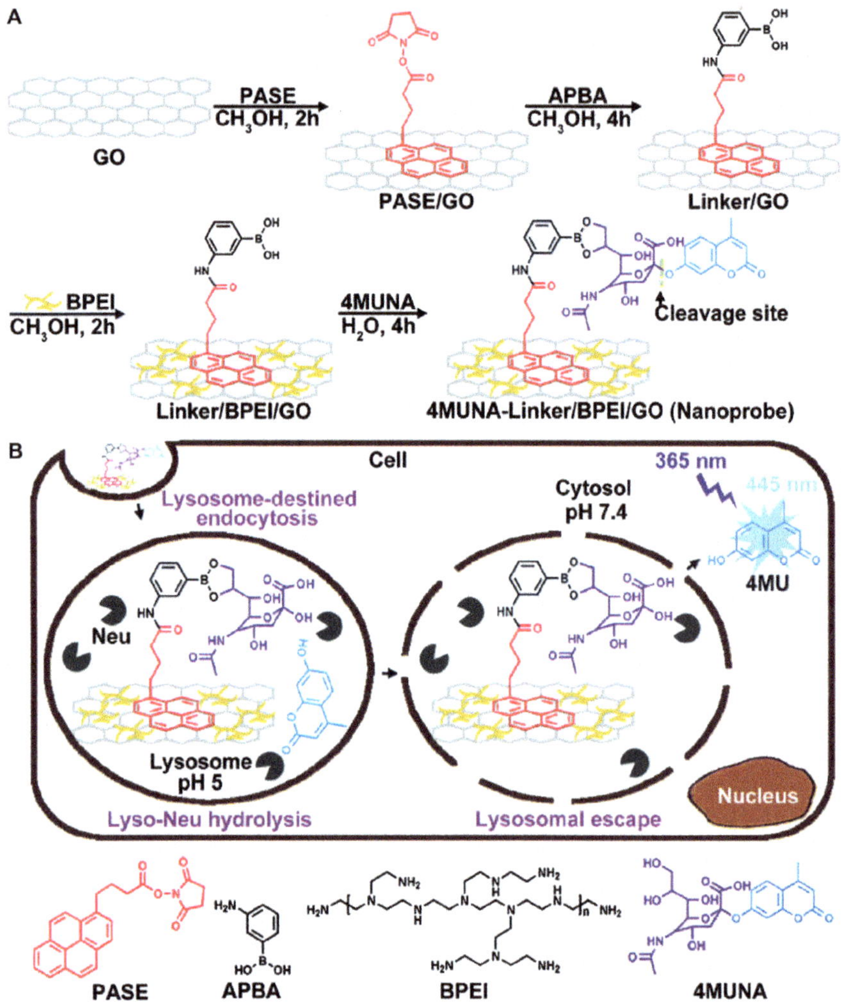

Figure 6.18 Schematic illustration of (A) construction of the nanoprobes and (B) the lighting-up imaging strategy for Lyso-Neu activity in living cells.[78] Reproduced from ref. 78 with permission from the Royal Society of Chemistry.

graphene oxide (GO) as the "carrier and locator" owing to its high loading capability and endocytosis-mediated cellular internalization.[79–81] 4-Methylumbelliferyl-*N*-acetylneuraminic acid (4MUNA) was assembled on the GO surface *via* a linker and could be specifically cleaved by Neus in acidic lysosomes (pH 5)[82] to produce a free 4-methylumbelliferone (4MU) moiety. The 4MU is pH responsive and can yield strong fluorescence at neutral pH.[83] Therefore, to facilitate the escape of the cleaved 4MU into the neutral cytosol for producing a strong signal, a cationic polymer, branched polyethylenimine (BPEI), was co-assembled on the GO surface,[84] which caused lysosomal osmotic swelling to form holes for escape of 4MU into the cytosol due to the unique proton-sponge effect.[85]

Bao *et al.*[78] prepared a linker/GO composite by covalently conjugating 3-aminophenylboronic acid (APBA) to the succinimidyl end of 1-pyrenebutyric acid *N*-hydroxysuccinimide ester (PASE) that was pre-assembled on GO by π–π stacking.[86] The flexible alkyl chain in PASE was favourable for reducing the steric effect for recognition of the enzyme substrate. After BPEI had been loaded onto the linker/GO by electrostatic interaction, 4MUNA was introduced by covalent binding of the *N*-acetylneuraminic acid (NA) moiety to the phenylboronic acid group of the linker to obtain the nanoprobe (Figure 6.18A).[87] Using human colon tumour (HCT-15) cells as a model, the nanoprobe was efficiently taken up through clathrin-mediated endocytosis and transported to lysosomes.[81] The Lyso-Neu-specific cleavage of 4MU followed by BPEI-assisted escape into the cytosol lit up the fluorescence of 4MU for visual detection of Lyso-Neu activity (Figure 6.18B). Moreover, the Neu associated with the cytosol did not interfere with the signal owing to the trapping of the nanoprobes in lysosomes by the endocytic pathway.

Bao *et al.* then investigated the specificity of the proposed nanoprobes to Neu by *in vitro* fluorescence measurements.[78] The Neu activity at pH 5.0 was reported to be fivefold that at pH 7.4.[82] After incubating the nanoprobe with Neu at pH 5.0, the FI was detected at pH 7.4. With increasing incubation time, the FI increased gradually and tended to a plateau after 90 min (Figure 6.19A), indicating complete cleavage of 4MU from the nanoprobe. The GO composite displayed a small effect (~11%) on the fluorescence signal of released 4MU (Figure 6.19C). When the incubation mixture contained *N*-acetyl-2,3-dehydro-2-deoxyneuraminic acid (NADNA), a Neu inhibitor,[88] incubation for 2 h did not produce the fluorescent signal of 4MU (Figure 6.19B), indicating complete suppression of Neu activity. Hence the signal could be attributed to the specific cleavage activity of Neu. As a control, the linker/BPEI/GO and the nanoprobe did not show any fluorescent signal (Figure 6.19B), indicating a low background for highly sensitive detection of Neu activity. The hydrolysed product on the nanoprobe was confirmed by matrix-assisted laser desorption/ionization time-of-flight (MALDI-TOF) mass spectrometry to verify the cleavage of 4MU. The mass spectrum showed a peak at *m*/*z* 703.74 (Figure 6.19D), which was attributed to $[\text{NA-linker} + \text{Na}]^+$, considering that the molecular weight of the NA-linker ($C_{37}H_{37}BN_2O_{10}$) is 680.51. This result confirmed the release of 4MU from the nanoprobe.

Bao *et al.* verified the entry of the nanoprobe into lysosomes by a colocalization experiment.[78] After incubation for 4 h, the distribution of the tracker overlapped well with lysosomes stained with LysoTracker Green (Figure 6.20A), suggesting that GO was a favourable carrier and locator for lysosomal sensing. After the living HCT-15 cells had been incubated with the nanoprobe for 4 h, a strong fluorescent signal was observed in the cytoplasmic region, which overlapped with the stained lysosomes with only 6%

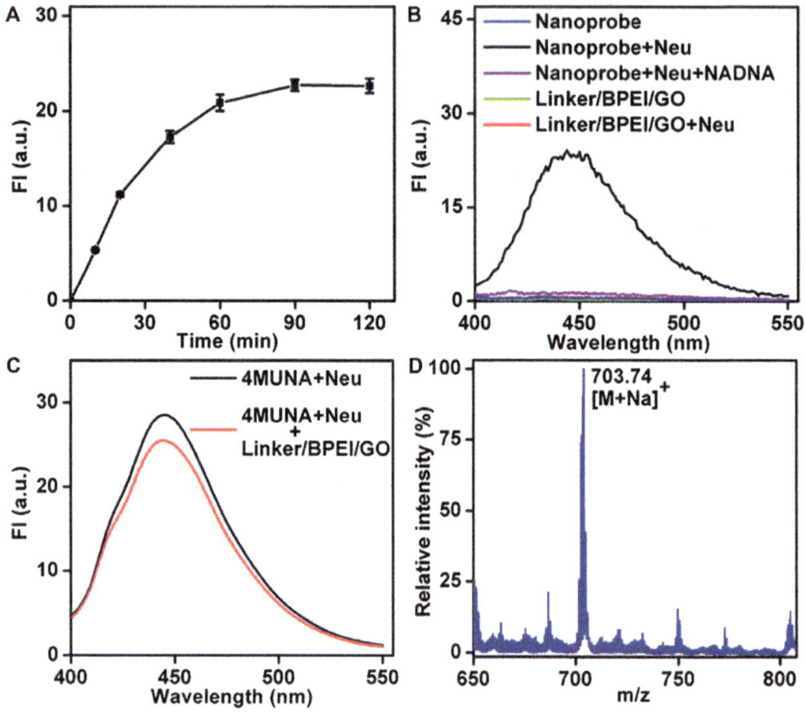

Figure 6.19 (A) Fluorescence intensity of the nanoprobe (1 mg mL^{-1}, 158 nM 4MUNA equiv.) after being incubated with Neu (2 mU mL^{-1}) for different times at pH 5.0 and transferred to pH 7.4 buffer. (B) Fluorescence responses of linker/BPEI/GO (1 mg mL^{-1}) and the nanoprobe (1 mg mL^{-1}, 158 nM 4MUNA equiv.) towards Neu, respectively. NADNA: a Neu inhibitor. $\lambda_{ex} = 365$ nm. (C) Fluorescence spectra of 4MUNA (158 nM, final concentration) pre-hydrolysed with Neu completely by incubation for 2 h and then mixed with glycine–NaOH buffer in the absence and presence of linker/BPEI/GO (1 μg mL^{-1}, final concentration) at pH 7.4 for 3 h. (D) MALDI-TOF mass spectrum of the hydrolysis product on the nanoprobe.[78]
Reproduced from ref. 78 with permission from the Royal Society of Chemistry.

of pixels (Figure 6.20B), indicating the escape of released 4MU from acidic lysosomes to the neutral cytosol due to the proton-sponge effect of BPEI. This demonstrated the feasibility of the proposed light-up imaging strategy. For the practical application of the imaging protocol, the incubation time of the nanoprobe was optimized with NaBu-treated HCT-15 cells, which induced the upregulation of Lyso-Neu.[89,90] The fluorescence of 4MU appeared unevenly in the cytoplasmic region upon incubation for up to 1.5 h, which showed some bright domains outside lysosomes (Figure 6.20C). With increasing incubation time, the fluorescence expanded and became uniform in the cytoplasmic region, while the signal became brighter, indicating that more 4MU molecules were cleaved from the nanoprobe and escaped to the

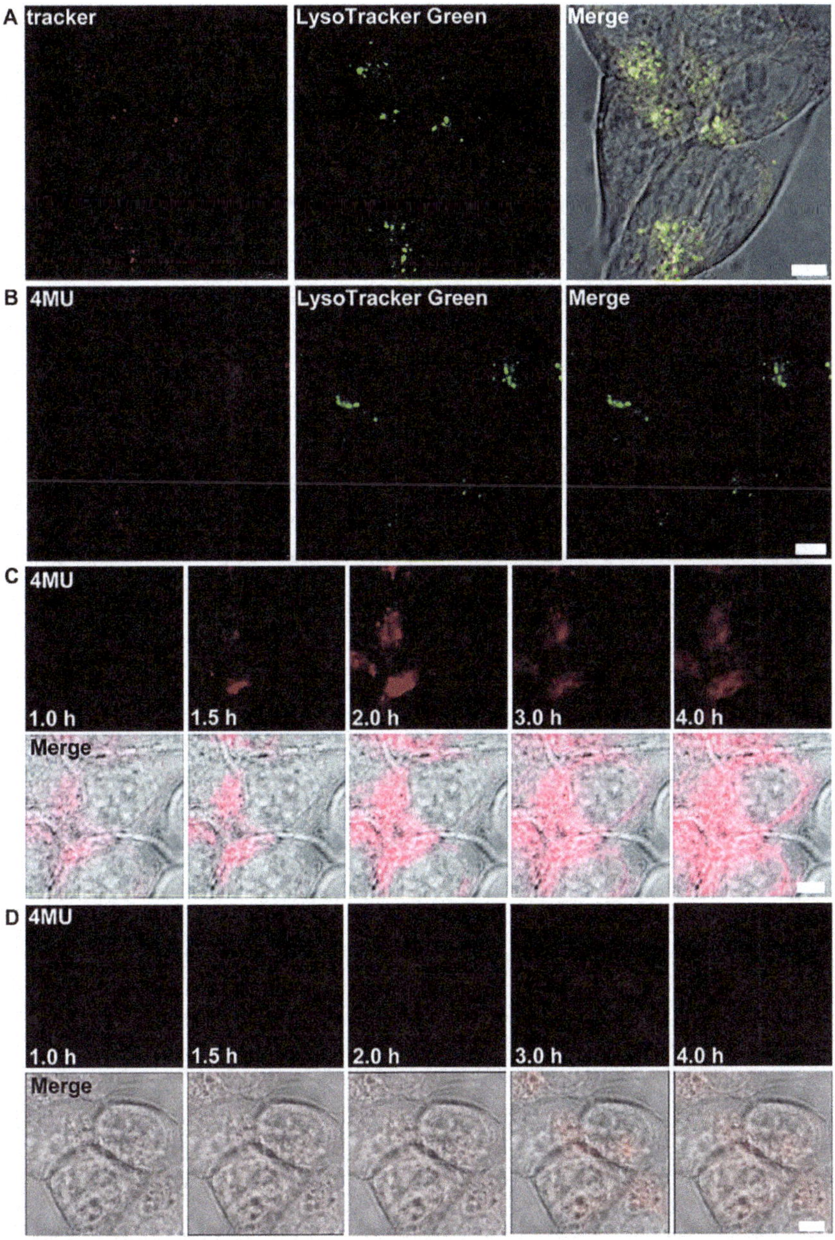

Figure 6.20 Confocal images of HCT-15 cells after incubation with the (A) tracker and (B) nanoprobe for 4 h, respectively, followed by LysoTracker Green staining. (C) Fluorescence and bright-field merged images of HCT-15 cells pretreated with NaBu for 48 h followed by incubation with a 10 mg mL^{-1} nanoprobe for the indicated times. Scale bars: 5 μm. (D) Fluorescent and bright-field merged images of HCT-15 cells after NADNA treatment for 48 h and then incubation with 10 μg mL^{-1} nanoprobe for the indicated times. Scale bar: 5 μm.[78]
Reproduced from ref. 78 with permission from the Royal Society of Chemistry.

cytosol. The maximum expansion and brightness occurred at an incubation time of 4 h. As a control, the cells with Lyso-Neus inhibited by NADNA showed much weaker fluorescence during the same incubation time (Figure 6.20D). These results further demonstrated that the nanoprobe was a powerful tool for the visualization of Lyso-Neu activity in living cells.

Bao *et al.* finally used this method to dynamically track the variation of Lyso-Neu activity in living cells during the apoptosis process and drug treatment by recording the mean FI of the cytoplasmic regions (Figure 6.21A).[78] The incubation of HCT-15 cells with fetal bovine serum (FBS)-free culture medium that induced cell apoptosis and NaBu-containing culture medium increased the Lyso-Neu activity 1.5- and 1.7-fold, respectively, at 24 h and 1.8- and 2.4-fold, respectively, at 48 h. These results were in

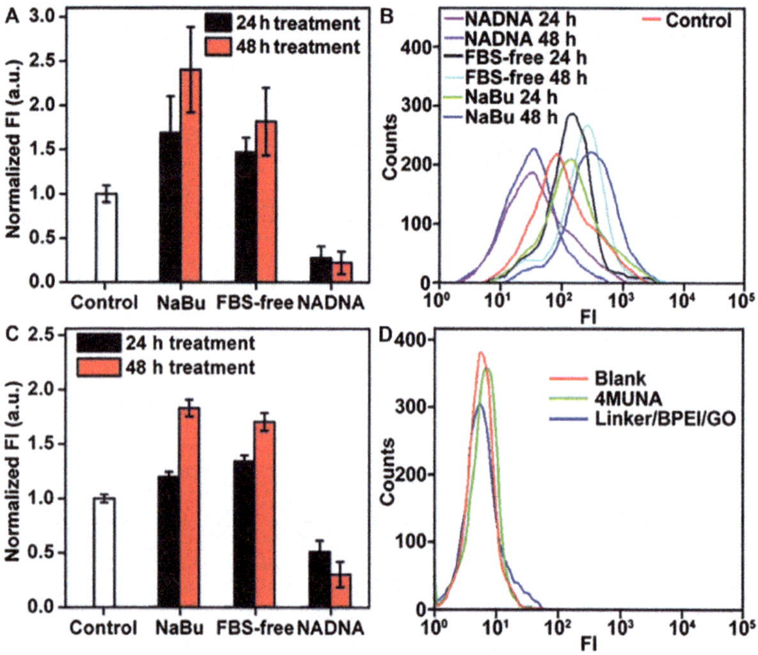

Figure 6.21 Normalized mean FI of the cytoplasmic region (A) and flow cytometric analysis (B) of HCT-15 cells treated with NaBu-containing, FBS-free and NADNA-containing culture media for 24 or 48 h, respectively, followed by incubation with the nanoprobe (10 mg mL^{-1}) for 4 h. Cells without treatment served as the control. (C) Normalized FIs of crude lysosomal lysates from HCT-15 cells pretreated with FBS-free, NaBu and NADNA for 24 or 48 h after incubation with nanoprobe at 37 °C for 10 min. The cells without pretreatment served as the control. (D) Flow cytometric analysis of HCT-15 cells after incubation with 4MUNA (1.58 μM) and linker/BPEI/GO (10 μg mL^{-1}) for 4 h. The cells without incubation served as the blank.[78]

good agreement with the fact that the Lyso-Neu activity could be upregulated in the early stage of apoptosis induced by NaBu or FBS-free treatment.[90] In response to the Neu inhibitor NADNA, the Lyso-Neu activity decreased to 27 and 22% after being treated for 24 and 48 h, respectively, suggesting obvious inhibition effects. These results were further verified by flow cytometric assays (Figure 6.21B) and were consistent with the data obtained from crude lysosomal lysates (Figure 6.21C). As a control, a flow cytometric experiment was also performed by monitoring the fluorescence of cells after being incubated with only 4MUNA or linker/BPEI/GO. These cells did not show any increase in the fluorescence signal (Figure 6.21D), suggesting a negligible background and eliminating the interference from cell surface Neus. These data also demonstrated the necessity for using linker/BPEI/GO to deliver 4MUNA towards lysosomes[76] for highly sensitive imaging of Lyso-Neu activity in living cells.

6.4 Perspective

Although some recent methods have achieved the visualization of intracellular glycosylation, the accuracy and quality of those methods need to be improved. Methods based on metabolic labelling still suffer from low labelling efficiency and long incubation times. FRET also faces many problems, such as photobleaching and low sensitivity. The developed nanoprobe-based methods bring complicated issues regarding the cellular uptake process. Further, none of these methods can give quantitative results. Future developments should be focused on the time–space-resolved quantitative analysis of intracellular glycosylation, which may require novel target-responsive molecular probes, such as derivatives of functional carbohydrates and analogues of glycan-related enzymatic substrates. Glycans in subcellular organelles, such as mitochondria, the Golgi apparatus and nucleus, need to be analysed independently. It is also necessary to track the dynamic and interactional processes of intracellular glycosylation, which is important for the exploration of mechanisms in biological and medical science.

Acknowledgement

Dr Yunlong Chen is thanked for expert assistance with the writing of this chapter.

References

1. A. Varki, R. D. Cummings, J. D. Esko, H. H. Freeze, P. Stanley, C. R. Bertozzi, G. W. Hart and M. E. Etzler, *Essentials of Glycobiology*, Cold Spring Harbor Laboratory Press, Cold Spring Harbor, NY, 2nd edn, 2009.
2. L. Krishnamoorthy and L. K. Mahal, *ACS Chem. Biol.*, 2009, **4**(9), 715.
3. L. Wells, K. Vosseller and G. W. Hart, *Science*, 2001, **291**(5512), 2376.

4. G. W. Hart, M. P. Housley and C. Slawson, *Nature*, 2007, **446**(7139), 1017.
5. L. K. Kreppel, M. A. Blomberg and G. W. Hart, *J. Biol. Chem.*, 1997, **272**, 9308.
6. Y. Gao, L. Wells, F. I. Comer, G. J. Parker and G. W. Hart, *J. Biol. Chem.*, 2001, **276**, 9838.
7. M. K. Tarrant, H.-S. Rho, Z. Xie, Y. L. Jiang, C. Gross, J. C. Culhane, G. Yan, J. Qian, Y. Ichikawa, T. Matsuoka, N. Zachara, F. A. Etzkorn, G. W. Hart, J. S. Jeong, S. Blackshaw, H. Zhu and P. A. Cole, *Nat. Chem. Biol.*, 2012, **8**, 262.
8. M. S. Macauley and D. J. Vocadlo, *Biochim. Biophys. Acta*, 2010, **1800**, 107.
9. G. W. Hart, M. P. Housley and C. Slawson, *Nature*, 2007, **446**, 1017.
10. E. J. Kim, *Molecules*, 2011, **16**, 1987.
11. Y. H. Chen, Y. Zhang and P. Y. Yang, *Curr. Proteomics*, 2010, **7**(3), 158.
12. D. L. Y. Dong and G. W. Hart, *J. Biol. Chem.*, 1994, **269**(30), 19321.
13. L. Wells, K. Vosseller, R. N. Cole, J. M. Cronshaw, M. J. Matunis and G. W. Hart, *Mol. Cell. Proteomics*, 2002, **1**(10), 791.
14. F. I. Comer, K. Vosseller, L. Wells, M. A. Accavitti and G. W. Hart, *Anal. Biochem.*, 2001, **293**(2), 169.
15. P. D. Bos, X. H. Zhang, C. Nadal, W. Shu, R. R. Gomis, D. X. Nguyen, A. J. Minn, M. J. van de Vijver, W. L. Gerald, J. A. Foekens and J. Massagué, *Nature*, 2009, **459**(7249), 1005.
16. F. Dall'Olio and M. Chiricolo, *Glycoconjugate J.*, 2001, **18**(11–12), 841.
17. A. Harduin-Lepers, M. A. Krzewinski-Recchi, F. Colomb, F. Foulquier, S. Groux-Degroote and P. Delannoy, *Front. Biosci., Elite Ed.*, 2012, **4**, 499.
18. M. Fukuda and X. Bao, *Nat. Chem. Biol.*, 2008, **4**(12), 721–722.
19. O. M. Pearce and H. Laubli, *Glycobiology*, 2016, **26**(2), 111–128.
20. T. Miyagi, K. Takahashi, K. Shiozaki and K. Yamaguchi, in *Sugar Chains: Decoding the Functions of Glycans*, ed. T. Suzuki, K. Ohtsubo and N. Taniguchi, Springer, Japan, Tokyo, 2015, pp. 159–176.
21. W. J. Grimes, *Biochemistry*, 1970, **9**, 5083.
22. H. Yu, H. Chokhawala, R. Karpel, H. Yu, B. Y. Wu, J. B. Zhang, Y. X. Zhang, Q. Jia and X. Chen, *J. Am. Chem. Soc.*, 2005, **127**(50), 17618.
23. K. Washiya, T. Furuike, F. Nakajima, Y. C. Lee and S.-I. Nishimura, *Anal. Biochem.*, 2000, **283**(1), 39.
24. M. Potier, L. Mameli, M. Bélisle, L. Dallaire and S. B. Melançon, *Anal. Biochem.*, 1979, **94**(2), 287.
25. C. Y. Zamora, M. d'Alarcao and K. Kumar, *Bioorg. Med. Chem. Lett.*, 2013, **23**(11), 3406.
26. Y. Kurebayashi, T. Takahashi, T. Otsubo, K. Ikeda, S. Takahashi, M. Takano, T. Agarikuchi, T. Sato, Y. Matsuda, A. Minami, H. Kanazawa, Y. Uchida, T. Saito, Y. Kawaoka, T. Yamada, F. Kawamori, R. Thomson, M. von Itzstein and T. Suzuki, *Sci. Rep.*, 2014, **4**, 4877.
27. G. Y. Yang, C. Li, M. Fischer, C. W. Cairo, Y. Feng and S. G. Withers, *Angew. Chem., Int. Ed.*, 2015, **54**(18), 5389.
28. W. Lin, L. Gao and X. Chen, *ChemBioChem*, 2015, **16**(18), 2571.

29. S. T. Laughlin and C. R. Bertozzi, *Proc. Natl. Acad. Sci. U. S. A.*, 2009, **106**(1), 12.
30. M. Boyce, I. S. Carrico, A. S. Ganguli, S.-H. Yu, M. J. Hangauer, S. C. Hubbard, J. J. Kohler and C. R. Bertozzi, *Proc. Natl. Acad. Sci. U. S. A.*, 2011, **108**(8), 3141.
31. C. Besanceney-Webler, H. Jiang, T. Zheng, L. Feng, D. Soriano Del Amo, W. Wang, L. M. Klivansky, F. L. Marlow, Y. Liu and P. Wu, *Angew. Chem., Int. Ed.*, 2011, **50**(35), 8051.
32. H. Wallrabe and A. Periasamy, *Curr. Opin. Biotechnol.*, 2005, **16**(1), 19.
33. H. Braak, E. Braak, I. Grundke-Iqbal and K. Iqbal, *Neurosci. Lett.*, 1986, **65**(3), 351.
34. M. Boyce, I. S. Carrico, A. S. Ganguli, S.-H. Yu, M. J. Hangauer, S. C. Hubbard, J. J. Kohler and C. R. Bertozzi, *Proc. Natl. Acad. Sci. U. S. A.*, 2011, **108**(8), 3141.
35. T. M. Gloster, W. F. Zandberg, J. E. Heinonen, D. L. Shen, L. Deng and D. J. Vocadlo, *Nat. Chem. Biol.*, 2011, **7**, 174.
36. S. A. Yuzwa, M. S. Macauley, J. E. Heinonen, X. Shan, R. J. Dennis, Y. He, G. E. Whitworth, K. A. Stubbs, E. J. McEachern, G. J. Davies and D. J. Vocadlo, *Nat. Chem. Biol.*, 2008, **4**, 483.
37. S. A. Yuzwa, X. Shan, M. S. Macauley, T. Clark, Y. Skorobogatko, K. Vosseller and D. J. Vocadlo, *Nat. Chem. Biol.*, 2012, **8**, 393.
38. R. Sayat, B. Leber, V. Grubac, L. Wiltshire and S. Persad, *Exp. Cell Res.*, 2008, **314**(15), 2774.
39. F. Doll, A. Buntz, A.-K. Späte, V. F. Schart, A. Timper, W. Schrimpf, C. R. Hauck, A. Zumbusch and V. Wittmann, *Angew. Chem., Int. Ed.*, 2016, **55**(6), 2262.
40. D. M. Patterson, L. A. Nazarova, B. Xie, D. N. Kamber and J. A. Prescher, *J. Am. Chem. Soc.*, 2012, **134**(45), 18638.
41. A.-C. Knall and C. Slugovc, *Chem. Soc. Rev.*, 2013, **42**(12), 5131.
42. J. Yang, J. Šečkutė, C. M. Cole and N. K. Devaraj, *Angew. Chem., Int. Ed.*, 2012, **51**(30), 7476.
43. R. J. Konrad, J. F. Tolar, J. E. Hale, M. D. Knierman, G. W. Becker and J. E. Kudlow, *Biochem. Biophys. Res. Commun.*, 2001, **288**(5), 1136.
44. J. Ma and G. W. Hart, *Clin. Proteomics*, 2014, **11**(1), 8.
45. F. I. Comer and G. W. Hart, *J. Biol. Chem.*, 2000, **275**, 29179.
46. Y. L. Chen, H. P. Liu, L. Ding and H. X. Ju, *Anal. Chem.*, 2018, **90**(5), 3073.
47. M. Monsigny, C. Sene, A. Obrenovitch, A.-C. Roche, F. Delmotte and E. Boschetti, *Eur. J. Biochem.*, 1979, **98**(1), 39.
48. C. S. Wright, *J. Mol. Biol.*, 1989, **209**, 475.
49. K. Harata, H. Nagahora and Y. Jigami, *Acta Crystallogr., Sect. D: Biol. Crystallogr.*, 1995, **51**, 1013.
50. D. Schwefel, C. Maierhofer, J. G. Beck, S. Seeberger, K. Diederichs, H. M. Möller, W. Welte and V. Wittmann, *J. Am. Chem. Soc.*, 2010, **132**(25), 8704.
51. C. S. Wright and G. E. Kellogg, *Protein Sci.*, 1996, **5**, 1466.
52. R. H. Rice and M. E. Etzler, *Biochemistry*, 1975, **14**(18), 4093.

53. S. A. Caldwell, S. R. Jackson, K. S. Shahriari, T. P. Lynch, G. Sethi, S. Walker, K. Vosseller and M. J. Reginato, *Oncogene*, 2010, **29**(19), 2831.
54. S. Lenzen and U. Panten, *Diabetologia*, 1988, **31**(6), 337.
55. R. S. Haltiwanger, K. Grove and G. A. Philipsberg, *J. Biol. Chem.*, 1998, **273**, 3611.
56. D. L. Dong and G. W. Hart, *J. Biol. Chem.*, 1994, **269**, 19321.
57. M. Horsch, L. Hoesch, A. Vasella and D. M. Rast, *Eur. J. Biochem.*, 1991, **197**(3), 815.
58. K. Vleminckx, L. Vakaet Jr, M. Mareel, W. Fiers and F. Van Roy, *Cell*, 1991, **66**(1), 107.
59. M. Münz, C. Kieu, B. Mack, B. Schmitt, R. Zeidler and O. Gires, *Oncogene*, 2004, **23**(34), 5748.
60. P. P. Hao, M. J. Lee, G. R. Yu, I. H. Kim, Y. G. Cho and D. G. Kim, *Mol. Cells*, 2013, **36**(5), 424.
61. F. Foglietta, L. Serpe, R. Canaparo, N. Vivenza, G. Riccio, E. Imbalzano, P. Gasco and G. P. Zara, *J. Pharm. Pharm. Sci.*, 2014, **17**(2), 231.
62. A. Krześlak, E. Forma, M. Bernaciak, H. Romanowicz and M. Bryś, *Clin. Exp. Med.*, 2012, **12**(1), 61.
63. L. Bao, L. Ding, M. Yang and H. X. Ju, *Sci. Rep.*, 2015, **5**, 10947.
64. R. G. Spiro, *J. Biol. Chem.*, 1960, **235**(10), 2860.
65. D. J. Lichlyter, S. A. Grant and O. Soykan, *Biosens. Bioelectron.*, 2003, **19**(3), 219.
66. A. Matsumoto, N. Sato, K. Kataoka and Y. Miyahara, *J. Am. Chem. Soc.*, 2009, **131**(34), 12022.
67. J. S. Reichner, S. W. Whiteheart and G. W. Hart, *J. Biol. Chem.*, 1988, **263**, 16316.
68. C. Oetke, S. Hinderlich, R. Brossmer, W. Reutter, M. Pawlita and O. T. Keppler, *Eur. J. Biochem.*, 2001, **268**(16), 4553.
69. Z. Rehman, I. S. Zuhorn and D. Hoekstra, *J. Controlled Release*, 2013, **166**(1), 46.
70. M. Rao and C. R. Alving, *Adv. Drug Delivery Rev.*, 2000, **41**(2), 171.
71. A. K. Varkouhi, M. Scholte, G. Storm and H. J. Haisma, *J. Controlled Release*, 2011, **151**(3), 220.
72. C. L. Bacon, E. O'Driscoll and C. M. Regan, *Int. J. Dev. Neurosci.*, 1997, **15**(6), 777.
73. S. Ishii and J. J. Volpe, *J. Neurosci. Res.*, 1990, **26**(4), 419.
74. X. Wang, L.-H. Zhang and X.-S. Ye, *Med. Res. Rev.*, 2003, **23**(1), 32.
75. S. Hinderlich, W. Weidemann, T. Yardeni, R. Horstkorte and M. Huizing, *Topics in Current Chemistry*, Springer, Berlin Heidelberg, 2013.
76. C. Y. Zamora, M. d'Alarcao and K. Kumar, *Bioorg. Med. Chem. Lett.*, 2013, **23**(11), 3406–3410.
77. G. Y. Yang, C. Li, M. Fischer, C. W. Cairo, Y. Feng and S. G. Withers, *Angew. Chem., Int. Ed.*, 2015, **54**(18), 5389.
78. L. Bao, L. Ding, J. J. Hui and H. X. Ju, *Chem. Commun.*, 2016, **52**(87), 12897.

79. Y. Liu, X. Dong and P. Chen, *Chem. Soc. Rev.*, 2012, **41**(6), 2283.
80. J. Huang, C. Zong, H. Shen, M. Liu, B. Chen, B. Ren and Z. Zhang, *Small*, 2012, **8**(16), 2577.
81. Q. Mu, G. Su, L. Li, B. O. Gilbertson, L. H. Yu, Q. Zhang, Y. P. Sun and B. Yan, *ACS Appl. Mater. Interfaces*, 2012, **4**(4), 2259.
82. T. Miyagi and S. Tsuiki, *Eur. J. Biochem.*, 1984, **141**(1), 75.
83. J. A. R. Mead, J. N. Smith and R. T. Williams, *Biochem. J.*, 1955, **61**(4), 569.
84. L. Feng, S. Zhang and Z. Liu, *Nanoscale*, 2011, **3**(33), 1252.
85. A. E. Nel, L. Madler, D. Velegol, T. Xia, E. M. Hoek, P. Somasundaran, F. Klaessig, V. Castranova and M. Thompson, *Nat. Mater.*, 2009, **8**(7), 543.
86. L. Zhang, Y. Cheng, J. P. Lei, Y. T. Liu, Q. Hao and H. X. Ju, *Anal. Chem.*, 2013, **85**(16), 8001–8007.
87. H. Otsuka, E. Uchimura, H. Koshino, T. Okano and K. Kataoka, *J. Am. Chem. Soc.*, 2003, **125**(12), 3493.
88. C. W. Cairo, *MedChemComm*, 2014, **5**, 1067.
89. C. Augeron and C. L. Laboisse, *Cancer Res.*, 1984, **44**(9), 3961.
90. Y. Kakugawa, T. Wada, K. Yamaguchi, H. Yamanami, K. Ouchi, I. Sato and T. Miyagi, *Proc. Natl. Acad. Sci. U. S. A.*, 2002, **99**(16), 10718.

CHAPTER 7

Real-time Monitoring of Intracellular Reactive Oxygen Species

7.1 Fluorescent Probes for Superoxide Anion

Superoxide anion ($O_2^{\bullet -}$) is the main reactive oxygen species (ROS) produced in organisms and is essential to life, being involved in the regulation of cell functions, namely signal transduction, neurotransmission, immune system control, production of energy, regulation of cellular growth, the synthesis of important biological compounds and the metabolism of xenobiotics.[1,2] However, when $O_2^{\bullet -}$ over-generation occurs, the reactive species become very harmful, causing oxidative stress through the oxidation of biomolecules, such as the lipids of cellular membranes, tissue proteins or enzymes, carbohydrates and DNA, leading to cellular damage and a variety of diseases.[3] It is therefore essential to develop methodologies representing new sensitive and specific tools for the purpose of ascertaining the pathological roles of $O_2^{\bullet -}$ in various diseases.

7.1.1 Phenol–Quinone Interconversion

Based on the intrinsic oxidation of $O_2^{\bullet -}$, a series of dynamic and reversible fluorescent probes based on the phenol structure were constructed (Figure 7.1). Using two-photon (TP) fluorescence microscopy, the highly selective and sensitive real-time detection of $O_2^{\bullet -}$ is possible.

As illustrated in probes TCA and PY-CA (probes **1** and **2**, Figure 7.1),[4,5] based on the scavenging activity of caffeic acid (or its derivatives) towards $O_2^{\bullet -}$, the probe *N*,*N*′-di[(2*E*)-3-(3,4-dihydroxyphenyl)acrylic

Detection Science Series No. 16
In Situ Analysis of Cellular Functional Molecules
By Huangxian Ju, Bo Tang and Lin Ding

Published by the Royal Society of Chemistry, www.rsc.org

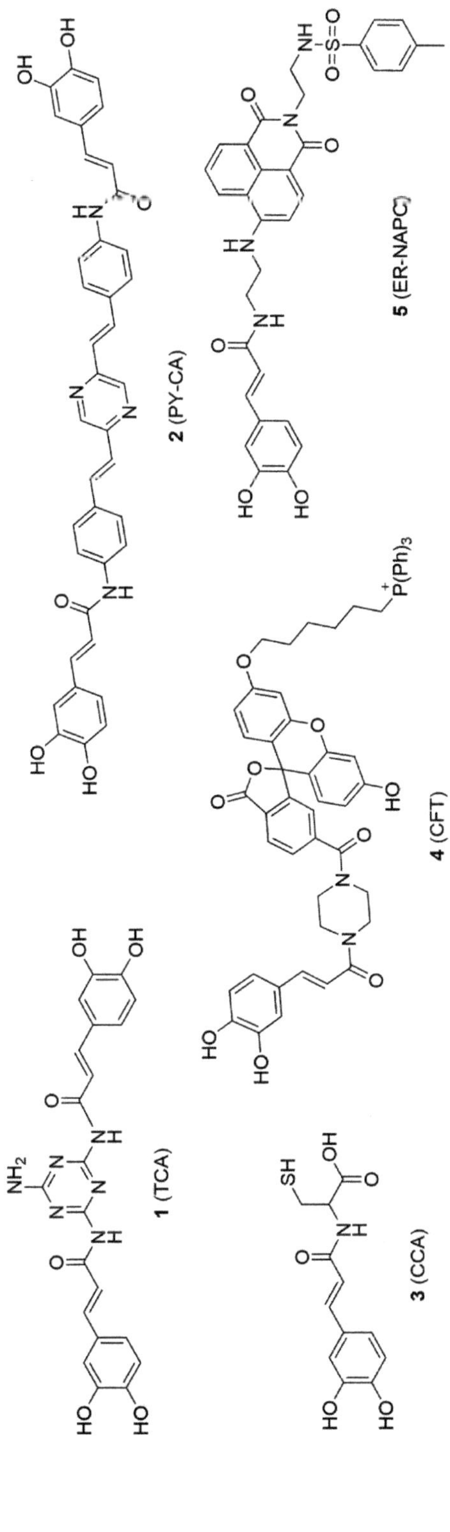

Figure 7.1 Structures of $O_2^{\bullet-}$ probes based on phenol–quinone interconversion.

acyl]-1,3,5-triazine-2,4,6-triamine (TCA), conjugated with two caffeic acid molecules and a symmetric tripolycyanamide, shows high specificity for $O_2^{\bullet-}$, instantaneous response and reversible interaction. After interaction with $O_2^{\bullet-}$, the generated TCAO can emit a fluorescence maximum at around 515 nm and the excitation wavelength is 800 nm for TP. This probe can be applied to image $O_2^{\bullet-}$ in hepatocytes, zebrafish and mice with mimicked ischaemia-reperfusion (I-R) injury and revealed that over-produced $O_2^{\bullet-}$ was closely related to reperfusion injury. To improve the depth of images further, 2,5-di(4′-caffeic acid amidestyrene)pyrazine (PY-CA) was developed, which contained symmetric styrylpyrazine as TP fluorophore, and showed a higher TP absorption cross-section. This probe can be applied to image $O_2^{\bullet-}$ in living tissues and intact organisms. The $O_2^{\bullet-}$ images can reach 900 μm into tumour tissues.

The fluctuations of $O_2^{\bullet-}$ in subcellular organelles of living cells can be analysed *via* covalently linking targeting groups to the phenol structure. The probe CCA is a new two-photon fluorescent probe for the instantaneous analysis of the Golgi apparatus $O_2^{\bullet-}$ (probe **3**, Figure 7.1).[6] The targeting strategy of CCA depends on an L-cysteine group for targeting the Golgi apparatus. CCA could respond to $O_2^{\bullet-}$ fluctuations instantaneously and reversibly. The Golgi $O_2^{\bullet-}$ level in I-R mice was detected with two-photon fluorescence microscopy and related molecular mechanisms regulated by the Golgi apparatus $O_2^{\bullet-}$ were illuminated during the hepatic I-R process.

The cross-talk between $O_2^{\bullet-}$ and other biologically active molecules or microenvironments can be detected based on phenol–quinone interconversion. The probe CFT monitors simultaneous changes in $O_2^{\bullet-}$ and pH levels in living cells and *in vivo* (probe **4**, Figure 7.1).[7] CFT contains caffeoyl, fluorescein and triphenylphosphonium moieties. The caffeoyl moiety is a satisfactory candidate for the $O_2^{\bullet-}$ recognition group, the fluorescein moiety serves as the pH response group and cationic triphenylphosphonium is used to target mitochondria. CFT was applied to the specific, instantaneous and dynamic detection of mitochondrial $O_2^{\bullet-}$ and pH fluctuations *via* blue–green channel fluorescence imaging based on two-photon single excitation. The results showed that both the $O_2^{\bullet-}$ and OH^- levels increased in hyperfused mitochondria of breast cancer cells and a mice tumour model. Further, the fluorescent probe ER-NAPC was developed for the simultaneous detection of $O_2^{\bullet-}$ and polarity in the endoplasmic reticulum (ER) (probe **5**, Figure 7.1).[8] The probe ER-NAPC contains 1,8-naphthalimide as the fluorophore, a caffeic acid group as the $O_2^{\bullet-}$ recognition unit and methylsulfonamide as an ER-targetable moiety. After reaction with $O_2^{\bullet-}$, the caffeic acid group of ER-NAPC is oxidized from pyrocatechol to benzoquinone, which results in an intense fluorescence resonance energy transfer (FRET) and emission of strong fluorescence of the naphthalimide moiety. Combined with a polarity probe, the simultaneous fluorescence imaging of $O_2^{\bullet-}$ and polarity in the ER of live cells and diabetic myocardial tissue was achieved. The results showed that $O_2^{\bullet-}$ and polarity will increase synergistically under ER stress.

7.1.2 Privileged Hydrogen Abstraction Reaction of Benzothiazoline

Based on the exclusive dehydrogenation reaction between benzothiazoline and $O_2^{\bullet -}$, a class of probes have been developed for the specific recognition of $O_2^{\bullet -}$ (Figure 7.2). Probe **6** (2-chloro-1,3-dibenzothiazolinecyclohexene) shows high selectivity and sensitivity to $O_2^{\bullet -}$.[9] It has a fluorescence response to $O_2^{\bullet -}$ over a large concentration range, with satisfactory detection limits at the nanomole level. By utilizing probe **6**, the release of $O_2^{\bullet -}$ in live RAW264.7 macrophages stimulated with 4β-phorbol 12-myristate-13-acetate (PMA) was successfully visualized.

The reaction-based TP fluorescent probe MF-DBZH[10] contains fluorine as TP fluorophore, benzothiazoline as an $O_2^{\bullet -}$-responsive group and a triphenylphosphonium salt as a mitochondrial-targeted site (probe **7**, Figure 7.2). $O_2^{\bullet -}$ dehydrogenates the benzothiazoline and expands the conjugated system of MF-DBZH to emit bright fluorescence at 512 nm. Applying MF-DBZH, mitochondrial $O_2^{\bullet -}$ fluctuations in inflamed cells and *in vivo* were successfully visualized by TP fluorescence imaging.

Based on benzothiazoline for $O_2^{\bullet -}$ recognition, a new ER-targeted two-photon fluorescent probe denoted ER-BZT was designed and synthesized for imaging $O_2^{\bullet -}$ (probe **8**, Figure 7.2).[11] ER-BZT contains 1,8-naphthalimide as a potent two-photon fluorophore, benzothiazoline as the $O_2^{\bullet -}$ recognition unit and methylsulfonamide as the ER-targetable moiety. ER-BZT shows high sensitivity, selectivity and stability and low cytotoxicity. In view of its advantageous features, ER-BZT was successfully used to detect $O_2^{\bullet -}$ release under conditions of ER stress. The results confirmed that $O_2^{\bullet -}$ is inextricably linked to ER-associated disorders.

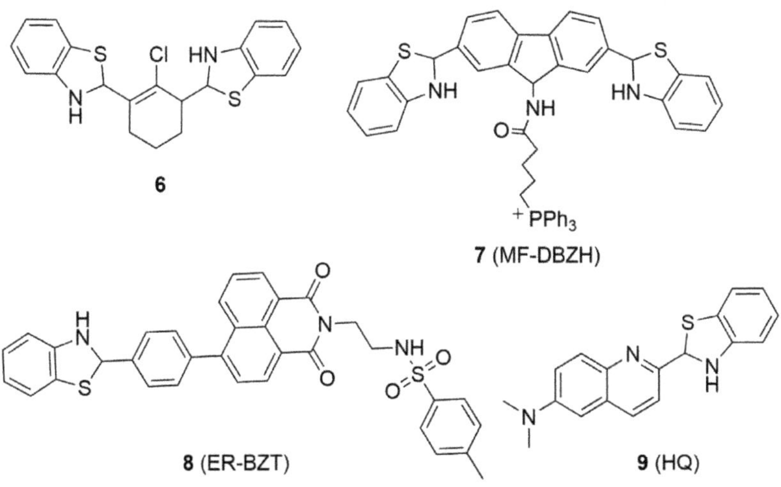

Figure 7.2 Structures of $O_2^{\bullet -}$ probes based on benzothiazoline.

A novel quinolone derivative-based two-photon fluorescent probe for $O_2^{\bullet-}$ detection, 6-(dimethylamino)quinoline-2-benzothiazoline (HQ), was reported (probe **9**, Figure 7.2).[12] HQ could recognize $O_2^{\bullet-}$ with a specific "turn-on" fluorescence response based on the extension of π-conjugations and a moderate intramolecular charge-transfer (ICT) process. Furthermore, HQ shows a favourable linear response towards $O_2^{\bullet-}$ in the range 0–500 μM in PBS buffer. With the assistance of two-photon microscopy, HQ was utilized to image endogenous $O_2^{\bullet-}$ in model animals with lung inflammation for the first time.

7.1.3 Chemiluminescence Method

Chemiluminescence (CL) methods are a promising approach for biological analysis owing to the obvious advantages of fast response, significantly reduced background and high signal-to-noise ratio (Figure 7.3). Therefore, a dual sensing platform with simultaneous turn-on signals of fluorescence and triggered by $O_2^{\bullet-}$ was fabricated (probe **10**, Figure 7.3).[13] The structure of Probe **10**, TPE-CLA, contains two moieties: the fluorescence moiety is a tetraphenylethene group with excellent aggregation-induced emission properties and the CL moiety is *Cypridina* luciferin analogues for a good CL response to $O_2^{\bullet-}$. TPE-CLA could monitor $O_2^{\bullet-}$ temporarily and continuously *in vitro* and *in vivo*. This fluorescence–CL system is expected to extend the application of dual detection probes in the life sciences. In addition, a novel visualization strategy based on CL resonance energy transfer (CRET) for $O_2^{\bullet-}$ detection was developed (probe **11**, Figure 7.3).[14] The CRET-based probe **11** contains a luminol group as the $O_2^{\bullet-}$ recognition site and an energy donor and BODIPY (a boron dipyrromethene) as the energy receptor. With the assistance of probe **11**, the $O_2^{\bullet-}$ increase in PMA-activated splenocytes was detected.

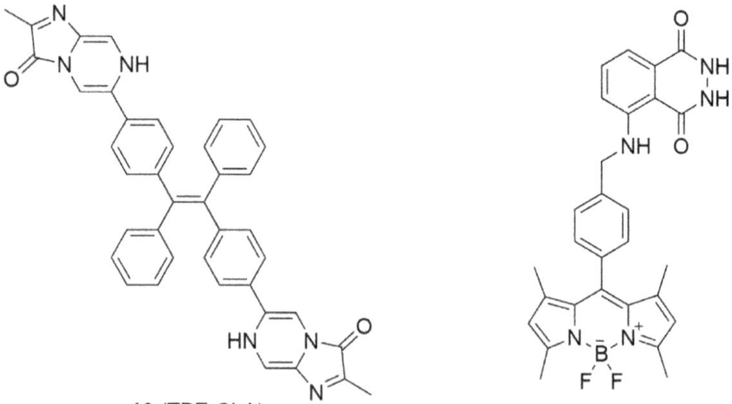

10 (TPE-CLA)

11

Figure 7.3 Structures of $O_2^{\bullet-}$ probes based on chemiluminescence methods.

7.1.4 Nucleophilic Substitution Reaction

In view of the nucleophilicity of $O_2^{\bullet-}$, a new type of $O_2^{\bullet-}$ probe was developed based on the nucleophilic substitution reaction between fluorescent probes and $O_2^{\bullet-}$. For example, probe **12** (Figure 7.4) was constructed as a highly sensitive and selective probe for $O_2^{\bullet-}$ detection.[15] Based on the fact that $O_2^{\bullet-}$ could cleave an aryl trifluoromethanesulfonate group to yield a free phenol, three new fluorescent probes, HKSOX-1 and HKSOX-1r for cellular retention and HKSOX-1m for mitochondria targeting, were prepared and used for the detection of $O_2^{\bullet-}$ in living cells. The results indicated that the $O_2^{\bullet-}$ level increased in multiple cellular models, such as inflammation and mitochondrial stress. These probes can also be efficiently applied to detect $O_2^{\bullet-}$ in intact live zebrafish embryos.

A two-channel fluorescent probe for the selective detection of $O_2^{\bullet-}$ in living cells was reported (probe **13**, Figure 7.4).[16] In the probe structure, the fluorescence reporter unit is a pyridinium-modified tetraphenylethene derivative, the reactive group is a diphenylphosphinyl moiety and the linkage unit is a cresol-like structure. The reactive group can react specifically with $O_2^{\bullet-}$ and results in the fracture of the linkage. Owing to the changes in the donor–acceptor structure, the aggregation-induced emission shifts from red to green. The probe exhibits favourable photophysical properties and could detect $O_2^{\bullet-}$ in living cells with different treatments. With the same reactive group, a novel near-infrared (NIR) turn-on fluorescent probe (CyR) for the selective detection of $O_2^{\bullet-}$ was developed (probe **14**, Figure 7.4).[17] The NIR fluorophore of CyR is an indole xanthene derivative and the emission peak is at 704 nm. CyR has a low limit of detection of 9.9 nM and good cell-membrane permeability. Further, CyR was successfully applied to the detection of $O_2^{\bullet-}$ fluctuations in living cells, zebrafish and mice.

With 2,4-dinitrobenzenesulfonyl as the recognition site, a fluorescent probe (probe **15**, Figure 7.4) was designed and synthesized for the detection of mitochondrial $O_2^{\bullet-}$ by coupling a mitochondrial locating signal peptide.[18] Probe **15** can detect the $O_2^{\bullet-}$ elevation in RAW264.7 cells treated with PMA and 2-methoxyestradiol (2-ME). In addition, $O_2^{\bullet-}$ over-generation during single-walled carbon nanotube invasion was monitored in live cells.

Trifluoromethylsulfonate can also be applied as an $O_2^{\bullet-}$recognition site, as illustrated in probe **16** (Figure 7.4).[19] Outstanding merits such as a large fluorescence enhancement (~235-fold) and a low detection limit (1.0 nM) for $O_2^{\bullet-}$, rapid response speed and high stability and selectivity endow this probe with the ability to detect $O_2^{\bullet-}$ in living cells and tissues. The $O_2^{\bullet-}$ elevation in RAW264.7 cells and rat liver tissues stimulated with PMA was detected by two-photon fluorescence imaging with probe **16**.

7.2 Fluorescent Probes for Hydrogen Peroxide

H_2O_2 as a reactive oxygen species (ROS) plays a crucial role in oxidative stress and signal transduction of organisms. Fluorescent probes have been

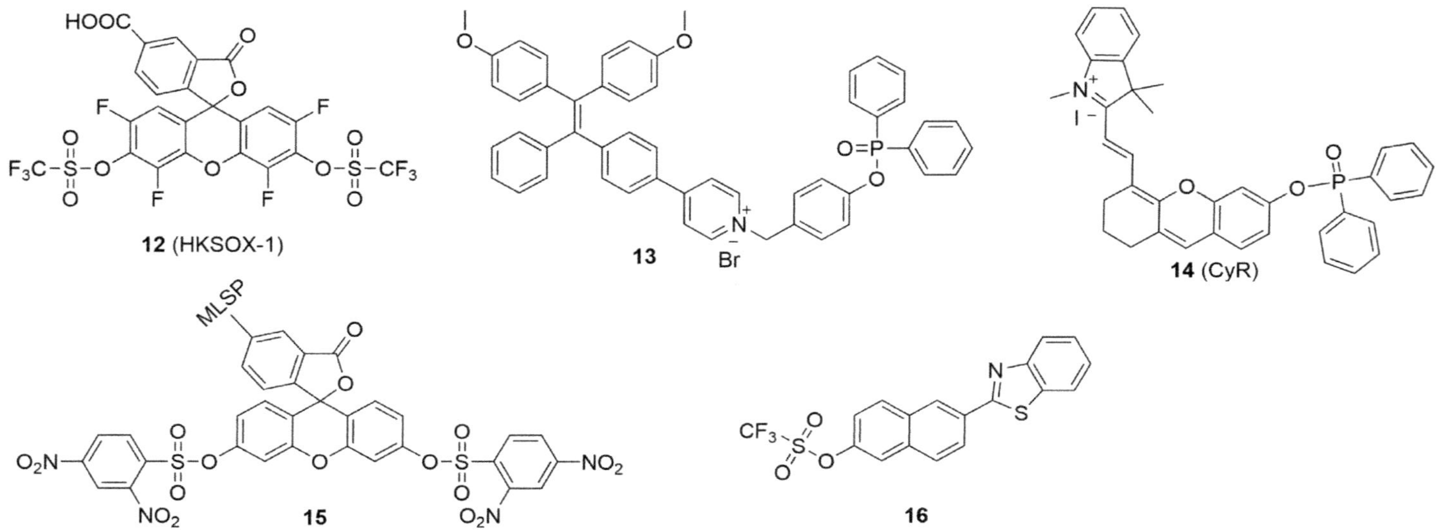

Figure 7.4 Structures of $O_2^{\bullet -}$ probes based on nucleophilic substitution reactions.

demonstrated to be a powerful tool for H_2O_2 analysis. However, a common problem is the slow response, leading to difficulty in tracking H_2O_2 *in situ.*

To address this problem, Tang and co-workers designed and synthesized a novel tetraphenylethene (TPE)-based fluorescent probe, TPE-BO, employing the activity of increased polarity in C–B bonds (probe **17**, Figure 7.5).[20] In order to confirm the reaction mechanism of the probe, a computational study was performed and the results indicated that the polarity of C–B bonds was increased, leading to their more facile replacement. The probe was successfully applied to detect H_2O_2 levels in living cells, providing a

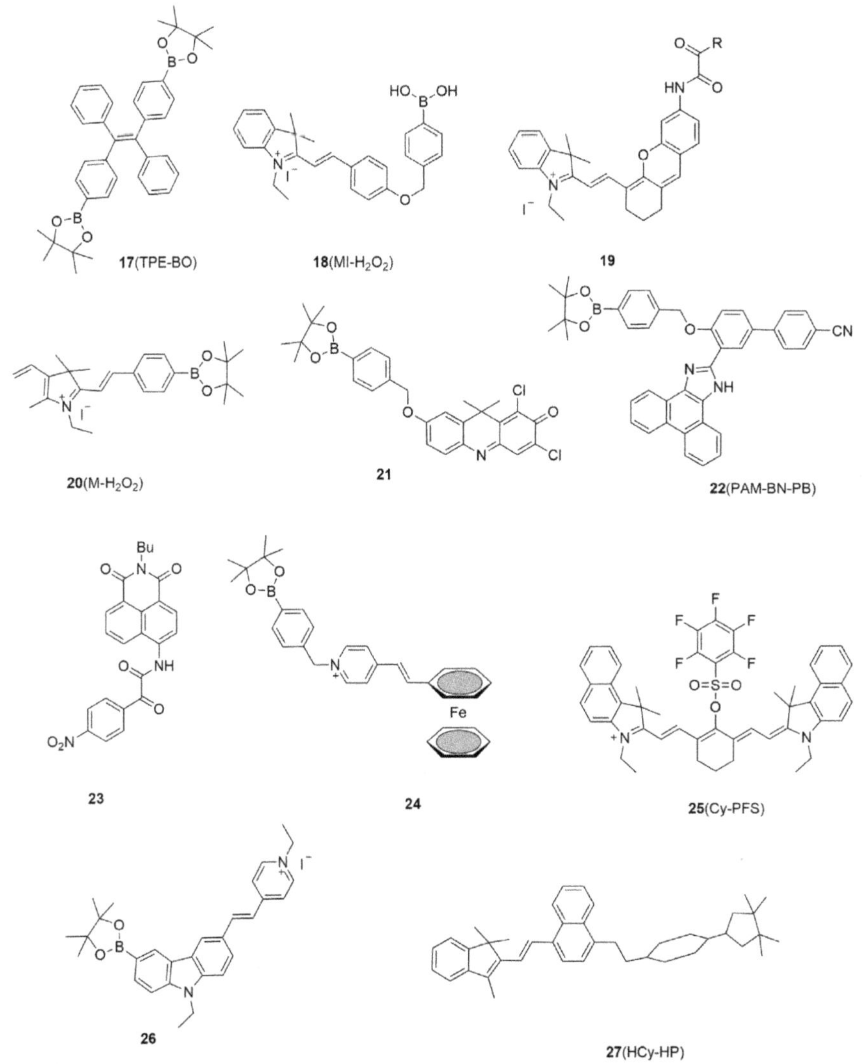

Figure 7.5 Structures of fluorescent probes for H_2O_2.

rapid-response and highly selective fluorescence tool for monitoring H_2O_2 levels in biological processes.

Tang and co-workers subsequently reported the synthesis and application of two H_2O_2-selective fluorescent probes, denoted MI-H_2O_2 (probe **18**, Figure 7.5) and ER-H_2O_2, based on a boronic acid/ester deprotection mechanism.[21] MI-H_2O_2 and ER-H_2O_2 can preferentially accumulate in the mitochondria and endoplasmic reticulum (ER), respectively, because of the lipophilic cationic and methylsulfonamide moieties, respectively. Additionally, they display obviously discriminable spectra, which benefits live cell multicolour imaging. MI-H_2O_2 contains merocyanine as a fluorescence reporter and boronic acid as a specific masking group. The boronic acid group can be removed upon reaction with H_2O_2 and converted to an oxygen anion, forming an intense push–pull conjugated system that fluoresces strongly. ER-H_2O_2 contains a 1,8-naphthalimide fluorophore and a boronic ester recognition group. After reaction with H_2O_2, electron donor and electron acceptor conformations were obtained, and these have been used in ratiometric fluorescent probes owing to the outstanding intramolecular charge-transfer (ICT) structure. As expected, the results indicated that MI-H_2O_2 and ER-H_2O_2 can visualize exogenous or endogenous H_2O_2 in mitochondria and the ER, respectively. Also, simultaneous multicolour fluorescence imaging utilizing these two probes revealed that the H_2O_2 levels changed differentially in the mitochondria and the ER during different apoptotic stimuli.

Tang and co-workers further reported the rational design, characterization and biological application of a mitochondria-targetable NIR fluorescent sensor, Mito-NIRHP, for H_2O_2 visualization (probe **19**, Figure 7.5).[22] Mito-NIRHP utilizes a unique reaction switch, an α-ketoamide moiety, to turn on a highly specific, sensitive and rapid fluorescence response towards H_2O_2 coupled with an intramolecular charge-transfer strategy. Mito-NIRHP is efficient in tracking endogenously produced H_2O_2 in both living cells and living animals. In addition, utilizing Mito-NIRHP, over-generation of H_2O_2 during I-R injury was directly visualized at both the cell and organ levels.

Tang and co-workers designed two organic fluorescent probes, denoted M-H_2O_2 (probe **20**, Figure 7.5) and M-Zn, for the simultaneous imaging of the two signalling molecules H_2O_2 and Zn^{2+}, respectively, inside mitochondria in living cells and *in vivo*.[23] M-H_2O_2 and M-Zn can preferentially target mitochondria because of the introduction of the lipophilic cation. They display distinct dual excitation–emission spectra, which favours live cell multicolour imaging. M-H_2O_2 contains merocyanine as the fluorescence reporter and a boronic ester as the specific H_2O_2 response moiety. Removal of the boronic ester upon reaction with H_2O_2 converts M-H_2O_2 into an intense push–pull conjugated system with an oxygen anion that fluoresces strongly. M-Zn contains Cy7 as the fluorescence indicator and a dipicolylamine (DPA) moiety as the Zn^{2+} reporter. The fluorescence of M-Zn is quenched by the photoinduced electron transfer (PET) mechanism. However, upon addition of Zn^{2+}, the Zn^{2+}–DPA complex formed blocks the PET process, which recovers the fluorescence of Cy7. M-H_2O_2 and M-Zn are highly sensitive and selective towards H_2O_2 and Zn^{2+}, respectively. They can detect H_2O_2 and Zn^{2+} in the mitochondria of live cells and zebrafish by confocal fluorescence imaging. Moreover, simultaneous multicolour fluorescence imaging utilizing these two probes revealed that the H_2O_2 and Zn^{2+} levels changed synergistically.

Liu and Song's group reported the synthesis of an NIR fluorescent probe (AB1) for H_2O_2 by modulating the intramolecular charge-transfer (ICT) process of the dye 9*H*-1,3-dichloro-7-hydroxy-9,9-dimethylacridin-2-one (DDAO) (probe **21**, Figure 7.5).[24] The probe AB1 exhibits both a large NIR fluorescence turn-on and a ratiometric response to H_2O_2 with high sensitivity and specificity. The fluorescence response of AB1 has a good linear relationship with H_2O_2 concentration over a wide range from 1 to 100 mM, affording a detection limit of 0.42 mM. Confocal microscopic experiments demonstrated that AB1 could ratiometrically detect exogenous and endogenous H_2O_2 in living cells. Moreover, owing to the NIR emission of DDAO, the probe was also utilized to detect endogenous H_2O_2 from the peritoneal cavity in a mouse model of lipopolysaccharide-induced acute inflammation, based on the fluorescence turn-on mode.

Wang and co-workers reported a new probe, PAM-BN-PB, based on an ICT process, with three components: phenanthroimidazole, benzonitrile and phenylboronate (probe **22**, Figure 7.5).[25] By modulation of the ICT process of PAM-BN-PB, H_2O_2 in solution systems can be detected with good selectivity. The exogenous and endogenous H_2O_2 in normal living cells, I-R injury cells and animals can all be detected by PAM-BN-PB.

Zhang and co-workers reported a ratiometric fluorescent probe, α-Naph, capable of determining basal H_2O_2 in the ER (probe **23**, Figure 7.5).[26] The probe specifically responds to H_2O_2. The limit of detection of the probe is as low as 38 nM, making it a feasible sensor for detecting intracellular basal H_2O_2. In addition, utilizing its ratiometric property, the probe can measure the concentration of H_2O_2 in the ER quantitatively, eliminating the errors caused by the probe concentration and environment. The intracellular concentration of H_2O_2 in the ER was calculated to be 0.692 μM under normal conditions and 1.26 μM under stimulation with PMA.

Shao and Xu's group reported a novel multichannel signalling fluorescence–electrochemistry combined probe (FE-H_2O_2) and applied it to the imaging and detection of H_2O_2 in living cell systems (probe **24**, Figure 7.5).[27] In this design, the probe FE-H_2O_2 contains an H_2O_2 reaction site and a 4-ferrocenyl(vinyl)pyridine unit that affords chromogenic, fluorescent and electrochemical signals. These structural motifs yield a combined chromogenic, fluorescent and redox sensor in a single molecule. The FE-H_2O_2 probe showed a turn-on fluorescence response to H_2O_2, which can be used for monitoring intracellular H_2O_2 *in vivo*. Furthermore, the electrochemical response of the FE-H_2O_2 probe decreased after the addition of H_2O_2, which can be applied to the accurate detection of H_2O_2 released from living cells. When the fluorescence imaging method is combined with electrochemical analysis technology, it is hoped that the well-designed multimodule probe can serve as a practical tool for understanding the metabolism and homeostasis of H_2O_2 in complex biological systems.

Yu's group developed an NIR ratiometric fluorescent probe, Cy-PFS, for specifically detecting endogenous H_2O_2 in cells and *in vivo* (probe **25**, Figure 7.5).[28] Employing the Cy-PFS probe, they examined the critical effects of endogenous H_2O_2 on the proliferation of cells in live hippocampal neuron cells and the results provided strong evidence for H_2O_2 signalling in cell mitosis through growth factor signalling. Furthermore, they successfully demonstrated the close association of endogenous H_2O_2 level changes with the brain development at various stages. It is envisioned that this probe has potential as a promising chemical imaging tool for exploring the roles of H_2O_2 in cell mitosis.

Lin and co-workers developed a novel fluorescent probe, CAI, which showed a large Stokes shift in various solutions (probe **26**, Figure 7.5).[29] They found that CAI was capable of detecting H_2O_2 in aqueous solution. Furthermore, the CAI probe can sense exogenous and endogenous H_2O_2 in the mitochondria of living RAW264.7 cells. In addition, the CAI probe exhibited ideal properties such as excellent photostability and high sensitivity and selectivity. This fluorescent dye with a large Stokes shift was expected to have a broad range of applications for developing various fluorescent probes with large Stokes shifts.

Tang's group reported a colorimetric and fluorescent selective turn-on probe (HCyHP) for H_2O_2 detection (probe **27**, Figure 7.5).[30] The HCyHP probe displayed very weak fluorescence in aqueous solution owing to the PET and weak ICT effect of the phenylboronate ester moiety. However, after H_2O_2-promoted cleavage of the boronate ester, a phenolate intermediate was generated, which can spontaneously undergo 1,6-elimination of *p*-quinonemethide to generate HCyOH with an inhibited PET and strong ICT effect, thus boosting strong red emission. HCyHP showed a capability for the turn-on detection of exogenous and endogenous H_2O_2 in live cells. Moreover, HCyHP could selectively light up endogenous H_2O_2 in hypoxic cancer cells with good biocompatibility. Its application to the *in vivo* imaging of H_2O_2 in tumour-bearing live mice and diagnosis of diabetes based on H_2O_2 detection was also evaluated.

7.3 Fluorescent Probes for Hydroxyl Radical

The hydroxyl radical ($^{\bullet}OH$), as one of the most reactive and oxidative ROS, can cause damage to biomacromolecules. Moreover, excess $^{\bullet}OH$ production leads to irreparable damage to cells, accelerating cellular senescence and death and potentially even leading to diseases. Therefore, there is an urgent need to develop efficient methods for tracing $^{\bullet}OH$ in living biosystems with high selectivity and sensitivity in order to define the role that $^{\bullet}OH$ plays in diseases.

Inspired by the specific one-electron oxidation reaction between $^{\bullet}OH$ and 3-methylpyrazolone in the neuroprotective agent edaravone, Tang's group created a novel two-photon fluorescent probe (MD-B) based on ICT that can detect •OH with outstanding selectivity and an instantaneous response (probe **28**, Figure 7.6).[31] The probe achieved the *in situ* detection of the release of •OH in human astrocytes under stimulation by glutamate and successfully traced •OH in the brain of mice with depression phenotype. The authors further demonstrated that •OH causes oxidative damage to the active sites of SIRT1, indicating that the inactivation of SIRT1 by •OH is responsible for depression phenotypes.

By incorporating a strong electron-donating methoxy group into the central benzene group of a cyanine fluorochrome, Ma and co-workers designed a sensitive fluorescence •OH probe to monitor the generation of trace •OH in iron autoxidation based on the unique aromatic hydroxylation and the electrophilicity of •OH (probe **29**, Figure 7.6).[32] After reaction with •OH, the probe was transformed into a larger π-conjugated structure and elicited bright NIR fluorescence. •OH generated in living cells under iron autoxidation and also various stimuli was detected.

Yang's group reported fluorescent probes for •OH detection with high sensitivity and selectivity.[33] Based on the electron-transfer reaction of •OH with diiodophenol to form a phenoxyl radical and steric hindrance of the two iodine atoms, new fluorescent probes HKOH-1 (probe **30**, Figure 7.6) for quantifying the scavenging capacities of the reported scavengers and HKOH-1r for better cellular uptake and retention were designed for detecting •OH in living cells. The authors monitored the •OH increase in HeLa cells in response to UV irradiation.

Xu and co-workers presented a novel mitochondrial-targetable probe (MPT-Cy2) based on a hybrid phenothiazine platform, which can be applied to detect the mitochondrial •OH in live cells and its distribution in living organisms (probe **31**, Figure 7.6).[34] Owing to the oxidation of the sulfur atom in the cyanine by •OH, MPT-Cy2 can measure the concentration of •OH. The probe can detect the endogenous •OH in living cells and various types of bacteria. In addition, it also possesses the capability to detect •OH generated following treatment with TiO_2NPs in zebra fish.

Based on a hybrid carbazole–cyanine platform, Lin's group designed and synthesized a turn-on fluorescent probe (CCy-OH) for detecting •OH in live cells with a fast response and high selectivity (probe **32**, Figure 7.6).[35] After reaction with •OH, a fluorescence enhancement of about 60-fold was observed. Further, the probe was applied to imaging endogenic and exogenous •OH in living HeLa cells.

By connecting the 4-position of the naphthalimide group and the 2-position of the naphthyridine group using ethylenediamine, Yi and co-workers designed a ratiometric fluorescent probe for the selective and sensitive detection of •OH (probe **33**, Figure 7.6).[36] It can distinguish •OH with high selectivity and a short response time. Further, the probe effectively detected intracellular •OH in living RAW264.7 cells under PMA stimulation.

Zhang's group reported a single fluorescent probe (FHZ) to achieve the imaging of •OH and hypochlorous acid (HClO) in living organisms

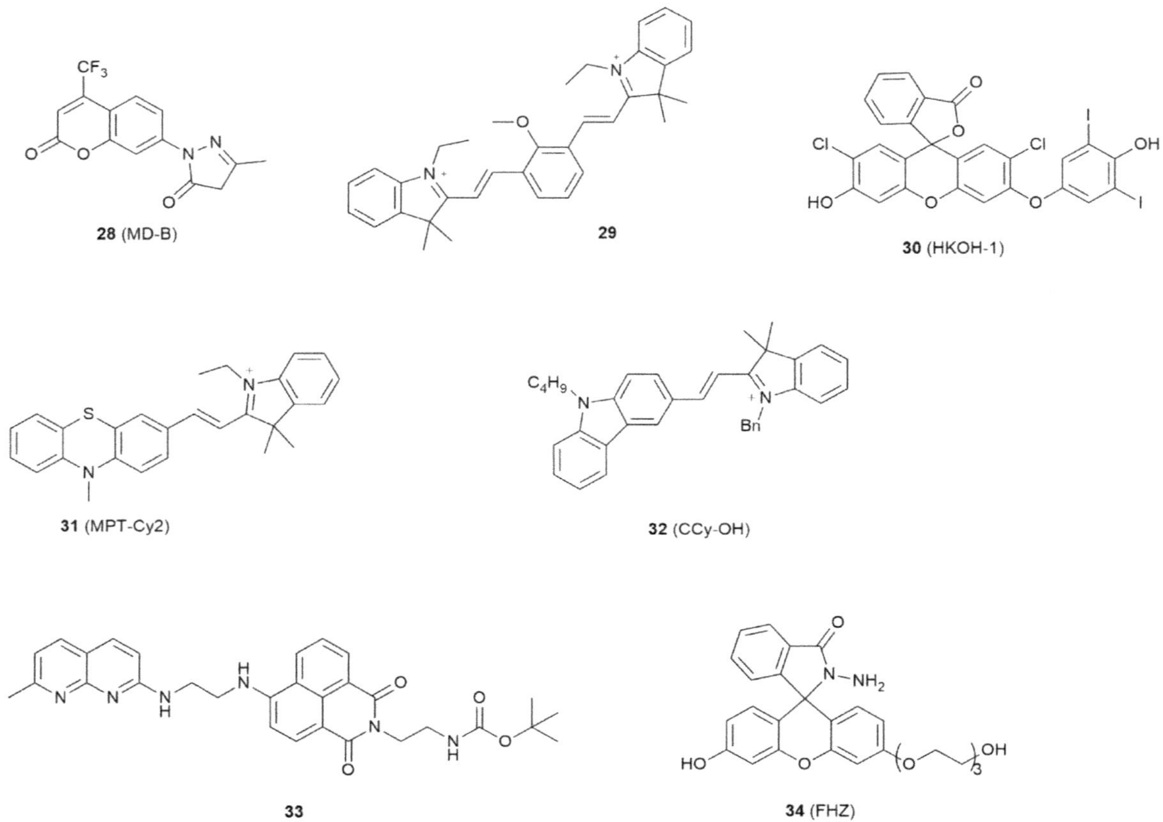

Figure 7.6 Structures of fluorescent probes for •OH.

(probe **34**, Figure 7.6).[37] Owing to the ultrastrong oxidizability, •OH can simultaneously break the bottom-left aromatic ring in the probe FHZ. The fluorescence of FHZ undergoes a hypsochromic shift from green to cyan due to the decreased size of the conjugated system. Utilizing the probe, they clearly presented the profiling of spontaneous •OH and HClO in cells and •OH and HClO in the organs of living zebrafish. Moreover, they successfully visualized for the first time the generation of •OH in a fresh wound in zebrafish.

7.4 Fluorescent Probes for Lipid Peroxy Radical

Lipids are major targets for ROS. Polyunsaturated fatty acids (PUFAs), especially arachidonic acid, are vulnerable to suffer oxidative modifications, causing lipid peroxidation.[38] This lipid peroxidative damage disrupts the lipid membrane structure and changes the fluidity of the membrane, leading to severe cellular injury.[39] Upon attack of ROS, the allylic hydrogen of PUFAs was abstracted, forming a carbon-centred radical (L•) through a molecular rearrangement. L• can quickly react with oxygen to form a more active lipid peroxy radical (LOO•) and subsequently lipid hydroperoxides (LOOH).[40] These products attack amino acids and nucleobases and lead to major protein disruption or other deadly genetic mutations, causing diseases such as cancer, atherosclerosis and neurodegenerative diseases.[41] In addition, these lipid peroxidation products could rapidly decompose to the secondary cytotoxic reactive carbonyl species.[42] Hence it is important to develop tools to capture these early-stage lipid peroxidation products at the site where they are produced, which is of benefit for exploring relevant physiopathology mechanisms. Fluorescent probes, combined with state-of-the-art fluorescence microscopy methods, can provide high levels of sensitivity and specificity in conjunction with exquisite temporal and spatial information in studies of living organisms.[43] Therefore, this section highlights recent research on fluorescent probes that selectively respond to LOO• and LOOH in the lipid membrane of living cells.

Considering the hydrophobicity of lipid membranes, the available fluorescent probes developed for monitoring lipid peroxidation are generally lipophilic, mimicking the unsaturated structure of PUFAs. Hudson and co-workers developed the first fluorescent probe for lipid peroxidation in 1975 (probe **35**, Figure 7.7). They chose the naturally polyunsaturated fatty acid *cis*-parinaric acid, acting as a fluorescent indicator of lipid peroxidation.[44] On oxidation, the fluorescence of this compound decreased owing to breakdown of the conjugation of the double bonds. However, its absorption in the UV region and photoinstability hindered its use in living systems.

Subsequently, based on the similar oxidation mechanism of *cis*-parinaric acid to lipid peroxidation, numerous probes were developed, and here we take C11-BODIPY$^{581/591}$ as an example (probe **36**, Figure 7.7).[45] This probe has a highly conjugated structure and red-shifted emission. After lipid

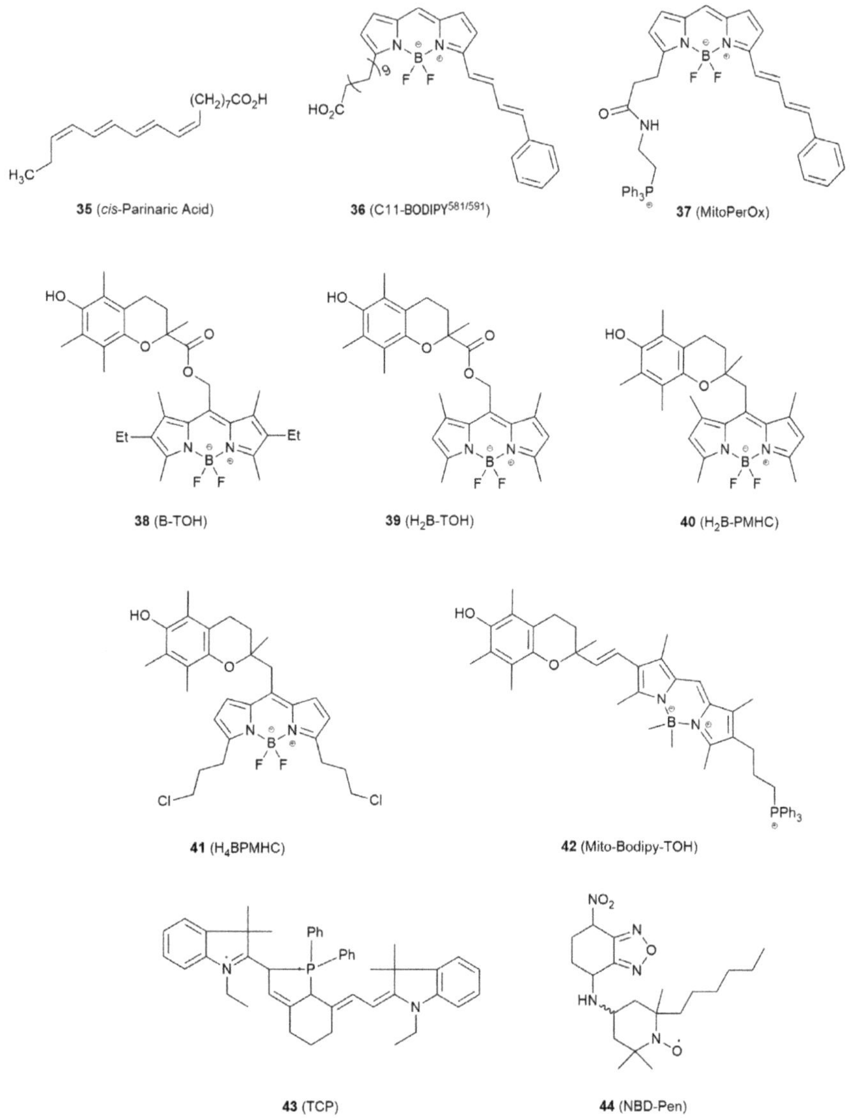

Figure 7.7 Structures of fluorescent probes for lipid peroxy radical.

peroxidation, the conjugation was broken and the emission shifted to lower wavelengths. This ratiometric probe realized the quantitative sensing of lipid peroxidation products by measuring the ratio of the emissions at red and green wavelengths, effectively reducing the false signal and instrumental factors (*e.g.* illumination stability). More recently, Murphy and co-workers modified C11-BODIPY$^{581/591}$ by incorporating the triphenylphosphonium lipophilic cation into the structure and named it MitoPerOx (probe **37**, Figure 7.7).[46] MitoPerOx allowed sensitive ratiometric assays of mitochondrial lipid peroxidation. However, this kind of probe still had a response to other free radicals, lacking selectivity. Additionally, the photostability of the probes was not perfect.

Vitamin E molecules, especially α-tocopherol, act as scavengers of LOO•, effectively terminating lipid chain autoxidation. Based on this, Cosa's group developed a series of fluorescent probes for LOO•. The first-generated probe was called B-TOH (probe **38**, Figure 7.7),[47] and mimicked the structure and activity of α-tocopherol and identified LOO• *via* an intramolecular PET mechanism. The probe was prepared by coupling a free radical scavenger segment (the chromanol compound trolox) to a fluorophore segment (BODIPY). The trolox, possessing similar activity to α-tocopherol, contributed as the recognition group towards LOO• and ensured the selectivity of the probe. The lipophilic BODIPY fluorophore ensured it was close to the lipid membrane and enhanced the sensitivity of the probe. The fluorescence of B-TOH was in the quenched state, and after reaction with LOO• the phenolic hydroxyl was converted into quinone and the fluorescence of BODIPY was recovered. However, compared with the homogeneous solution, the sensitivity of B-TOH decreased in lipid membranes.

In order to modulate the sensitivity of the first-generation probe, Cosa and co-workers made improvements to B-TOH and the second-generation probe H_2B-TOH was prepared (probe **39**, Figure 7.7).[48] In H_2B-TOH, by changing the substituents at positions C2 and C6 of the BODIPY core to electron-withdrawing groups, the HOMO energy level of the fluorophore was lower and more exergonic electron transfer would be accomplished. Therefore, the emission enhancement of H_2B-TOH was up to 40-fold and the sensitivity of H_2B-TOH was increased.

To improve further the sensitivity of the probe towards LOO•, Cosa and co-workers constructed a third-generation probe, called H_2B-PMHC (probe **40**, Figure 7.7). In H_2B-PMHC, by changing the ester linker to methylene, the phenolic hydroxyl was strengthened owing to the electron-withdrawing effect. Hence the reactivity of the chromanol moiety towards LOO• was improved. Compared with B-TOH, the emission enhancement of H_2B-PMHC increased to 100-fold, along with higher sensitivity.

Subsequently, a fourth-generation probe was developed by Cosa's group, called H_4BPMHC (probe **41**, Figure 7.7).[49] In H_4BPMHC, the methyl groups at the 3- and 5-positions of the BODIPY core in H_2B-PMHC were replaced by propyl chloride, mimicking α-tocopherol's lipid solubility to a great extent. The emission enhancement of H_4BPMHC increased to 1000-fold, and it was extremely sensitive to LOO• at the cell level. Taking advantage of this exquisitely designed H_4BPMHC, they realized real-time measurements to indicate the production of endogenous LOO• in lipid membranes and in a non-disruptive manner. The authors continued to modify the properties of above probes, and Mito-Bodipy-TOH was prepared (probe **42**, Figure 7.7).[50] Through the addition of a triphenylphosphonium cation, Mito-Bodipy-TOH could specifically target the inner mitochondrial lipid membrane (IMM) and track mitochondrial LOO•.

Tang and co-workers developed an NIR fluorescent probe, TCP (tricarbocyanine diphenylphosphine), for detecting lipid hydroperoxides in living cells (probe **43**, Figure 7.7). The probe was almost non-fluorescent because of its non-conjugation, but on contacting LOOH conjugated TCP=O was formed and the fluorescence signal was strongly restored. TCP exhibited specific selectivity, high sensitivity (detection limit 38 pM) and a rapid response to LOOH and effectively avoided background fluorescence interference in biological systems owing to the NIR excitation and emission. This low cytotoxic probe provides a useful tool for detecting lipid hydroperoxides in living biological systems.

Yamada *et al.* designed a selective and sensitive fluorescent probe, NBD-Pen (probe **44**, Figure 7.7), for $L^•$ for the first time.[51] This α-substituted nitroxide had a high affinity for $L^•$, acting as the recognized group and also the fluorescence quencher of NBD. After the α-substituted nitroxide had reacted with lipid radicals, the fluorescence of NBD rapidly recovered. NBD-Pen could directly track $L^•$ in living cells and in an animal model. In addition, the nitroxide portion of NBD-Pen could significantly relieve inflammation in the early stage of carcinogenesis.

7.5 Fluorescent Probes for Other Reactive Oxygen Species

7.5.1 Fluorescent Probes for HClO

As an endogenous ROS, hypochlorous acid (HClO) can be generated from the peroxidation reaction of H_2O_2 and chloride ions catalysed by myeloperoxidase (MPO) in cells. HClO is an unstable weak acid that can partially dissociate into hypochlorite (ClO^-) under physiological conditions. HClO is an extremely important oxidant *in vivo* that can react with a variety of biomolecules in living systems. HClO plays an important role in the immune system. However, excessive intracellular levels of HClO contribute to cardiovascular disease, inflammation, atherosclerosis, kidney damage and even cancers. Therefore, identifying and detecting cellular HClO fluctuations accurately are of vital significance for disease diagnosis.

Many fluorescent probes have been reported for detecting HClO *in vivo*. The reaction mechanisms basically use the strong oxidizing property of HClO. Specific functional groups in the fluorescent probe are oxidized by HClO. The fluorescence signal of the probe changes owing to the changes in the structure of the fluorescent probe. Quantitative detection of HClO levels can be realized based on oxidative desulfurization reactions, oxidative unsaturated double bond reactions, oxidation of *p*-methoxyphenols, oxidation of thioethers, organic sulfur, selenium or antimony compounds and other newly discovered mechanisms. A number of organic molecular fluorescent probes, such as coumarins, cyanine dyes, rhodamines, fluorescein, BODIPY, naphthalimide, benzothiazole, *etc.*, have been reported (Figure 7.8).

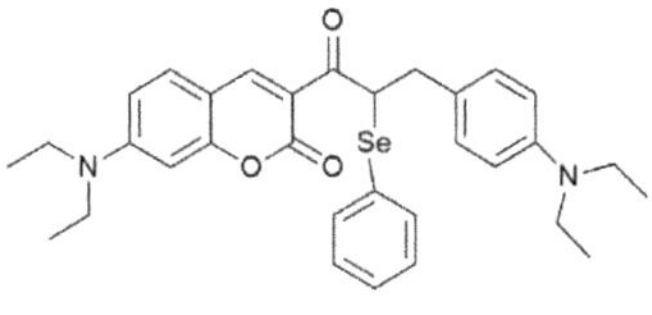

45 (Coum-Se)

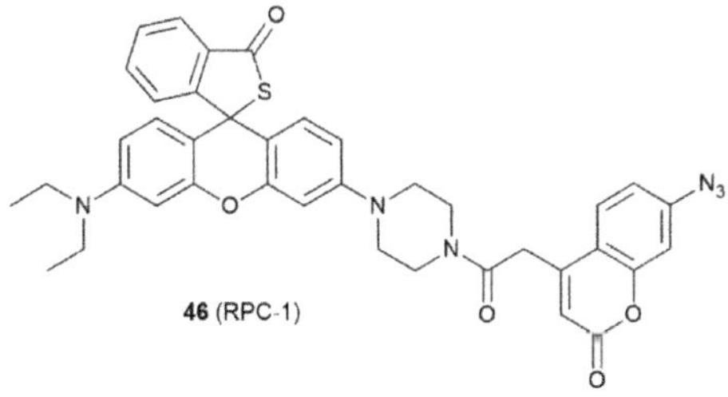

46 (RPC-1)

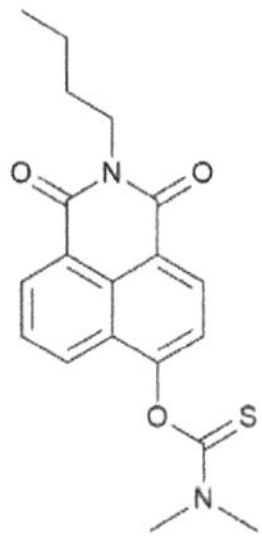

47 (NDMTC)

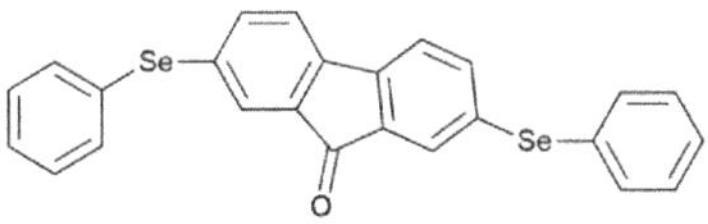

48 (FO-PSe)

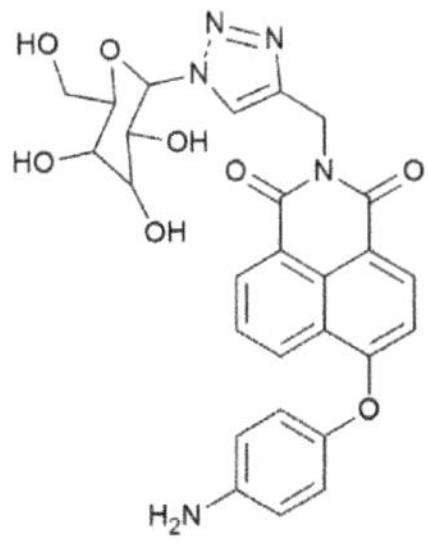

49 (Gal-NPA)

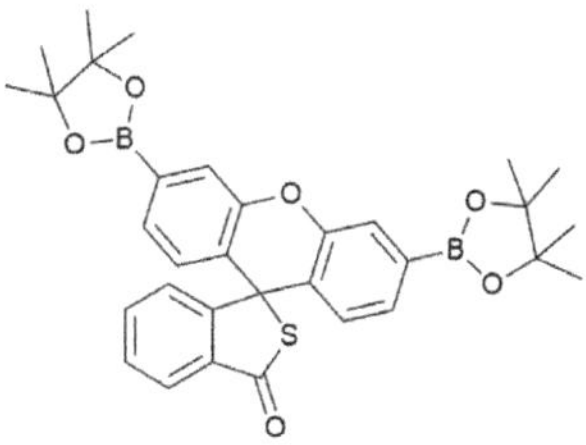

50 (FBS)

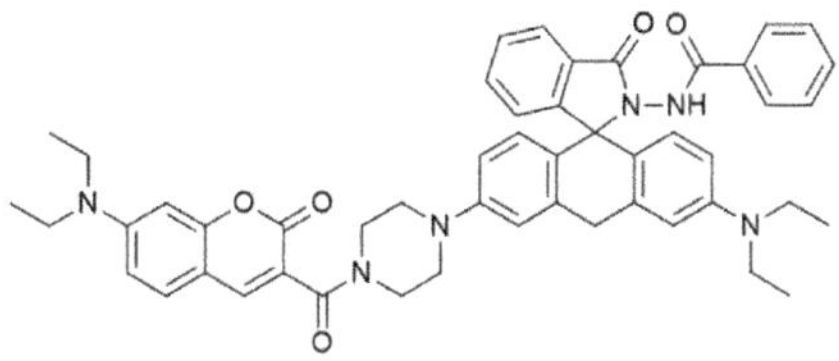

51

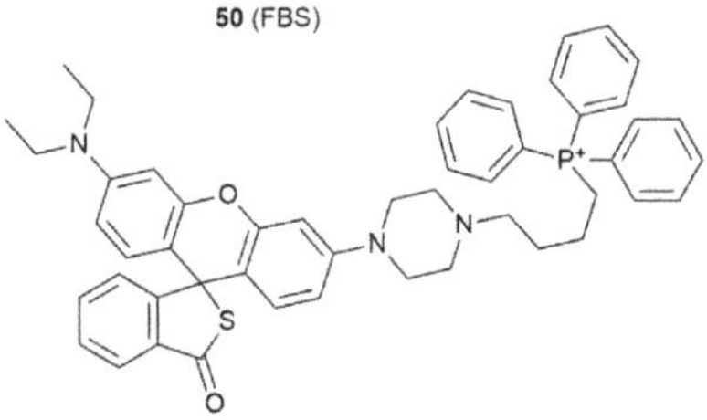

52 (RSTPP)

Figure 7.8 Structures of fluorescent probes for HClO.

Tang and co-workers constructed a coumarin derivative possessing α-phenylselenocarbonyl moiety that is expected to be HClO reactive and release an α,β-unsaturated ketone (probe **45**, Figure 7.8).[52] In the presence of HClO, the green fluorescence of the coumarin fluorophore is reduced and a red-shifted fluorescence emission is released due to the extended conjugation structure. The probe yields a ratiometric fluorescent signal towards HClO. Additionally, the probe displays remarkable selectivity and sensitivity to detect both exogenous and endogenous HClO in living cells.

A novel two-photon fluorescent probe, RPC-1, was constructed by Tang and co-workers for the simultaneous discrimination and imaging of H_2S and HClO (probe **46**, Figure 7.8).[53] They chose 7-aminocoumarin and rhodamine B as suitable fluorophores, which were linked by a rigid piperazine, to constitute the probe skeleton. Thiolactone and azido groups were employed to mask rhodamine B and 7-aminocoumarin with the aim of detecting HClO and H_2S, respectively. Upon titration of HClO and H_2S, gradual fluorescence increases at 445 and 580 nm were detected. Moreover, H_2S could not interfere with the signal of HClO. With the excellent properties of RPC-1, a large fluorescence enhancement response for HClO levels in drug-induced liver injury was detected, which indicates that HClO could be an efficient indicator for antidepressant-induced hepatotoxicity. Applying the probe in living cells, the use of H_2S as an antidotal molecule participating in the protective effect of *N*-acetylcysteine against drug-induced liver injury was exploited and verified.

Tang and co-workers first designed and synthesized an ICT-based two-photon fluorescent probe, NDMTC, for the detection of picomolar hypochlorous acid (probe **47**, Figure 7.8). They employed 4-hydroxy-1,8-naphthalimide fluorophore with a strong ICT structure and an *N*,*N*-dimethylthiocarbamate (DMTC) group as the recognition receptor of HClO.[54] The free probe is almost non-fluorescent in PBS–ethanol buffer solution as an expected consequence of the presence of the DMTC, which prevents fluorescence of the 4-hydroxy-1,8-naphthalimide. Addition of HClO to a solution of the probe results in a large fluorescence enhancement at 547 nm. Owing to the oxidative cleavage of the DMTC moiety *via* the initiation of an electrophilic addition of Cl^+, NDMTC exhibits excellent selectivity towards HClO over other ROS, including ClO^-. Subsequently, to track the HClO levels of lysosome in living cells, a lysosome-targetable fluorescent probe, Lyso-NDMTC, was constructed by introducing an alkylmorpholine group to the NDMTC platform. The two probes can monitor native HClO levels in living cells and *in vivo*.

A 9-fluorenone-based fluorescent probe was developed to monitor endogenous HClO in zebrafish and mice *via* two-photon imaging (probe **48**, Figure 7.8). The probe possesses two selenium compounds as selective recognition groups for HClO.[55] HClO could react with the selenium of the probe, leading to the oxidation of selenide. The two-photon fluorescent probe has unique advantages, such as deeper tissue penetration and minimum photo-damage to biological samples. After titration of glutathione, the fluorescence intensity of the probe response for HClO decreases instantaneously, which indicates that the probe is an ideal tool to image HClO fluctuations *in vitro* and *in vivo*. More importantly, the probe was successfully applied to imaging HClO in mice and zebrafish.

Zhu and co-workers reported a hepatoma-specific turn-on fluorescent probe (Gal-NPA) based on a *p*-aminophenyl ether derivative, which can quickly detect HClO levels. A galactose moiety was employed as a hepatoma-targeting unit to trace HClO in living HepG2 cells. With addition of HClO, a stronger ICT structure was produced, resulting in remarkable enhancement at 558 nm (probe **49**, Figure 7.8).[56]

H_2O_2 and $ONOO^-$ show strong interference when applied as fluorescent probes for detecting HClO, owing to similar characteristics. Yoon and co-workers therefore developed FBS as a fluorescent probe for detecting HClO exclusively (probe **50**, Figure 7.8).[57] Compared with other probes, the "dual-lock" structure of FBS displayed excellent selectivity and a fast response to HClO. To investigate whether FBS can monitor HClO levels in biological systems, the probe was applied to *Drosophila*. Bacteria induced HClO was imaged successfully, indicating that the probe is a promising tool for tracing HClO.

Zhao and co-workers developed a ratiometric fluorescent probe for tracing HClO fluctuations based on coumarin. Rhodamine acid was incorporated as the detection moiety for HClO (probe **51**, Figure 7.8).[58] Before reaction with HClO, the probe displayed negligible fluorescence of rhodamine due to the off-FRET. Upon addition of HClO, a new fluorescence emission (580 nm) of rhodamine could be detected *via* FRET.

Ma and co-workers designed a mitochondrial-targeting fluorescent probe, RSTPP, for detecting HClO. To monitor mitochondrial HClO in macrophages, a mitochondria-targeting moiety and recognition unit for HClO were employed. In the presence of HClO, remarkable fluorescence enhancement was detected. The high Pearson's coefficient of 0.92 indicates that the probe can target mitochondria specifically (probe **52**, Figure 7.8).[59]

7.5.2 Fluorescent Probes for HOBr

HOBr, a type of ROS, plays important roles in the formation of sulfilimine cross-links in collagen IV, which is generated from the peroxidation reaction of the bromide anion (Br^-) with H_2O_2. The low concentration of Br^- in living systems makes it difficult to monitor HOBr fluctuations. Many fluorescent probes have been designed based on the high eletrophilicity of Br^+ (Figure 7.9).

Tang and co-workers developed an ultrasensitive sulfilimine-based fluorescent probe, BPP, for detecting HOBr exclusively in biosystems (probe **53**, Figure 7.9).[60] Bromide ions (Br^+) trigger an electrophilic addition reaction with the *S*-methyl group of the probe. Subsequently, the amino group reacts with the *S*-methyl group to form an S=N bond. The cyclization-based biphenyl product responds to HOBr with a red-shifted emission. The probe possesses unique specificity and ultrasensitivity (picomole level). Endogenous HOBr was detected by applying the probe in HepG2 cells and zebrafish. More importantly, native HOBr levels could be monitored in HepG2 cells without bromine anion stimulation.

Inspired by the ingenious strategy of this cyclization reaction triggered by Br^+, Tang and co-workers subsequently designed a rhodamine 110-based fluorescent probe, RhSN-mito, to track HOBr fluctuations in HepG2 cells and zebrafish (probe **54**, Figure 7.9). Rhodamine 110, as a high quantum yield fluorophore, was employed to target mitochondria. With addition of HOBr, an electrophilic addition reaction and a Suzuki cross-coupling

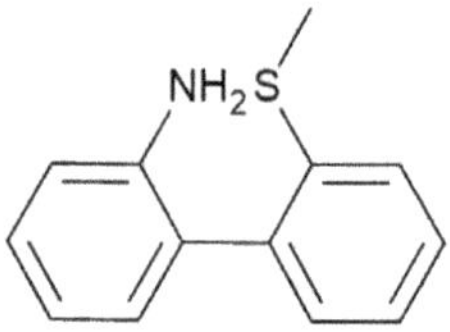

53 (BPP)

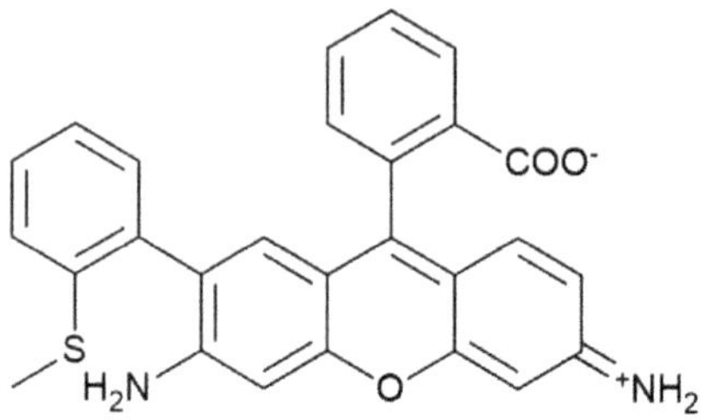

54 (RhSN-mito)

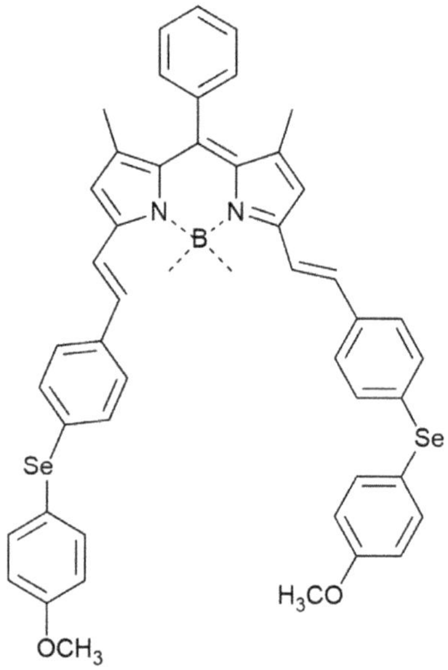

55 (diMPhSe-BOD)

Figure 7.9 Structures of fluorescent probes for HOBr.

reaction occur.[61] In particular, the probe has the following advantages: (a) a high fluorescence quantum yield ($\Phi_F = 0.68$), (b) an NIR emission with large red shift (~100 nm) between RhSN-mito and the product RhSN, (c) targets mitochondria with high specificity and (d) fast response to HOBr *in vitro*.

Han and co-workers developed an NIR fluorescent probe based on BODIPY dye for detecting HOBr in living systems. BODIPY dye shows a high fluorescence quantum yield and strong absorption. 4-Methyoxyphenylselenide and a styrene bridge were incorporated for monitoring HOBr and extending the conjugation system, respectively. The selenide group of diMPhSe-BOD was oxidized to selenoxide by HOBr, resulting in a blue-shift fluorescence. With addition of H_2S, the selenoxide of diMPhSeO-BOD was reduced to selenide, recovering its original fluorescence with a red shift. Applying diMPhSe-BOD in macrophage cells, real-time imaging of the HOBr–H_2S redox cycle was realized (probe **55**, Figure 7.9).[62]

7.5.3 Fluorescent Probes for $^1O_2$

As a highly reactive oxygen species (hROS), singlet oxygen ($^1O_2$) plays important roles in pathological and physiological processes of living organisms. $^1O_2$ can be produced endogenously through enzyme-mediated catalysed oxidations, such as with peroxidase, lipoxygenase or bleomycin. Owing to its high reactivity and oxidizability, $^1O_2$ can oxidize various cell components, including proteins, DNA and lipids. Furthermore, $^1O_2$ can damage tumour cells in the photodynamic therapy (PDT) of cancer. A range of photosensitizers have been designed for enhancing $^1O_2$ generation for the photodynamic therapy of cancer.

Tang and co-workers synthesized a mitochondria-targeted NIR probe, His-Cy, for the detection of $^1O_2$ (probe **56**, Figure 7.10).[63] Histidine and tricarbocyanine are employed as the identification group of $^1O_2$ and an NIR fluorophore, respectively. Histidine, as a $^1O_2$ scavenger, can react with $^1O_2$ specifically *via* a 1,4-cycloaddition of an imidazole ring. The fluorescence of His-Cy is weak owing to the electron transfer based on the PET mechanism. After reaction with $^1O_2$, the fluorescence of the cyanine fluorophore is recovered, resulting from the reduced electron-donating ability of oxidized histidine. A series of experiments indicated that His-Cy possesses a fast response and is highly sensitive towards $^1O_2$. His-Cy shows low cytotoxicity and photobleaching in cell culture. Further, this probe could be applied to detecting endogenous $^1O_2$ in living cells and tissues.

7.5.4 Fluorescent Probes for O_3

As an ROS, O_3 is closely related to depression. To illustrate the relationship between O_3 and depression, Tang and co-workers developed an NIR fluorescent probe (ACy7) for the specific detection of O_3 in living brains (probe **57**, Figure 7.11).[64] A Cy7-like molecule and 3-butenyl are employed as a fluorophore precursor and recognition group, respectively. O_3 triggers a

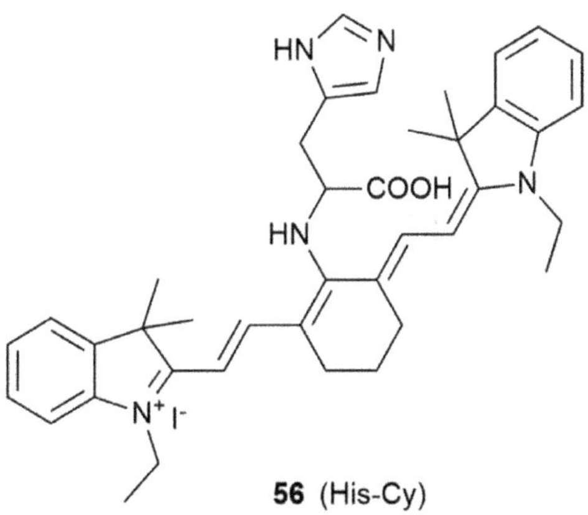

56 (His-Cy)

Figure 7.10 Structure of a fluorescent probe for $^1O_2$.

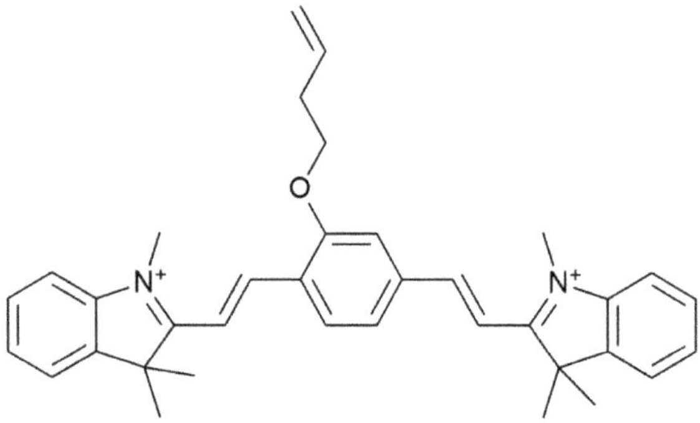

57 (ACy7)

Figure 7.11 Structure of a fluorescent probe for O_3.

cycloaddition reaction with the terminal alkene of the 3-butenyl moiety. After treatment with O_3, a remarkable fluorescence at 690 nm was observed. Importantly, *in situ* imaging observation of O_3 in the brain tissue of mice with depression was realized for the first time (Figure 7.11).

Acknowledgement

Dr Ping Li is thanked for expert assistance with the writing of this chapter.

References

1. A. Gomes, E. Fernandes and J. L. Lima, *J. Biochem. Biophys. Methods*, 2005, **65**, 45.
2. B. Halliwell and M. B. Whiteman, *Br. J. Pharmacol.*, 2004, **142**, 231.
3. X. Chen, X. Tian, I. Shin and J. Yoon, *Chem. Soc. Rev.*, 2011, **40**, 4783.
4. W. Zhang, P. Li, F. Yang, X. F. Hu, C. Z. Sun, W. Zhang, D. Z. Chen and B. Tang, *J. Am. Chem. Soc.*, 2013, **135**, 14956.
5. W. Zhang, X. Wang, P. Li, F. Huang, H. Wang, W. Zhang and B. Tang, *Chem. Commun.*, 2015, **51**, 9710.
6. W. Zhang, J. Zhang, P. Li, J. Liu, D. Su and B. Tang, *Chem. Sci.*, 2019, **10**, 879.
7. W. Zhang, X. Wang, P. Li, H. B. Xiao, W. Zhang, H. Wang and B. Tang, *Anal. Chem.*, 2017, **89**, 6840.
8. H. Xiao, C. Wu, P. Li and B. Tang, *Anal. Chem.*, 2018, **90**, 6081.
9. J. J. Gao, K. H. Xu, B. Tang, L. L. Yin, G. W. Yang and L. G. An, *FEBS J.*, 2007, **274**, 1725.
10. P. Li, W. Zhang, K. Li, X. Liu, H. Xiao, W. Zhang and B. Tang, *Anal. Chem.*, 2013, **85**, 9877.
11. H. Xiao, X. Liu, C. Wu, Y. Wu, P. Li, X. Guo and B. Tang, *Biosens. Bioelectron.*, 2017, **91**, 449.

12. R. Q. Li, Z. Q. Mao, L. Rong, N. Wu, Q. Lei, J. Y. Zhu, L. Zhuang, X. Z. Zhang and Z. H. Liu, *Biosens. Bioelectron.*, 2017, **87**, 73.
13. J. Niu, J. Fan, X. Wang, Y. Xiao, X. Xie, X. Jiao, C. Sun and B. Tang, *Anal. Chem.*, 2017, **89**, 7210.
14. S. Bag, J. C. Tseng and J. Rochford, *Org. Biomol. Chem.*, 2015, **13**, 1763.
15. J. J. Hu, N.-K. Wong, S. Ye, X. Chen, M.-Y. Lu, A. Q. Zhao, Y. Guo, A. C. H. Ma, A. Y.-H. Leung, J. Shen and D. Yang, *J. Am. Chem. Soc.*, 2015, **137**, 6837.
16. X. Gao, G. Feng, P. N. Manghnani, F. Hu, N. Jiang, J. Liu, B. Liu, J. Z. Sun and B. Z. Tang, *Chem. Commun.*, 2017, **53**, 1653.
17. J. Zhang, C. Li, R. Zhang, F. Zhang, W. Liu, X. Liu, S. M.-Y. Lee and H. Zhang, *Chem. Commun.*, 2016, **52**, 2679.
18. F. Si, Y. Liu, K. Yan and W. Zhong, *Chem. Commun.*, 2015, **51**, 7931.
19. D. Q. Lu, L. Y. Zhou, R. W. Wang, X. B. Zhang, L. He, J. Zhang, X. X. Hu and W. H. Tan, *Sens. Actuators, B*, 2017, **250**, 259.
20. W. Zhang, W. Liu, P. Li, F. Huang, H. Wang and B. Tang, *Anal. Chem.*, 2015, **87**, 9825–9828.
21. H. Xiao, P. Li, X. Hu, X. Shi, W. Zhang and B. Tang, *Chem. Sci.*, 2016, **7**, 6153–6159.
22. X. Xie, X. Yang, T. Wu, Y. Li, M. Li, Q. Tan, X. Wang and B. Tang, *Anal. Chem.*, 2016, **88**, 8019–8025.
23. H. Xiao, P. Li, S. Zhang, W. Zhang, W. Zhang and B. Tang, *Chem. Commun.*, 2016, **52**, 12741–12744.
24. J. Hou, M. Qian, H. Zhao, Y. Li, Y. Liao, G. Han, Z. Xu, F. Wang, Y. Song and Y. Liu, *Anal. Chim. Acta*, 2018, **1024**, 169–176.
25. Y. Chen, X. Shi, Z. Lu, X. Wang and Z. Wang, *Anal. Chem.*, 2017, **89**, 5278–5284.
26. C. Gao, Y. Tian, R. Zhang, J. Jing and X. Zhang, *Anal. Chem.*, 2017, **89**, 12945–12950.
27. Y. Ni, H. Liu, D. Dai, X. Mu, J. Xu and S. Shao, *Anal. Chem.*, 2018, **90**, 10152–10158.
28. H. Guo, G. Chen, M. Gao, R. Wang, Y. Liu and F. Yu, *Anal. Chem.*, 2019, **91**, 1203–1210.
29. Y. Liu, J. Niu, J. Nie, F. Meng and W. Lin, *New J. Chem.*, 2017, **41**, 3320–3325.
30. J. Wang, W. Zhu, G. Niu, G. Jiang, Q. Chen, P. Gao, Y. Li, G. Zhang, X. Fan and B. Z. Tang, *Chem. Commun.*, 2018, **54**, 13957–13960.
31. X. Wang, P. Li, Q. Ding, C. Wu, W. Zhang and B. Tang, *Angew. Chem., Int. Ed.*, 2019, **58**, 4674–4678.
32. H. Li, X. Li, W. Shi, Y. Xu and H. Ma, *Angew. Chem., Int. Ed.*, 2018, **57**, 12830–12834.
33. X. Bai, Y. Huang, M. Lu and D. Yang, *Angew. Chem., Int. Ed.*, 2017, **56**, 12873–12877.
34. F. Liu, J. Du, D. Song, M. Xu and G. Sun, *Chem. Commun.*, 2016, **52**, 4636–4639.
35. L. Meng, Y. Wu and T. Yi, *Chem. Commun.*, 2014, **50**, 4843–4845.

36. R. Zhang, J. Zhao, G. Han, Z. Liu, C. Liu, C. Zhang, B. Liu, C. Jiang, R. Liu, T. Zhao, M. Y. Han and Z. Zhang, *J. Am. Chem. Soc.*, 2016, **138**, 3769–3778.
37. J. Y. Wang, Z. R. Liu, M. Ren, X. Kong, K. Liu, B. Deng and W. Lin, *Sens. Actuators, B*, 2016, **236**, 60–66.
38. K. J. Barnham, C. L. Masters and A. I. Bush, *Nat. Rev. Drug Discovery*, 2004, **3**, 205–214.
39. C. Chatgilialoglu and C. Ferreri, *Acc. Chem. Res.*, 2005, **38**, 441–448.
40. B. Kalyanaraman, *Redox Biol.*, 2013, **1**, 244–257.
41. P. Li, B. Tang, Y. Xing, P. Li, G. Yang and L. Zhang, *Analyst*, 2008, 133.
42. K. Krumova and G. Cosa, *Photochemistry*, 2013, **41**, 279.
43. X. Wang, P. Li, Q. Ding, C. Wu, W. Zhang and B. Tang, *J. Am. Chem. Soc.*, 2019, **141**, 2061–2068.
44. L. A. Sklar, B. S. Hudson and R. D. Simoni, *Proc. Natl. Acad. Sci. U. S. A.*, 1975, **72**, 1649–1653.
45. G. P. Drummen, L. C. van Liebergen, J. A. Op den Kamp and J. A. Post, *Free Radical Biol. Med.*, 2002, **33**, 473–490.
46. T. A. Prime, M. Forkink, A. Logan, P. G. Finichiu, J. McLachlan, P. B. Li Pun, W. J. Koopman, L. Larsen, M. J. Latter, R. A. Smith and M. P. Murphy, *Free Radical Biol. Med.*, 2012, **53**, 544–553.
47. P. Oleynik, Y. Ishihara and G. Cosa, *J. Am. Chem. Soc.*, 2007, **129**, 1842.
48. K. Krumova, S. Friedland and G. Cosa, *J. Am. Chem. Soc.*, 2012, **134**, 10102–10113.
49. L. E. Greene, R. Lincoln and G. Cosa, *J. Am. Chem. Soc.*, 2017, **139**, 15801–15811.
50. K. Krumova, L. E. Greene and G. Cosa, *J. Am. Chem. Soc.*, 2013, **135**, 17135–17143.
51. K. Yamada, F. Mito, Y. Matsuoka, S. Ide, K. Shikimachi, A. Fujiki, D. Kusakabe, Y. Ishida, M. Enoki, A. Tada, M. Ariyoshi, T. Yamasaki and M. Yamato, *Nat. Chem. Biol.*, 2016, **12**, 608–613.
52. X. Xie, T. Wu, X. Wang, Y. Li, K. Wang, Z. Zhao, X. Jiao and B. Tang, *Chem. Commun.*, 2018, **54**, 11965.
53. X. Jiao, Y. Xiao, Y. Li, M. Liang, X. Xie, X. Wang and B. Tang, *Anal. Chem.*, 2018, **90**, 7510.
54. B. Zhu, P. Li, W. Shu, X. Wang, C. Liu, Y. Wang, Z. Wang, Y. Wang and B. Tang, *Anal. Chem.*, 2016, **88**, 12532.
55. W. Zhang, W. Liu, P. Li, J. Kang, J. Wang, H. Wang and B. Tang, *Chem. Commun.*, 2015, **51**, 10150.
56. Q. Duan, P. Jia, Z. Zhuang, C. Liu, X. Zhang, Z. Wang, W. Sheng, Z. Li, H. Zhu, B. Zhu and X. Zhang, *Anal. Chem.*, 2019, **91**, 2163.
57. Q. Xu, K.-A. Lee, S. Lee, K. M. Lee, W.-J. Lee and J. Yoon, *J. Am. Chem. Soc.*, 2013, **135**, 9944.
58. Y.-R. Zhang, X.-P. Chen, S. Jing, J.-Y. Zhang, Q. Yuan, J.-Y. Miao and B.-X. Zhao, *Chem. Commun.*, 2014, **50**, 14241.
59. J. Zhou, L. Li, W. Shi, X. Gao, X. Li and H. Ma, *Chem. Sci.*, 2015, **6**, 4884–4888.

60. K. Xu, D. Luan, X. Wang, B. Hu, X. Liu, F. Kong and B. Tang, *Angew. Chem., Int. Ed.*, 2016, **55**, 12751.
61. X. Liu, A. Zheng, D. Luan, X. Wang, F. Kong, L. Tong, K. Xu and B. Tang, *Anal. Chem.*, 2017, **89**, 1787.
62. B. Wang, P. Li, F. Yu, J. Chen, Z. Qu and K. Han, *Chem. Commun.*, 2013, **49**, 5790.
63. K. Xu, L. Wang, M. Qiang, L. Wang, P. Li and B. Tang, *Chem. Commun.*, 2011, **47**, 7386.
64. P. Li, J. Wang, X. Wang, Q. Ding, X. Bai, Y. Zhang, D. Su, W. Zhang, W. Zhang and B. Tang, *Chem. Sci.*, 2019, **10**, 2805.

CHAPTER 8

Imaging of Intracellular Reactive Nitrogen Species and Reactive Sulfur Species

8.1 Introduction

Reactive nitrogen species (RNS) refer to a family of intracellular active molecules containing nitrogen and oxygen, such as nitric oxide (NO), nitrogen dioxide (NO_2), peroxynitrite ($ONOO^-$), *S*-nitrosothiols (RSNO) and nitroxyl (HNO).[1] RNS are highly reactive and can react with a wide range of intracellular substances, including nucleic acids, proteins, saccharides and lipids *via* oxidation, nitration and nitrosylation reactions, resulting in nitrative stress.[2] Moreover, RNS also have a signal transducing effect during physiological/pathophysiological processes.[3] Reactive sulfur species (RSS) are sulfur-containing compounds with the sulfur oxidation state ranging from −2 to +4. RSS mainly include biothiols (GSH, Cys, Hcy), H_2S, thiyl radical ($RS^•$), disulfide (RSSR), sulfenic acid (RSOH), sulfur dioxide/sulfite/bisulfite ($SO_2/SO_3^{2-}/HSO_3^-$), persulfides (R–S–SH/H_2S_2) and polysulfides (R_2S_n/H_2S_n, $n > 2$).[4] Normally, there is broad cross-talk between RNS and RSS in redox biochemistry, which plays an important role in the regulation of signal pathways.[5] The deep understanding or elucidation of the interaction between RNS and RSS will contribute to revealing the vital molecular mechanism regarding cell activity.

The *in situ* presentation of the dynamic behaviours of RNS and RSS can provide important information for the study of their biofunction. Fluorescence imaging has attracted much attention since it showed useful properties in terms of direct visualization and good temporal and spatial resolution.[6–8] Therefore, a number of fluorescent probes for RNS and RSS

Detection Science Series No. 16
In Situ Analysis of Cellular Functional Molecules
By Huangxian Ju, Bo Tang and Lin Ding

Published by the Royal Society of Chemistry, www.rsc.org

have been developed that provide good analytical tools for these compounds. In this chapter, we summarize representative small-molecule fluorescent probes for RNS and RSS in living cells. The design, recognition principle and response mechanism are mainly described. It is hoped that this survey will provide useful information for researchers in the relevant fields.

8.2 Imaging of Intracellular Reactive Nitrogen Species

As important class of chemically reactive substances, RNS, which include nitric oxide (NO), peroxynitrite ($ONOO^-$) and nitroxyl (HNO),[9] are essential for regulating normal physiological processes. In addition to cancer, inflammation, cardiovascular diseases and organ fibrosis, an abnormally high cellular level of RNS production has been reported in chronic wound healing and keloid scars. Because they have pathophysiological significance, RNS have been the subject of extensive investigations aimed at determining their biological roles. Many fluorescent probes have been developed as powerful tools to detect RNS in biological and environmental samples and to elucidate their biological functions.[10]

8.2.1 Probes for Nitroxyl (HNO)

Nitroxyl (HNO), the one-electron reduced and protonated form of the well-known signalling agent nitric oxide (NO), is a novel regulator of cardiovascular function and may provide a useful tool for treating cardiovascular diseases such as heart failure.[11] In addition, HNO exhibits myocardial antihypertrophic and superoxide-suppressing actions.[12] In order to advance insight into the physiological and pathological roles of HNO, numerous fluorescent probes for HNO have been reported based on the reaction of HNO with triphenylphosphine or the reduction of Cu(II) to Cu(I) by HNO[13] (Figure 8.1).

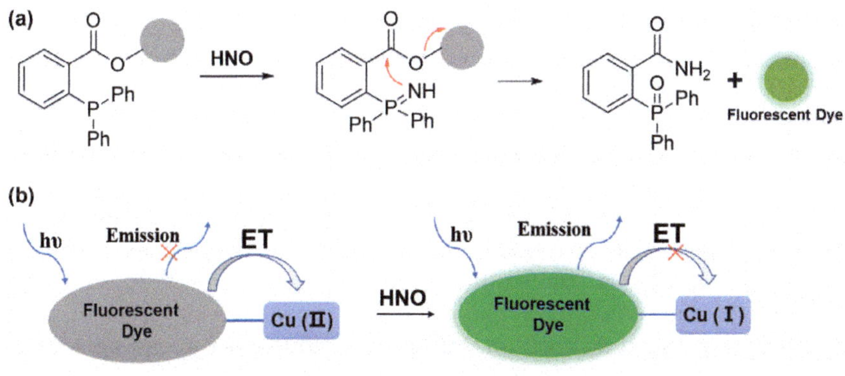

Figure 8.1 Detection mechanism of HNO probes.

For triphenylphosphine-based fluorescent probes, the reaction of HNO with 2 equiv. of phosphine nucleophiles produces phosphine oxides and the corresponding phosphine azaylides. Then, in the presence of an internal electrophilic ester, these azaylides undergo Staudinger ligation to yield the thermodynamically stable amide and the corresponding fluorescent dye (Figure 8.1). Based on this detection strategy, probes **1–12** were reported for the detection of HNO in different biological samples (Figure 8.2). Tan and co-workers[14] designed a coumarin-based fluorescent probe, P-CM (**1**), for the quantitative detection of nitroxyl (HNO). The probe affords high sensitivity to HNO in aqueous solutions with high selectivity for HNO over other biologically relevant species. Moreover, P-CM is unaffected by various biological reductants, in contrast to previously reported fluorescent probes. Further application to cellular imaging indicated that P-CM could detect HNO quantitatively in bovine serum.

Zhang and co-workers[15] developed a naphthalimide-based ratiometric fluorescent probe (**2**) for HNO. Probe **2** exhibits high HNO selectivity even in the presence of high concentrations of GSH and ascorbate and displays colour changes from colourless to yellow on addition of HNO due to the intramolecular charge-transfer (ICT) effect, which serves as a "naked-eye" probe for HNO. Probe **2** can detect HNO quantitatively by ratiometric fluorescence with excellent sensitivity and a fast response.

Based on an excited-state intramolecular proton-transfer (ESIPT) mechanism, Zhu and co-workers[16] designed and synthesized a fluorescent turn-on probe, FLA-1 (**3**), for the detection of nitroxyl. FLA-1 consists of a flavone derivative moiety as the fluorophore and a diphenylphosphinobenzoyl group as a recognition receptor. The probe exhibited a selective response to HNO over other related biologically active species and can quantitatively detect HNO with a detection limit of 1.28×10^{-7} M in aqueous solution. Furthermore, FLA-1 was applied for the quantitative detection of HNO in bovine serum. In a similar fashion, Zhang and co-workers[17] reported another ESIPT-based probe, HNO-HBT (**4**), using a 2-(2′-hydroxyphenyl)benzothiazole fluorophore. HNO-HBT could quantitatively detect HNO with high sensitivity with a detection limit of 50 nM and was applied to the imaging of the changes in HNO levels in living HeLa cells upon addition of Angeli's salt (AS), a commonly employed HNO donor. King's group[18] designed and synthesized the phosphine-based HNO probe **5** by conjugating triarylphosphine to fluorescein dye. Probe **5** is highly sensitive and selective for HNO detection under physiological conditions and can detect the release of HNO from AS in HeLa cells.

A two-photon fluorescent probe is suitable for monitoring RNS not only in living cells but also in living tissues, because it can facilitate three-dimensional imaging of living tissues, reduce photodamage to biosamples, increase tissue penetration and lower background fluorescence. Lin and co-workers[19] constructed the first two-photon fluorescent turn-on HNO probe **6**, based on GCTPOC dye, which showed a large cross-section. Probe **6** can detect HNO with high sensitivity, high selectivity and low cytotoxicity. It showed a 49-fold enhancement in the presence of AS (75 mM) with a detection limit of

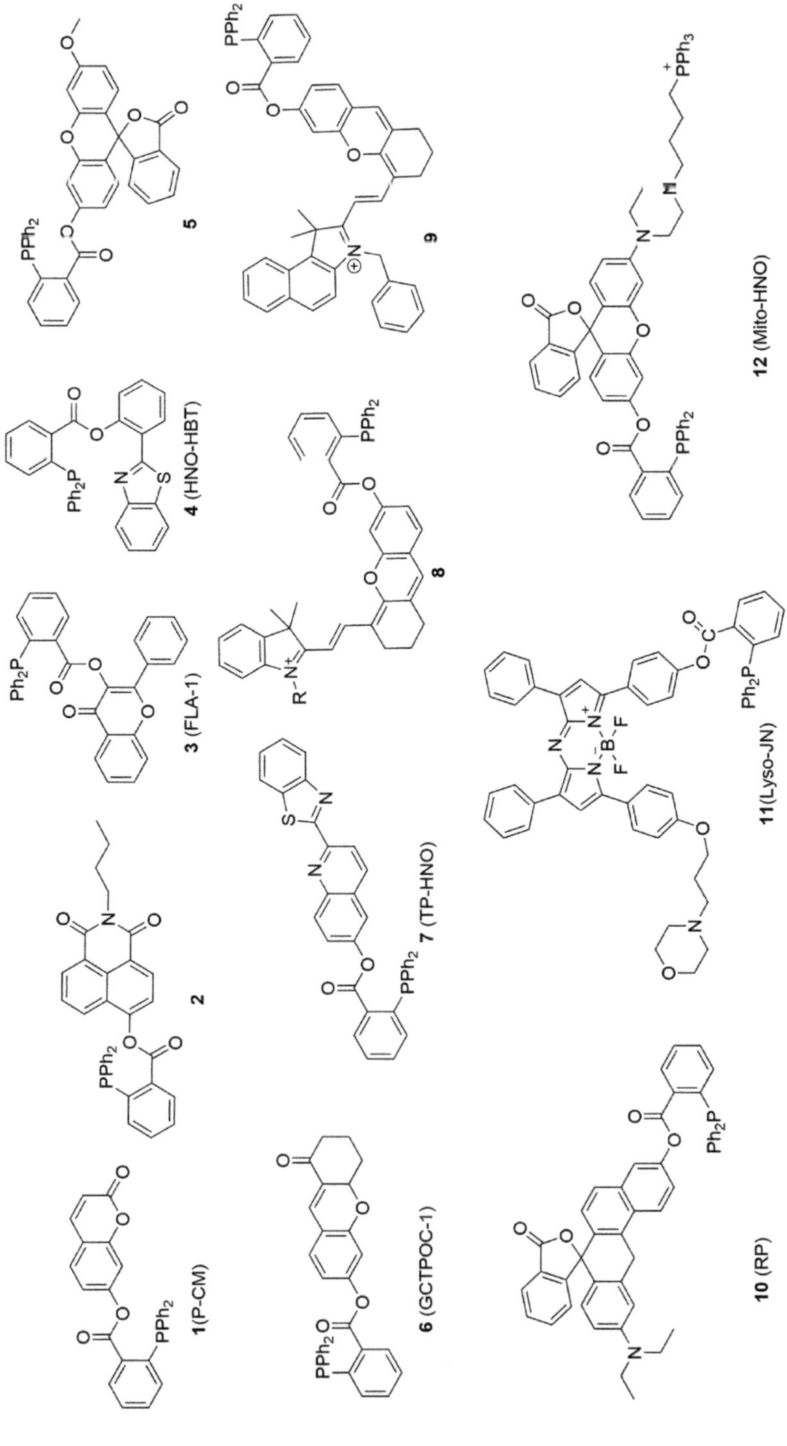

Figure 8.2 Structures of fluorescent probes for HNO based on triphenylphosphine.

5.90×10^{-7} M and was used for fluorescence imaging of HNO in living cells and rat liver slices by two-photon fluorescence microscopy. Peng and co-workers[20] developed a two-photon fluorescent probe, TP-HNO (7), for HNO based on 6-hydroxyquinonline-2-benzothiazole derivatives. Probe 7 demonstrated outstanding selectivity, excellent sensitivity and rapid response and was successfully used for the qualitative and bioimaging of exogenous and endogenous HNO concentration fluctuations in living HeLa cells using two-photon laser confocal fluorescence microscopy.

In contrast to classic fluorescent probes, near-infrared (NIR) probes boast distinct advantages, such as the low absorption of biological molecules in the NIR region and NIR light penetrates deeper into biological tissues than visible light. The groups of Sun[21] and Yang[22] independently reported NIR fluorescent probes (**8** and **9**) for HNO based on a merocyanine skeleton. These two probes display an intense fluorescence turn-on response to HNO *via* azaylide formation and its subsequent Staudinger ligation to release the NIR merocyanine fluorophore with outstanding selectivity and excellent sensitivity. Cell imaging and colocalization experiments demonstrated that probes **8** and **9** are suitable for visualizing nitroxyl in the mitochondria of living cells.

Based on extending the π system of xanthene, Lin and co-workers[23] developed a novel two-photon red emissive fluorescent probe, RP (**10**), for the detection of HNO. When responding to HNO, probe **10** displayed a maximum δ value (two-photon absorption cross-section) of about 90 GM and was successfully applied for two-photon imaging with red emission for the imaging of HNO in living HeLa cells and mouse liver tissues. A BF_2-chelated tetraarylazadipyrromethene fluorophore (Aza-BODIPY) is frequently used as the fluorophore for constructing fluorescent probes, because its absorption/emission wavelength is located in the NIR spectral region. Chen and co-workers[24] reported an NIR fluorescent probe, Lyso-JN (**11**), based-on Aza-BODIPY for the detection of HNO. This probe exhibits high selectivity, good sensitivity and low cytotoxicity in the detection of HNO. Moreover, the lysosomal locator alkylmorpholine group in probe **11** demonstrated perfect lysosomal localization in living RAW264.7 cells. In order to detect HNO precisely in mitochondria, Lin and co-workers[25] synthesized a mitochondria-targeted fluorescent HNO probe, Mito-HNO (**12**). Probe **12** was highly selective towards HNO over other reactive nitrogen-containing species and reducing species and was rapidly responsive and suitable for the visualization of HNO in the mitochondria of HeLa cells.

Reduction of a Cu(II) complex by HNO was another important strategy for designing HNO fluorescent probes that was devised by Rosenthal and Lippard[26] and used to monitor HNO in both aqueous solution and living cells. By introducing an *N*,*O*-chelated-Cu(II) reducing group to 7-diethylaminocoumarin, Yao and co-workers[27] developed a cell-permeable Cu(II)-based fluorescent probe, Cu(II)-COET (**13**), for HNO-specific detection (Figure 8.3). Probe **13** was successfully used for visualizing HNO in living NIH3T3 cells and zebrafish larvae without interference from thiols by using *N*-ethylmaleimide (NEM). The same group developed a Cu(II) coumarin conjugate probe (**14**)[28] that acted as a

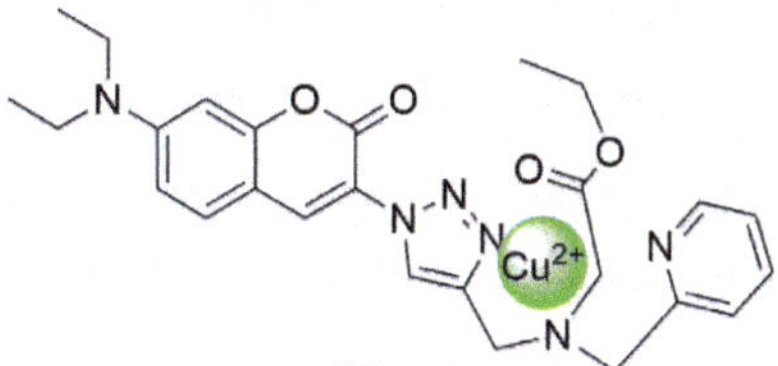

13

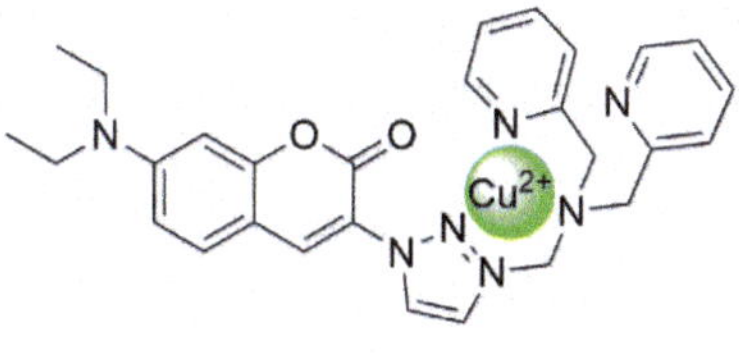

14

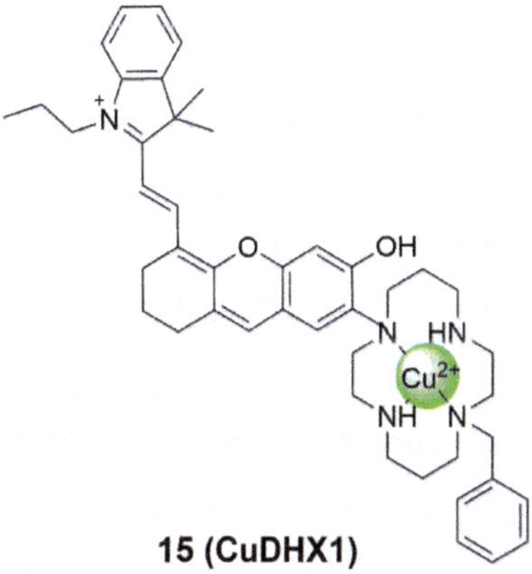

15 (CuDHX1)

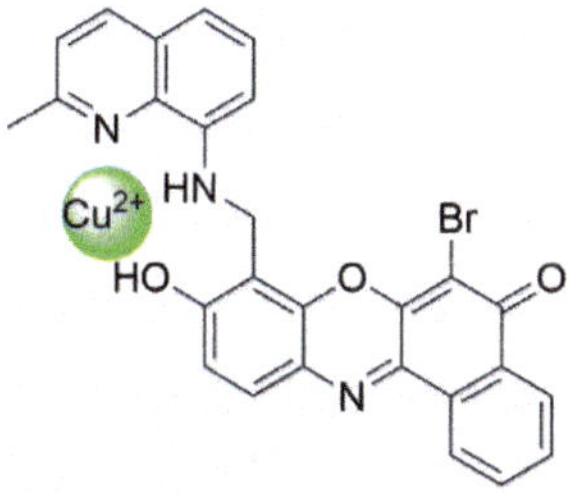

16 (BRNO1)

Figure 8.3 Structures of fluorescent probes for HNO based on Cu complexes.

dual-response probe to HNO for both fluorescence and electron paramagnetic resonance detection and could be used to detect HNO with high selectivity over other reactive oxygen species (ROS) and RNS. The large changes in the fluorescence colour make it a "naked-eye" probe. Probe **14** was also successfully used for the detection of changes in HNO levels in living A375 cells with low cytotoxicity. Lippard and co-workers[29] reported the first NIR fluorescent turn-on copper-based probe, CuDHX1 (**15**), for the detection of HNO; it contains a dihydroxanthene (DHX) fluorophore and a cyclam derivative as a Cu(II) binding site. Probe **15** was selective for HNO over thiols and RNS and ROS and can detect exogenously applied HNO in live mammalian cells. Further, when used in combination with the green-fluorescent sensor ZP1, probe **15** was successfully used for tracking the increase in intracellular levels of mobile zinc elicited by exogenously applied AS. Based on the benzoresorufin fluorophore, Lippard's group developed the red-fluorescent sensor **16** for HNO.[30] Probe **16** has been used to detect exogenously applied HNO in live HeLa cells.

8.2.2 Probes for Nitric Oxide (NO)

Nitric oxide (NO) is another important RNS and is also a signalling molecule implicated in numerous physiological and pathological conditions. Misregulation of NO production is implicated in a large number of pathologies. The biological events associated with NO are diverse and the effect of NO depends on the cell microenvironment and also its concentration and spatial and temporal constraints. Owing to its significance in both human health and disease treatment, numerous fluorescent probes for the determination of NO in biological samples have been explored.[31]

In 2010, Anslyn and co-workers developed a 2-dimethylaminophenyl-5-cyano-α-naphthylamine fluorescent probe, NO_{550} (**17**), for NO[32] (Figure 8.4). The addition of NO to probe **17** in aqueous buffer yielded AZO_{550} (**18**), resulting in a > 1500-fold fluorescence enhancement. An excellent linear relationship between the fluorescence response of the probe **17** and NO concentration was found, with a detection limit of about 30 nM. Further, probe **17** shows high selectivity for NO over competitive ROS or RNS and other analytes and has been used to visualize NO in astrocytes stimulated with interleukin-1β and interferon-γ. Based on the above principle, Song's group[33] developed a two-photon fluorescent probe, NO-QA5 (**19**), with a dimethylaminophenyl linking at the 6-position of 5-aminoquinoline. Probe **19** can detect NO with a fast response (within ~20 s), high specificity and high sensitivity (limit of detection 15 nM). Probe **19** achieved the visual detection of exogenous and endogenous NO in RAW264.7 cells and in tissue slices under two-photon excitation. To extend this method into a versatile design platform, Guo and co-workers[34] developed its modified version for constructing the fluorescent probe **20** through integration of an NO-binding group, 2-amino-3′-dimethylaminobiphenyl (AD), and BODIPY (a boron dipyrromethene) fluorophore. The probe **20** can selectively sense NO over ROS and RNS, and also ascorbic acid (AA), dehydroascorbic acid (DHA) and methylglyoxal (MGO) with a detection limit of 30 nM. Cell imaging experiments indicated its potential to probe NO chemistry in living HL-7702 cells.

The *o*-diamine functionality has been widely used to develop NO-responsive organic-based probes (Figure 8.5). Nagano's group[35] first developed a series of molecules containing *o*-diamine functional groups. The electron-rich vicinal amines are thought to transfer energy to the aromatic group with photoinduced electron transfer (PET) to quench the fluorescence. The reaction between *o*-diamine functional groups and an oxidized form of NO yields an electron-poor fluorescent triazole derivative, which dissipates the PET pathway and restores the fluorescence. Nagano's group designed an NIR fluorescent probe, DAC-P (**21**),[36] for NO based on the PET mechanism. The reaction rate of probe **21** with NO was fast and the observation of the NIR fluorescence was less subject to interference by biological substances. Furthermore, the probe was applicable to imaging NO in rat kidneys. Based on the rigid conjugate plane of the BODIPY chromophore, a far-red fluorescent probe, DANPBO-M (**22**),[37] was designed and synthesized for NO imaging in cells and tissues. The detection sensitivity of NO imaging was considerably improved owing to the high and photostable turn-on fluorescence, low background interference and excellent intracellular retention of probe **22**. This probe was used to visualize NO in ECV-304 cells and frozen liver slices.

The traditional off–on type of NO probes based on a PET mechanism showed some limitations, such as the fluorescence quantum yield of the

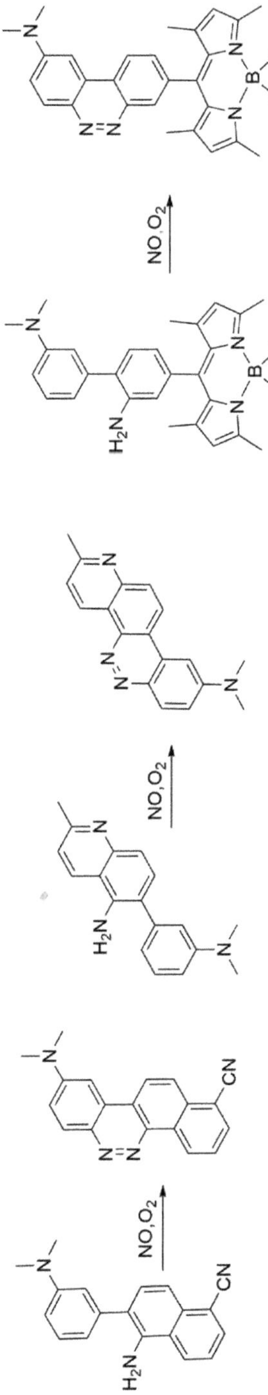

17 (NO_{550}) **18 (AZO_{550})** **19 (NO-QA5)** **20**

Figure 8.4 Structures of fluorescent probes for HNO based on diazotization.

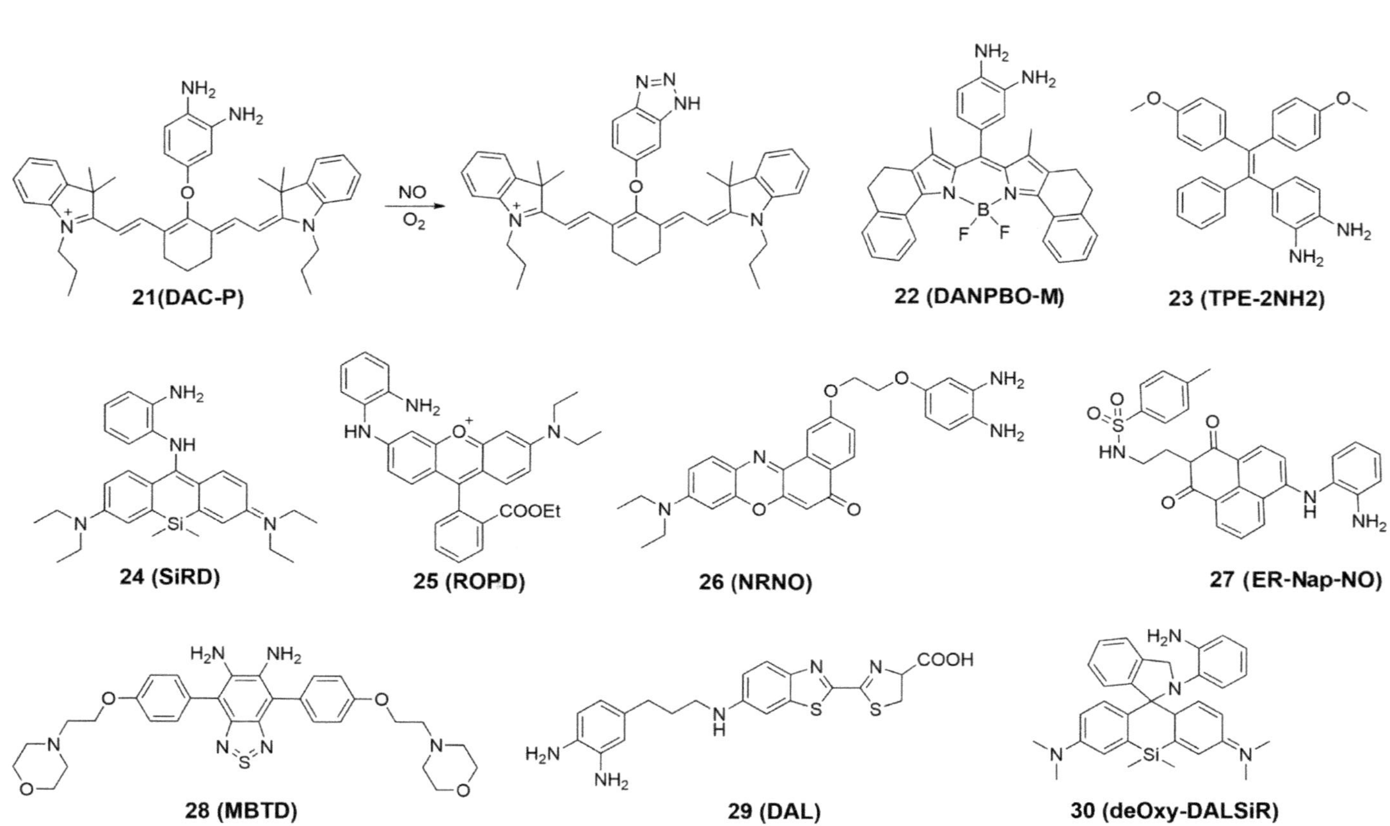

Figure 8.5 Structures of fluorescent probes for HNO based on an *o*-diamine functionality.

probes is not enhanced substantially after the reaction with NO due to weak suppression of the PET effect or aggregation-caused quenching of the fluorescence, or possible fluorescence interference from the biological matrix makes the fluorescence of the background and probes indistinguishable. In order to solve this problem, Tian and co-workers[38] designed and synthesized an aggregation-induced emission (AIE) fluorescent probe, TPE-2NH2 (**23**), for the highly sensitive ratiometric detection of NO. Probe **23** was applied to the highly selective and sensitive ratiometric detection of NO through an ICT process and responded highly sensitively and selectively to NO produced in MCF-7 cancer cells activated by lipopolysaccharides (LPS).

In order to achieve real-time monitoring of NO at the subcellular level, based on silicon-substituted xanthene, Guo and co-workers developed a fast and selective NIR fluorescent probe, SiRD (**24**),[39] for the targeted tracing of endogenous NO. Probe **24** possessed important features for the real-time detection of intracellular NO, including a significant turn-on NIR response, high specificity and a fast response. Moreover, the probe was applicable to the real-time monitoring of endogenous NO in mitochondria of HeLa cells.

Based on the PET principle, Zeng and co-workers designed and prepared a rhodamine-based NO fluorescent probe, ROPD (**25**),[40] *via* integration of the NO-trapping *o*-phenylenediamine (OPD) unit on the rhodamine fluorophore. Probe **25** exhibited high selectivity and sensitivity towards NO with a fast response over a wide pH range and can be used for the real-time visualization of exogenous NO in both L929 and HeLa cells and endogenous NO in stressed RAW264.7 cells.

Based on the Nile Red fluorophore, Liu and co-workers developed a far-red-emissive two-photon (TP) excitable probe, NRNO (**26**),[41] for NO detection. Probe **26** showed a fast (within 180 s) and specific fluorescence response towards NO with a limit of detection as low as 46 nM. The probe was able to visualize the generation of NO in an LPS-mediated inflammation process in mice under a TP microscope. For imaging of NO during endoplasmic reticulum (ER) stress, a naphthalimide-based TP fluorescent probe, ER-Nap-NO (**27**),[42] was designed and synthesized by Li's group. Probe **27** showed a short response time (less than 6 min) and high selectivity towards NO over other potential interfering species in biological samples. In addition, this probe has low cytotoxicity and can be applied for TP imaging of exogenous and endogenous NO levels in the ER of living cells. In particular, the probe has been used for investigating the relationship between ER stress and the level of NO in HeLa cells and a mouse model. Based on a dual intramolecular charge-transfer (ICT) mechanism, Ding and co-workers developed the probe MBTD (**28**)[43] for acid-promoted and far-red fluorescence imaging of lysosomal NO. Probe **28** displays a large Stokes shift, a huge on–off ratio with minimal autofluorescence, high NO specificity and an acid-promoted response to NO with high acid tolerance. Furthermore, the probe was employed for ratiometric examination of the exogenous or endogenous NO of macrophages, and was able to detect the variation of lysosomal NO level in an acute liver injury mouse model *ex vivo*.

Traditional fluorescence imaging requires external illumination, whereas a bioluminescence signal generated deep inside the body can be detected with high signal-to-noise ratio and is unaffected by tissue absorption/scattering of excitation light or background autofluorescence of tissues. Hence numerous chemical probes utilizing bioluminescence have been used to visualize various molecular activities *in vivo*. Urano and co-workers[44] developed a luciferin-based bioluminescent probe, DAL (**29**), for NO. The results indicated that probe **29** permitted quantitative and sensitive detection even in luciferase transgenic (Luc-Tg) rats. Guo's group presented an Si-rhodamine-based NIR fluorescent probe, deOxy-DALSiR (**30**),[45] for the detection of NO with a rapid response rate, huge fluorescence off–on ratio and ultra-low detection limit. In addition, probe **30** was successfully used to image of endogenous NO in RAW264.7 macrophages, pancreatic β cells and endothelial EA.hy926 cells, and also in inflamed and diabetic mouse models.

Utilization of the nitrosation reactions of electron-rich aromatic secondary amines with NO was another strategy for designing NO fluorescent probes (Figure 8.6). Guo and co-workers developed an NO fluorescent probe (**31**)[46] based on the NO-triggered nitrosation of *N*-benzyl-4-hydroxyaniline reaction groups. Probe **31** displayed a fast fluorescence response, high sensitivity and excellent selectivity for NO. The images of the basal and stimulation-induced NO in macrophages RAW264.7 and also the endogenous NO in endothelial cells after oxygen–glucose deprivation (OGD) were successfully acquired by the probe. Based on the same principle, Liu and co-workers developed an "NIR-to-NIR" TP probe, SiRNO (**32**),[47] for NO. Probe **32** exhibited high sensitivity and specificity and a fast response for NO detection. The NIR excitation and emission enabled it to track the subtle variations of intracellular NO content in living HeLa cells. In addition, the probe permitted the detection of NO *in situ* in the xenograft tumour mouse model.

Photoacoustic (PA) imaging is a non-invasive imaging modality with deep tissue capabilities. Based on the nitrosation reaction of aromatic secondary amines with NO, Chan and co-workers reported five PA probes for NO.[48] Among these, the ratiometric probe APNO-5 (**33**) can rapidly respond to NO to form an *N*-nitroso product with a concomitant 91 nm hypsochromic shift. Moreover, probe **33** displayed the requisite photophysical characteristics for *in vivo* PA imaging and also high biocompatibility, stability and selectivity for NO over a variety of biologically relevant analytes. Importantly, probe **33** was successfully applied to the detection of endogenous NO in a murine LPS-induced inflammation model. To overcome the drawbacks of the OPD-based NO probe (interference from AA, DHA and MGO), Liu developed a red-emissive TP fluorescent probe, NCNO (**34**),[49] for NO based on the *N*-nitrosation of an aromatic secondary amine. Probe **34** could detect NO with high selectivity and sensitivity over ROS, RNS and reactive carbonyl species (RCS) and was validated by detecting NO in live cells and deep tissues owing to its TP excitation and red-light emission. Moreover, probe **34** was successfully applied to monitor NO in ischaemia reperfusion injury (IRI) in mice kidneys and the results clearly revealed the profile of NO generation *in situ* during the renal IRI process.

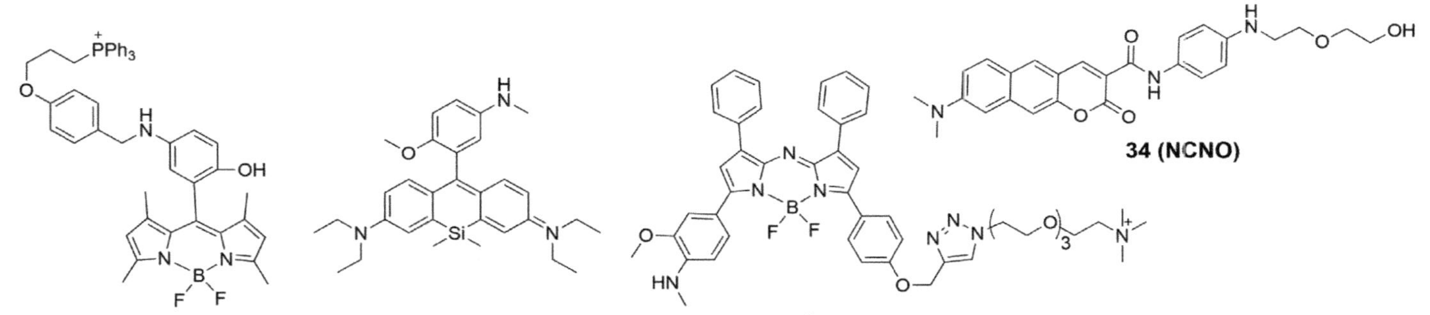

31

32 (SiRNO)

33 (APNO-5)

34 (NCNO)

Figure 8.6 Structures of fluorescent probes for HNO based on nitrosation reactions.

8.2.3 Probes for Peroxynitrite ($ONOO^-$)

Peroxynitrite ($ONOO^-$) is an extremely powerful oxidant with a short lifetime (~10 ms) in biological systems and can react with a wide variety of molecular targets, including proteins, lipids and nucleic acids, eventually resulting in a series of disease states such as diabetes, Alzheimer's disease, cancer, arthritis, autoimmune diseases and other disorders.[50] To appreciate better the diverse pathophysiological roles of $ONOO^-$ and exploit $ONOO^-$-related therapeutic strategies, it is important to develop simple, efficient and reliable methods to detect or image $ONOO^-$ in various biological systems. A number of small-molecule fluorescent probes have been designed and tested for the detection of $ONOO^-$, mainly based on peroxynitrite-triggered reactions such as the formation of dioxirane with activated ketones, boronates/boronic acids, oxidative cleavage, oxidative *N*-dearylation reactions and oxidation of organoselenium/-tellurium compounds (Figure 8.7).

Based on the ketone group, Tang and co-workers developed a two-photon-excitable and NIR-emissive fluorescent probe, TPNIR-FP (**35**), for the detection of $ONOO^-$.[51] Probe **35** displayed high specificity and sensitivity and a fast response towards $ONOO^-$ with mitochondria-targeting characteristics, which permitted the direct visualization of mitochondrial $ONOO^-$ upregulation during anthracycline-induced cardiotoxicity in both live cardiomyocytes and cardiac tissues. In particular, in combination with other bioanalytical methods, the direct correlation between elevated $ONOO^-$ levels and anthracycline cardiotoxicity has been well established. Based on a similar principle, Tang's group developed another TP fluorescent probe, TP-KA (**36**), for detecting $ONOO^-$ with a rapid response and high specificity and sensitivity.[52] Drug (acetaminophen and tolcapone)-related liver injury and the remedial effect of *N*-acetylcysteine (NAC) were investigated using the probe, and the results demonstrated that $ONOO^-$ is a good and appropriate biomarker of hepatotoxicity. Moreover, probe **36** can be employed as a powerful tool to pre-detect drug-induced organism injury and to study the effect of antidotes. Based on luciferase–luciferin pairs and the α-ketoamide group, Zhang and co-workers developed a bioluminescent probe, BP-PN (**37**), for the highly sensitive detection and imaging of endogenous $ONOO^-$ in living cells.[53] Attributed to the bioluminescence without external excitation, probe **37** exhibited a high signal-to-noise ratio with relatively low autofluorescence. Furthermore, it was used to image endogenous $ONOO^-$ in inflamed mice. To investigate how $ONOO^-$ is generated and metabolized during the hepatotoxicity process, Zhang and co-workers reported probe **38** for detecting $ONOO^-$ in mitochondria with superior selectivity and sensitivity.[54] Application of probe **38** achieved real-time imaging of peroxynitrite generation in LPS-stimulated live cells in one- or two-photon imaging modes and revealed that the CYP450/CYP2E1-mediated enzymatic metabolism of acetaminophen leads to $ONOO^-$ generation in liver cells.

Because boronates/boronic acids have a greater reactivity towards $ONOO^-$ than $HClO/ClO^-$ (minutes) and H_2O_2 (hours), the former would therefore

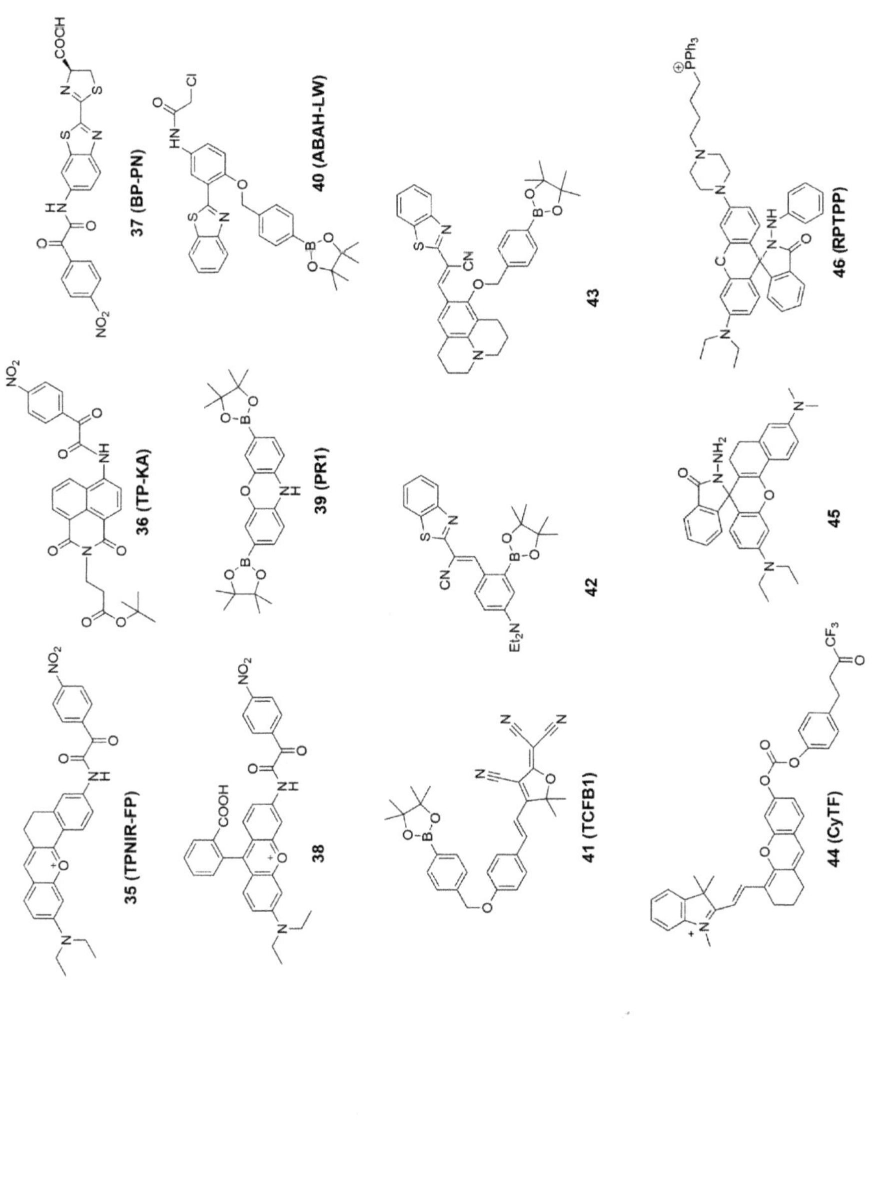

Figure 8.7 Structures of fluorescent probes for $ONOO^-$ based on ketones and boronates/boronic acids.

allow the selective detection of $ONOO^-$. A number of boronate/boronic acid-based fluorescent probes have been developed for recognizing $ONOO^-$, such as probes **39**, **40**, **41** and **42**. Using probe PR1 (**39**), first reported by Chang's group,[55] James and co-workers[56] investigated the changes in $ONOO^-$ production due to different factors, such as enhanced generation of $ONOO^-$ through LPS and interferon-gamma or diminished $ONOO^-$ production with the introduction of superoxide scavengers and nitric oxide synthase inhibitors. The results validated probe **39** as an effective tool for the detection of $ONOO^-$ in J774.2 murine macrophages and should allow the further elucidation of the biology and chemistry of ROS. James's group also developed an ER-directed ESIPT-based ratiometric fluorescent probe, ABAH-LW (**40**),[57] for detecting $ONOO^-$ with high sensitivity and selectivity. Probe **40** was further used for the ratiometric imaging of endogenous and exogenous $ONOO^-$ in HeLa cells. Based on 2-dicyanomethylene-3-cyano-4,5,5-trimethyl-2,5-dihydrofuran (TCF) fluorophores, the same group explored probe TCFB1 (**41**),[58] which demonstrated high sensitivity towards peroxynitrite and mitochondria localization with the ability to detect exogenous and endogenous peroxynitrite in live cells.

Based on $ONOO^-$-triggered covalent assembly of a bright iminocoumarin fluorophore, Guo and co-workers designed and synthesized the fluorescent probe **42**[59] for detecting $ONOO^-$, with a huge fluorescence off–on ratio, fast response rate, high sensitivity and excellent selectivity over a series of biologically relevant ROS. Probe **42** successfully imaged the endogenous $ONOO^-$ in endothelial cells after OGD. Using the same principle, Kim and co-workers developed the arylboronate-derived fluorescent probe **43** for monitoring of $ONOO^-$.

Pu's group developed an NIR fluorescent probe, CyTF (**44**), by linking a trifluoromethyl ketone group to an NIR hemicyanine dye.[60] Probe **44** can specifically distinguish keloid fibroblasts from normal dermal fibroblasts by detecting $ONOO^-$ and permitted the specific detection of implanted keloid fibroblasts in a xenograft live mouse model. Based on the hydrazide reactive group for $ONOO^-$, Yoon and co-workers explored the far-red-emitting fluorescent probe **45** for peroxynitrite detection and imaging.[61] Probe **45** showed high sensitivity and selectivity in the detection of peroxynitrite over other ROS/RNS and was successfully utilized to detect exogenous and endogenous peroxynitrite in HeLa and RAW264.7 cells. Similarly, Ma's group reported a mitochondria-targetable fluorescent probe, RPTPP (**46**), obtained by conjugating rhodamine with the triphenylphosphonium (TPP) cation and phenylhydrazine.[62] Probe **46** exhibited good mitochondria-targeting properties and high sensitivity and selectivity for the mitochondrial $ONOO^-$ over other ROS and was successfully utilized to image the exogenous and endogenous $ONOO^-$ in mitochondria.

Because of its good photostability, large Stokes shift and high quantum yield, the coumarin scaffold sas commonly been used to construct fluorescent probes. Yu and co-workers reported the coumarin-based probe C-Py-1 (**47**) (Figure 8.8) for the detection of $ONOO^-$.[63] Upon addition of $ONOO^-$, the C=C bond linking the coumarin and pyridinium moieties was broken *via* an

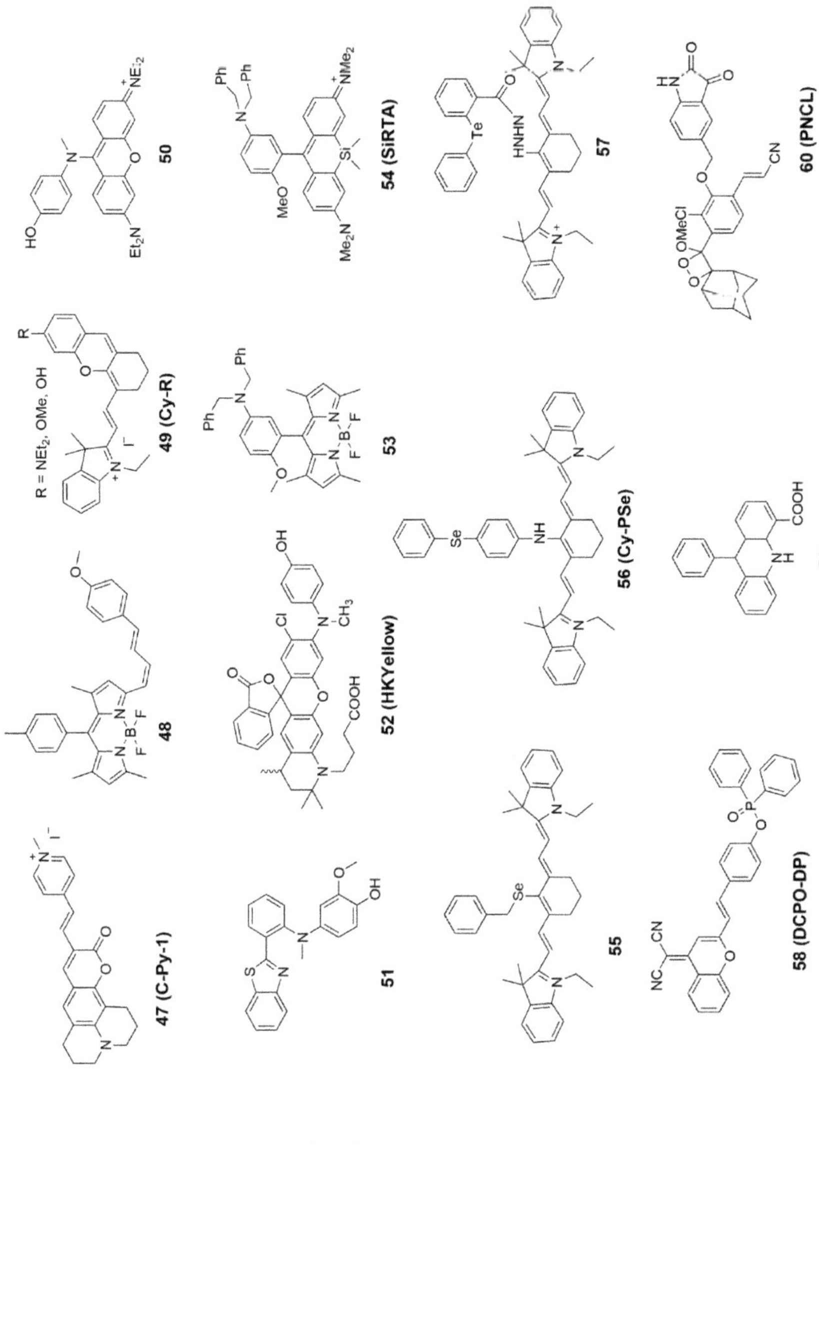

Figure 8.8 Structures of fluorescent probes for $ONOO^-$ based on other mechanisms.

oxidative cleavage. Probe **47** can detect $ONOO^-$ with a rapid response time, high sensitivity and excellent selectivity over other ROS and RNS and was successfully applied in the imaging of endogenous $ONOO^-$ in RAW264.7 cells with outstanding cell membrane penetrability and low cytotoxicity. Based on $ONOO^-$-induced diene oxidation, Wang and co-workers designed and synthesized probe **48**[64] by a Knoevenagel condensation reaction of the BODIPY scaffold with a *trans*-cinnamaldehyde molecule. Probe **48** functions as a specific ratiometric chemosensor that features a stable structure in live cells and shows a sensitive response towards $ONOO^-$ with a marked red-to-green emission change with a detection limit of 150.54 nM. Furthermore, probe **48** was applied to image basal $ONOO^-$ in immune-stimulated macrophage cells and the $ONOO^-$ generation in phagosomes. Based on hemicyanine dye, Li and co-workers developed a ratiometric fluorescent probe, Cy-R (**49**), for detecting $ONOO^-$.[65] The probe was successfully applied for monitoring the level of mitochondrial $ONOO^-$ and assessing the LPS-induced mitochondrial oxidative stress status in living HepG2 cells and in KM mice.

Guo and co-workers explored the *N*-phenylrhodol-based fluorescent probe **50** obtained by the direct conjugation of a methyl(4-hydroxyphenyl)amino group with a selected fluorophore.[66] Probe **50** sensed $ONOO^-$ not only with high selectivity and sensitivity, but also with mitochondria-targetable characteristics and could detect the exogenous and endogenous $ONOO^-$ in BIU-87 cells. Yoon and James's group developed the simple probe **51**[67] for the detection of peroxynitrite incorporating a benzylboronic ester "protecting" unit for 2-(benzo[*d*]thiazol-2-yl)aniline. The 2-(benzo[*d*]thiazol-2-yl)aniline unit of probe **51** was removed by peroxynitrite to turn-on ESIPT fluorescence. Furthermore, the probe was cell permeable and was used in cell imaging experiments, showing an off–on response towards peroxynitrite in HeLa and RAW264.7 cells. On the basis of the efficient fluorescence quenching effect of the *N*-phenyl group on the rhodamine core, Yang and co-workers reported a rhodamine-based fluorescent probe, HKYellow (**52**),[68] for the detection and molecular imaging of peroxynitrite. Probe **52** was thoroughly evaluated for the robust imaging of peroxynitrite in live SH-SY5Y cells and in mouse livers.

Based on $ONOO^-$-triggered *N*-oxidation and *N*-nitrosation reactions of aromatic tertiary amines, Guo and co-workers developed fluorescent probe **53**,[69] which could detect $ONOO^-$ with a fairly fast fluorescence off–on response (within seconds), ultrasensitivity (detection limit $<$2 nM) and excellent selectivity over a series of biologically relevant ROS and also metal cations. Probe **53** was successfully applied to detect the endogenous $ONOO^-$ in activated RAW264.7 murine macrophage, EA.hy926 endothelial cells and kidney tissue of diabetic rats. The same group also developed an NIR fluorescent probe, SiRTA (**54**),[70] containing an aromatic tertiary amine functional group and a Si–rhodamine fluorophore for endogenous $ONOO^-$ detection and imaging. Probe **54** exhibited a fast, sensitive and specific fluorescence off–on response towards $ONOO^-$ with excellent imaging ability for endogenous $ONOO^-$ in living INS-1 cells. In addition, with this probe the therapeutic effects of phenolic acid antioxidants in EA.hy926 endothelial cells suffering from ischaemia–reperfusion injury and the pathogenesis of diabetes and diabetic nephropathy in pancreatic β cells and diabetic rats were successfully evaluated.

Tang and co-workers reported the NIR organoselenium fluorescent turn-off probe **55** for $ONOO^-$ sensing.[71] Probe **55** was converted to its oxidized state upon addition of $ONOO^-$, preventing the PET effect and resulting in enhancement of the fluorescence intensity. The off–on fluorescence intensity could be reversed by the addition of the reducing agent ascorbate. Probe **55** was used for the real-time imaging of cellular redox cycles in living RAW264.7 cells. Based on the same mechanism, Han and co-workers devised the reversible NIR fluorescent turn-on probe Cy-PSe (**56**) for $ONOO^-$ sensing.[72] Probe **56** exists in a non-fluorescent state as a result of the action of PET between the donor and acceptor groups. Addition of $ONOO^-$ causes an increase in emission from the probe at 775 nm, a change that can be reversed by the addition of GSH. The probe was also successfully applied to the detection of peroxynitrite in live cells. Based on heptamethine cyanine, the same group developed the NIR fluorescent probe **57**[73] by integrating with telluroenzyme mimics for monitoring the changes in $ONOO^-$/GSH levels in cells and *in vivo*. Probe 57 can reversibly respond to $ONOO^-$ and GSH with high selectivity, sensitivity and mitochondrial localization function and was successfully applied to visualize the changes in redox cycles during the release of $ONOO^-$ and the antioxidant GSH repair in cells and animals.

By integrating a diphenylphosphinate group with a dye, dicyanomethylene-4*H*-pyran, Churchill and co-workers developed the fluorescent probe DCPO-DP (**58**) for $ONOO^-$ sensing.[74] The probe displayed excellent selectivity and sensitivity with a fast response for the detection of peroxynitrite. In addition, probe 58 was used to image $ONOO^-$ in HeLa cells. On the basis of the oxidative aromatization of non-fluorescent 9,10-dihydroacridine derivatives by peroxynitrite, Gong and co-workers developed the highly selective and rapid-response fluorescent probe **59**, which can detect $ONOO^-$ with high selectivity and sensitivity. Furthermore, the probe was successfully utilized to image the exogenous $ONOO^-$ in RAW264.7 cells. Lippert and co-workers reported the development, characterization and biological applications of a reaction-based chemiluminescent probe, PNCL (**60**),[75] for $ONOO^-$ detection. Probe **60** can react with $ONOO^-$ *via* an isatin moiety through an oxidative decarbonylation reaction to initiate light emission with high selectivity against other reactive species. Moreover, the probe was applied for $ONOO^-$ detection in aqueous solution and live cells and can be employed to detect cellular $ONOO^-$ generated in macrophages stimulated to mount an immune response with LPS.

8.3 Fluorescent Probes for GSH, Cys and Hcy

8.3.1 Fluorescent Probes for GSH

Glutathione (GSH) is the most abundant (1–10 mM) of the small non-protein thiols inside cells and is a reducing agent that can be converted to its oxidized form glutathione disulfide (GSSG). GSH plays an essential role in a variety of physiological and pathological events, such as inflammation, oxidative stress, gene expression, signal transduction and apoptosis. Abnormal biological levels of GSH are directly linked to many human diseases, including cancer, Alzheimer's disease and cardiovascular disease.[76,77] Therefore, the tracing of the fluctuations of GSH and its metabolism in biological specimens has attracted a great deal of attention. So far, fluorescent probes for GSH detection have mainly been constructed by utilizing the strong nucleophilicity of the sulfhydryl group, which enables reactions such as Michael addition and nucleophilic substitution. Some typical examples of fluorescent probes based on these reactions (Figure 8.9) are described in this section.

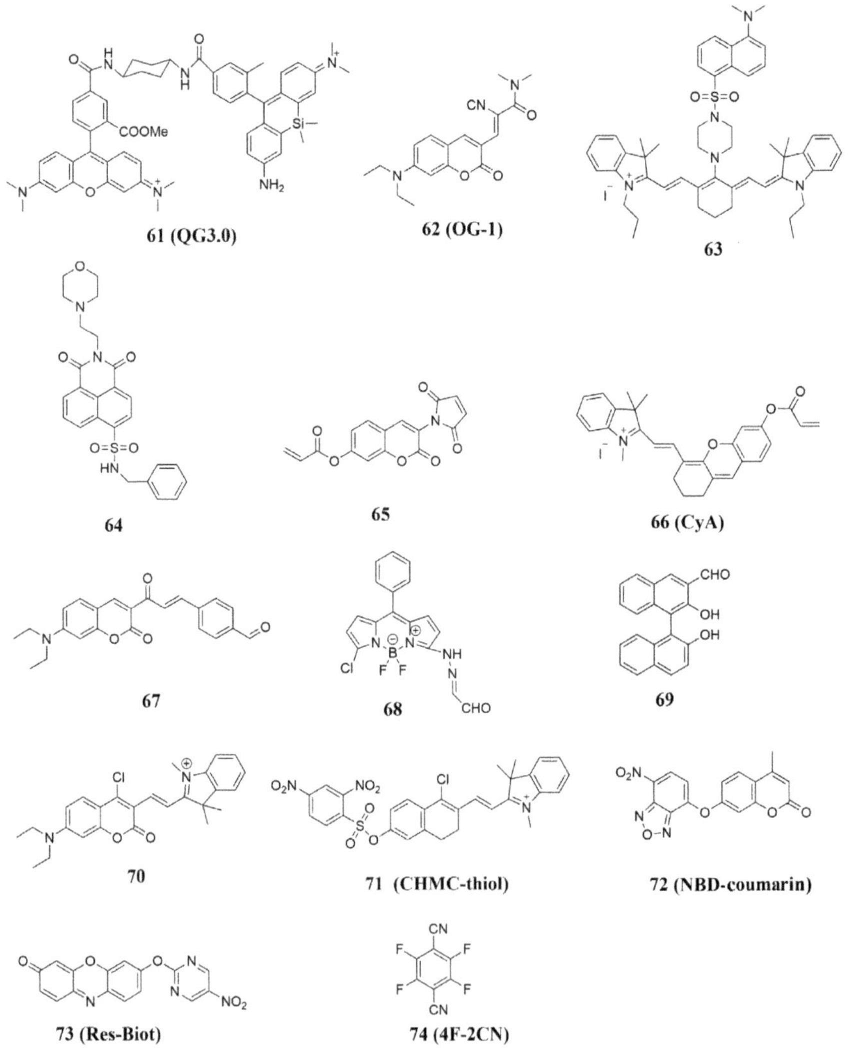

Figure 8.9 Structures of fluorescent probes for GSH, Cys and Hcy.

8.3.1.1 Probes Based on Michael Addition

The Michael reaction, involving the addition of a nucleophile to an α,β-unsaturated carbonyl moiety, is widely utilized as a basis of chemical recognition for fluorescent probes for GSH. To quantify intracellular GSH, two reversible fluorescent probes were developed, both of which displayed a ratiometric response towards GSH.[78,79] Utilizing silicon-substituted rhodamine as response moiety and energy acceptor, the fluorescence resonance energy transfer (FRET)-based fluorescent probe QG3.0 (**61**) was designed.[78] On exposure to GSH, the FRET process of **61** was interrupted and the fluorescence emission of the energy donor, tetramethylrhodamine, increased considerably, and the fluorescence signal of the energy acceptor recovered after removal of GSH, thus resulting in a ratiometric readout. Another reversible ratiometric probe for GSH, OG-1 (**62**), was implemented *via* controlling the length of the resonant system of the probe, thereby providing fluorescence emission at two different wavelengths ($\lambda_{em} = 488$ and 560 nm).[79] By integrating rhodamine dye into the cyanine skeleton, an NIR fluorescent turn-on probe for GSH based on Michael addition accompanied by the hydrolysis process was constructed. The probe underwent an addition reaction of the unsaturated bond followed by hydrolysis to release the fluorophore, thus representing a fluorescence turn-on event.[80]

8.3.1.2 Probes Based on Nucleophilic Substitution

The nucleophilic substitution capability of GSH has also been utilized for GSH probe fabrication. A ratiometric fluorescent probe based on chloro-BODIPY was developed.[81] The sensor operated through the rapid displacement of chloride atom with deprotonated thiols, in which intramolecular displacement to form a secondary amine can occur in the presence of cysteine (Cys) and homocysteine (Hcy), but not GSH, thus causing discrimination of GSH over Cys and Hcy. Utilizing sulfonamide as the response group and cyanine as reporter, the NIR fluorescent probe **63** for GSH was synthesized.[82] In the presence of GSH, the piperazine bridge between cyanine and sulfonamide underwent cleavage to release free piperazine-modified cyanine, accompanied by fluorescence enhancement. A labile nitroazo aryl ether group-based GSH sensor was also prepared by utilizing cyanine as fluorescent indicator.[83] The nitroazo ether group was introduced as both the GSH-selective capture group and the fluorescence quencher because of the PET effect. Once sulfonamide had been integrated into naphthalimide, the fluorescence of **64** was silenced because of the ICT effect.[84] The nucleophilic attack of GSH would relieve the electron-withdrawing capacity of sulfonamide and switch on fluorescence.

8.3.2 Fluorescent Probes for Cys

Cysteine (Cys), a semi-essential proteinogenic amino acid, plays an important role in living systems.[85] A deficiency of Cys causes multiple dysfunctions that lead to many human diseases, such as liver damage, loss of muscle and fat and skin lesions.[86] In view of the strong nucleophilicity of Cys, the Michael α,β-unsaturated addition reaction is a widely used design strategy for Cys probe preparation.

8.3.2.1 Probes Based on Addition of Maleimide

The maleimide group, one of the best known electrophiles, has been integrated into various fluorescent reporters, including coumarin (**65**),[87] chromenoquinoline[88] and naphthalimide,[89] to afford Cys probes featuring outstanding detection performance. Michael addition of Cys to maleimide converts the maleimide from an electron-withdrawing group to an electron-donating unit, which removes its fluorescent quenching effect and results in turn-on fluorescence.

8.3.2.2 Probes Based on Addition of Acrylate or Acrylate Derivatives

Acrylate and acrylate derivatives represent another widely used trigger for Cys fluorescent probe construction. The acrylate group can be easily integrated into various hydroxyl-containing dyes.[90,91] For example, the NIR fluorescent probe CyA (**66**) was prepared by conjugating an acrylate group to a hemicyanine skeleton. Both naked-eye and fluorescence changes of **66** can be observed upon Cys addition.

8.3.2.3 Probes Based on Addition of a Ketone or Aldehyde

By incorporating terephthalaldehyde into ethanoylcoumarin, a novel dual-site fluorescent sensor (**67**) for Cys and SO_2 was obtained.[92] The terephthalaldehyde moiety of the probe can be specifically attacked by Cys, thereby turning on fluorescence. This probe was also capable of tracing SO_2, a metabolite of Cys, which enabled Cys metabolism activity at the cellular level to be explored. Utilizing the reductive capacity of Cys, aldehyde-based fluorescent probes for selective Cys detection were formulated.[93–95] For example, the mode of action of **68** involved the formation of a five-membered ring of thiazolidine triggered by Cys, which interrupted the ICT effect of aldehyde to recover the fluorescence of BODIPY.[95]

8.3.3 Fluorescent Probes for Hcy

Homocysteine (Hcy), a homologue of the amino acid cysteine, is considered to be risk factor for atherothrombotic vascular disease.[96] Making use of the cyclization reaction between an aldehyde and Hcy, the 1,10-binaphthyl-based probe **69** was developed for Hcy detection in buffer solution.[97] Taking advantage of the reducibility of Hcy, an azide-functionalized dansyl derivative was prepared for the specific tracking of Hcy. In view of the high reactivity of the azide group towards H_2S, zinc chloride was added to precipitate out the sulfide in the solution through ZnS formation, thus avoiding H_2S interference.[98] By integrating propionyl chloride into 1-hydroxypyrene-2-carbaldehyde fluorophore, a novel fluorescent probe with distinctly different fluorescence responses towards Hcy and Cys was obtained. Theoretical calculations demonstrated that the different contributions of ICT and PET to the singlet excited states of the probe was responsible for high selectivity of the probe towards Hcy over Cys.[99]

8.3.4 Fluorescent Probes for Two or All of GSH, Cys and Hcy

8.3.4.1 Probes Based on α,β-Unsaturated Addition

Based on chlorinated coumarin hemicyanine, a sensor featuring three reaction sites was developed. This strategy can selectively track Cys and GSH on the basis of their different emission wavelengths. The probe **70** first underwent nucleophilic substitution between the chloro group and three thiols, followed by intramolecular unsaturated addition at different sites to afford the corresponding cyclization products based on the different lengths of thiols.[100] Similarly, through integrating a chlorine atom and an α,β-unsaturated malonitrile into coumarin, another probe based on different cascade reactions for differentiation amongst Cys, Hcy and GSH was obtained.[101]

8.3.4.2 Probes Based on S_NAr Substitution

Biothiols are capable of selectively cleaving the O–S bond to remove the 2,4-dinitrobenzenesulfonyl unit, which served as a fluorescence quencher owing to its strong electron-withdrawing effect, and to release the hydroxyl-free fluorophore, thus triggering a notable fluorescence light-up response.[102–104] By simultaneously incorporating a chlorine atom and a 2,4-dinitrobenzenesulfonyl unit in an NIR merocyanine dye, the probe CHMC-thiol (**71**) for total thiols was obtained, which shows distinct response pathways towards different concentration ranges of thiols *via* employing the 2,4-dinitrobenzenesulfonyl moiety and the chloro group as the high- and low-sensitivity site of thiols, respectively.[103] Attaching 7-nitro-2,1,3-benzoxadiazole (NBD) to a hydroxyl-containing red fluorophore through an ether linkage afforded the probe NBD-coumarin (**72**) for GSH and Cys/Hcy.[105] In the presence of GSH or Cys/Hcy, the ether bond was cleaved with the concomitant generation of red fluorescent dye. Further, production of a green fluorescence of NBD with Cys/Hcy can be achieved through an intramolecular Smiles rearrangement, hence this probe can be utilized to detect and differentiate GSH and Cys/Hcy. By implementing a combinatorial developmental programme, pyrimidine derivatives were introduced into various fluorescent platform to generate the biothiol-specific sensor Res-Biot (**73**) through an S_NAr pathway.[106] A remarkably simple molecule, tetrafluoroterephthalonitrile,

4F-2CN (**74**), was discovered to be an outstanding fluorescent probe featuring a distinct reaction performance for Cys, Hcy and GSH in the presence of the surfactant cetyltrimethylammonium bromide (CTAB).[107] The disulfide bond was proved to be an effective response group for the construction of thiol fluorescent probes.[108–111] For example, by incorporating a thiol-specific cleavable disulfide bond into a galactose-modified naphthalimide fluorophore, a hepatocyte-targeting fluorescent probe was constructed for thiol imaging. On exposure to thiols, the probe underwent disulfide bond cleavage *via* reaction with the thiols, which released the naphthalimide moiety and led to obvious fluorescence changes.[108]

8.4 Fluorescent Probes for H_2S

Hydrogen sulfide (H_2S), the simplest and smallest biothiol in living cells, is considered to be the third gaseous signal molecule after nitric oxide (NO) and carbon monoxide (CO).[112] The specific protective or deleterious physiological function of H_2S is dynamically controlled by its intracellular concentrations,[113] which may fluctuate considerably in a reported range of 30–160 μM with diverse types of cells and different subcellular regions.[114] At the molecular level, H_2S acts both as a good reducing agent that helps to maintain the balance of intracellular redox status like a cytoprotectant and as a highly toxic nucleophile that can bind to haem proteins such as cytochrome *c* oxidase and inhibit the electron transport chain at sufficient levels.[113] H_2S influences various physiological and pathological processes, including the modulation of blood vessel tone and cardioprotection,[115] the endogenous stimulation of angiogenesis[116] and mitochondrial bioenergetics.[117] Unregulated, abnormal levels of H_2S may contribute to disease, as observed in models of Huntington's,[118] Parkinson's[119] and Alzheimer's diseases.[120] Considering the importance of H_2S as a cytoprotectant and gasotransmitter, the *in situ* assay and visual monitoring of H_2S in living cells are important for gaining a better understanding and positive regulation of its basic cellular functions, especially the challenges in the quantitative detection of this fluctuating small molecule.

The principles and strategies employed for the enhancement of the detection sensitivity and specificity of intracellular H_2S are mainly based on its nucleophilic attack and reducing capacity endowed by the electronegativity of the sulfur atom and reducibility of divalent sulfur ions, respectively. With a pK_a of 7.0, H_2S exists predominantly as anionic HS^- at physiological pH, which means a higher nucleophilicity of H_2S compared with many other thiols found in living systems. Therefore, nucleophilic attack principles have been widely used to develop H_2S-specific probes. In view of the reducibility of H_2S, electron-withdrawing moieties such as the azide group are commonly used in the selective analysis of intracellular H_2S. Typical examples of fluorescent probes for the selective and sensitive detection of intracellular H_2S are illustrated in Figure 8.10.

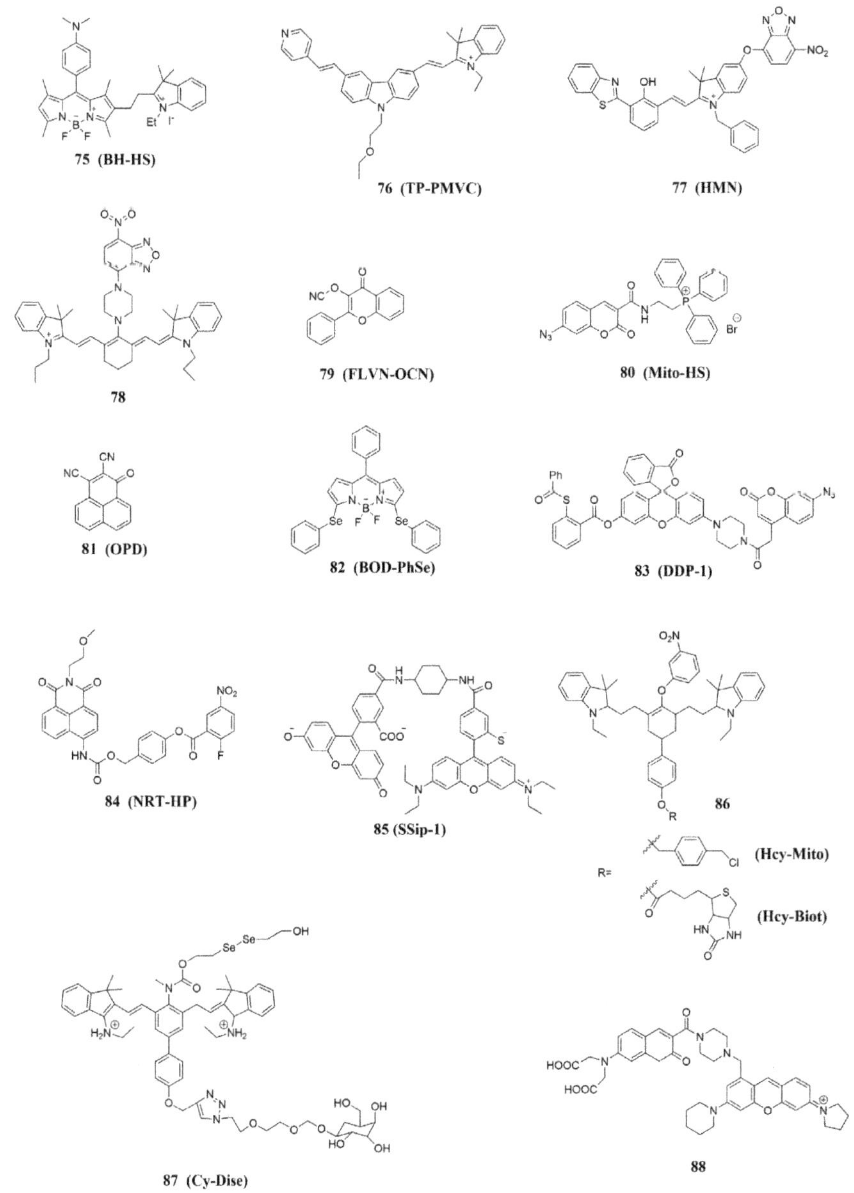

Figure 8.10 Structures of fluorescent probes for H_2S, H_2S_n and Cys-SSH.

8.4.1 Nucleophilic Addition of H_2S

The nucleophilicity of HS^- at physiological pH in living cells makes H_2S susceptible to attack by electrophilic centres in conjugated fluorescent molecules, including cyanine[121–124] and other dyes,[125] through H_2S-induced nucleophilic addition. Based on this reaction, typical fluorescent signal transduction mechanisms such as FRET,[121] intramolecular twisted internal charge transfer (TICT),[122] intramolecular charge transfer (ICT)[123,124] and PET[124,125] can be employed to develop highly selective fluorescent probes. For the enhancement of assay sensitivity, weakly fluorescent probes, such as structures characteristic of TICT, with a low background signal are the preferred options. Let us take probe **75** (BH-HS), the first reported TICT-based fluorescent probe for H_2S, as an example.[122] BODIPY is selected as the fluorescence reporting group and hemicyanine is used as recognition site for H_2S. In terms of design principles, the intramolecular TICT pathway is considered to be effectively blocked by the nucleophilic addition of H_2S to the indole unit of the probe, which enhances the fluorescence of the system within 2 min in the presence of 50 μM H_2S. Accordingly, rapid and specific detection, and also visualization of exogenous and endogenous H_2S in living cells, are achieved.

Furthermore, to obtain the subcellular localization of small active species in living cells, targeting groups are combined on the fluorescent probes, such as triphenylphosphine for mitochondria and pyridine for lysosomes. In view of the overlap between coumarin emission and merocyanine absorption, a FRET-based ratiometric fluorescent probe, CPC, composed of a coumarin–merocyanine dyad, was developed for mitochondrial H_2S.[121] Once the FRET process had been interrupted by the nucleophilic addition of H_2S to the indole unit, the emission ratio was considerably enhanced in a dose-dependent manner. A dual-site, two-photon fluorescent probe, TP-PMVC (**76**), was successfully developed for visualizing lysosomes and tracking lysosomal H_2S with two different sets of fluorescence signals in living cells and mouse liver tissues.[123] Specifically, a lysosome-targetable pyridine moiety responsive to H^+ and an indole unit recognizing H_2S are covalently bound on a carbazole platform as a whole imaging probe. The D–π–D–π–A structure (D = donor, A = acceptor) enables the probe to be almost non-fluorescent. Once targeted to an acidic lysosome, the pyridine moiety is protonated, which induces efficient ICT and a red-shifted fluorescence signal is observed. Meanwhile, the overall conjugation in the probe is blocked by the nucleophilic addition of lysosomal H_2S to the indolenium C2 atom, which inhibits the ICT efficiency and thus produces a blue-shifted fluorescence readout. Therefore, simultaneously tracking of lysosomal H_2S and visualization of lysosomes in living cells are achieved with two distinguishable sets of signals by dual-channel fluorescence imaging.

The enhancement of selectivity is another key challenge for the *in situ* assay of H_2S in living cells, because of the close network of generation and metabolic pathways within biothiols. The H_2S level is susceptible to the levels of Cys, Hcy and GSH. However, it is difficult to discriminate these small molecules, with very similar chemical structures and properties, from each other. Distinguishable multi-channel detection and imaging provide an effective solution, as illustrated in the probe HMN (**77**).[124] The hybrid multi-signal fluorescent probe **77**, containing a hydroxyphenylbenzothiazole merocyanine (HBTMC) dye and a nitrobenzoxadiazole (NBD) chromophore connected through an ether linker, was developed for simultaneously distinguishing and sequentially sensing Cys/Hcy, GSH and H_2S for the first time. The fluorescent behaviour of probe **77** can be flexibly modulated by the PET from the NBD moiety to the HBTMC moiety and from the HBT unit to the indole unit. Specifically, the fluorescence of **77** is effectively quenched by the PET process from the protected hydroxyl-NBD moiety to the HBTMC moiety and red emission ($\lambda_{ex} = 550$ nm, $\lambda_{em} = 609$ nm) of the released HBTMC unit can be obtained upon cleavage of the ether bond by Cys/Hcy, GSH or low-level H_2S, while another green emission ($\lambda_{ex} = 470$ nm, $\lambda_{em} = 546$ nm) of the amino-substituted NBD chromophore is also observed for Cys/Hcy. In addition, the C=N double bond of the HBTMC unit would be further added by HS^- to produce an interrupted HBTMC derivative (HBTMC-SH), which interrupts the ICT process from the electron-donating HBT group to the electron-withdrawing indole group, generating a new blue emission ($\lambda_{ex} = 410$ nm, $\lambda_{em} = 485$ nm). Accordingly, distinguishable fluorescent signals at three distinct emission bands (blue–green–red) with 61 and 63 nm distances between two adjacent bands, respectively, are obtained upon responding to the respective biothiols. Three-colour fluorescent imaging in living cells can be simultaneously achieved for the sequential detection of Cys/Hcy, GSH and H_2S.

8.4.2 H_2S-induced Thiolysis

As the predominantly existing component of intracellular H_2S at physiological pH, HS^- could be selectively added to a single electrophilic position on the probe scaffold, followed by the removal of the electrophilic functionalities and the subsequent production of turned-on fluorescence signals. Typical probes based on this principle usually have electron-withdrawing motifs incorporated that can be selectively cleaved by intracellular H_2S, such as 7-nitro-1,2,3-benzoxadiazole (NBD) amine. Naphthalimide,[126] styryl-BODIPY[127] and cyanine[128] are commonly used for the fluorescent skeleton, in view of their ability to undergo thiolysis by H_2S *via* a nucleophilic aromatic substitution (S_NAr) mechanism.

Let us consider probe **78**,[128] an NIR structure with a cyanine skeleton, as an example. The water-soluble probe **78** is highly specific to H_2S even in the presence of millimolar GSH, with dramatic fluorescence enhancement at 796 nm, which is employed to monitor endogenous H_2S in colorectal cancer cells. Notably, through intratumoural injection of this probe, endogenous H_2S in colorectal cancer cells *in vitro* and in a murine tumour model could be monitored selectively and rapidly. Furthermore, a targetable group can be flexibly incorporated for the *in situ* subcellular distribution assay of H_2S. A mitochondria-targeting fluorescent probe was developed for fluorescence imaging of H_2S in HeLa cells, based on the H_2S-induced thiolysis of NBD amine bonding on the naphthalimide scaffold with a triphenylphosphine (TPP) group.[126]

In addition to NBD, other electron-withdrawing ester moieties[129–132] that could be specifically removed by nucleophilic H_2S can also be incorporated as leaving groups to release fluorescent scaffolds. Thiophenecarboxylate was incorporated into a fluorescein platform to develop a sensitive probe for detecting H_2S endogenously produced by cystathionine beta-synthase (CBS) and cystathionine gamma-lyase (CSE) in living cells and living mice.[131] Through a cut-through mechanism confirmed by mass spectrometric analysis, thiophenecarboxylate is specifically removed by H_2S and thus the released fluorescein accounted for the pronounced increase in fluorescence intensity. Moreover, the activity of CBS and CSE in H9C2 cells and living mice was successfully monitored, owing to the high selectivity and sensitivity of this probe. Similarly, an ESIPT-based fluorescent probe, FLVN-OCN (**79**), was developed for detection of H_2S *in vitro* and imaging in living cells.[130] With an electrophilic cyanate as the recognition site masked on the 3-hydroxyflavone, probe **79** is characterized by a low detection limit (0.25 μM) and a relatively rapid response time (<60 s), and also excellent selectivity towards H_2S over other biothiols.

8.4.3 Azide Group Reduction

The azide group, well known as an electron-withdrawing moiety, has been reported for its selective reduction by H_2S to electron-donating amino products. The main principle of azide-based probes is based on the significant change in the electronic distribution of the whole molecule once installed onto fluorophores. Moreover, the compatibility of the azide motif with living cellular systems and the relative ease of reaction with amine-containing fluorophores encouraged the development of a wide variety of fluorescent azide-based H_2S probes.[133–135] Subcellular localization probes can be flexibly and conveniently developed with an azide motif, and this has been the main development trend for this kind of probe for the *in situ* assay of subcellular H_2S. Two positively charged fluorescent probes for H_2S detection in mitochondria and lysosomes were designed based on the selective reduction of azide.[133] A triphenylphosphonium (TPP) and a dimethylamino group were attached to two naphthalimide platforms as the targeting group of mitochondria and lysosomes and also to enhance the solubility of the probes. Both probes could detect H_2S *in situ* with good selectivity. In addition, the mitochondria-targeting probe successfully visualized H_2S in tumours of living mice. Notably, the probe Mito-HS (**80**) is capable of sensing endogenous H_2S in cancer cells without any external stimulators within 15 min.[134] Thanks to the high sensitivity of **80** towards H_2S, as low as 24.3 nM, cancer cells can be successfully distinguished from normal cells based on the level of the endogenous H_2S formation.

The potential interference on the inherent redox equilibrium caused by the introduction and irreversible reaction of organic molecules has been one of the controversial factors that restrict the practical application of these kinds of fluorescent probes. In order to detect H_2S without its irreversible consumption, an H_2S donation strategy coupled with a fluorescent probe was developed based on the thiocarbamate platform.[136] When the azide group is reduced by H_2S, a transient arylamine intermediate is formed, followed by a self-immolative cascade reaction to release the amine-bound fluorophore and carbonyl sulfide, which could be hydrolysed to H_2S and CO_2 in the biological environment. This process can be triggered by endogenous or biorthogonal triggers to maintain the inherent redox equilibrium.

8.4.4 Other Assay Principles

Some other representative assay principles that have been employed in recent years,[137–139] other than nucleophilic addition, H_2S-induced thiolysis and azide group reduction, are introduced in this section. The first involves nucleophilic aromatic substitution of hydrogen (S_NAr^H) reactions. Specifically, the sulfide is added to the double bond of the phenalene-based fluorescent chemodosimeter OPD (**81**),[137] followed by sequential isomerization, rearrangement, hydrolysis and elimination to afford the final product. The second is based on a familiar ESIPT fluorophore with AIE characteristics, characteristic of two-photon fluorescence imaging for the specific detection of H_2S. With an 80-fold fluorescence enhancement along with a large Stokes shift upon addition of H_2S, this probe could successfully detect H_2S in HeLa cells with both one- and two-photon microscopy. The third principle is illustrated by the fluorescent turn-off probe BOD-PhSe (**82**).[139] The leaving of the phenylselenide group at the 3- and 5-positions of the BODIPY core substituted by H_2S significantly decreases the fluorescence intensity of probe **82**, which could be employed to image intracellular H_2S in living cells.

8.5 Fluorescent Probes for H_2S_n

Compared with H_2S, the amount of attention paid to the *in situ* assay of higher-order reactive sulfur species (RSS) is much less, although these species are also found to exhibit important biological activities in the modulation of enzyme functions and the synthesis of sulfur-containing vitamins and cofactors.[140] As suggested in studies on H_2S redox biology, higher-order RSS are considered to be responsible for most biological mechanisms that were originally attributed to H_2S.[140–142] From the viewpoint of chemical reactivity, hydrogen polysulfides (H_2S_n, $n>1$) are expected to be stronger nucleophiles but weaker reductants than H_2S, and can be considered as redox partners or oxidized forms of H_2S. Therefore, H_2S_n may coexist in biological systems as a redox couple to collectively regulate the sulfur-containing redox balance. Despite an increasing recognition of the importance of H_2S_n in biological settings, their detection in living cells remains a challenge. A few typical fluorescent structures for the detection and *in situ* assay of intracellular H_2S_n are given in this section to illustrate the basic principles for the design this kind of probe.

Considering the stronger nucleophilicity and weaker reductibility of H_2S_n than H_2S, phenyl 2-(benzoylthio)benzoate[143,144] and 2-fluoro-5-nitrobenzoate[145–147] are the two most commonly used recognition units specific for H_2S_n. Furthermore, in order to understand better the symbiotic relationship and cellular cross-talk between H_2S_n and H_2S, it is highly desirable to develop single fluorescent probes that permit dual-channel discrimination between H_2S and H_2S_n. The first desirable probe capable of visualizing H_2S and H_2S_n with different fluorescence signals, DDP-1 (**83**),[143] deserves a systematic detailed introduction. In the presence of single H_2S_n, only rhodol could be released from **83** with green fluorescence at 542 nm. For single H_2S, both rhodol and coumarin could be simultaneously released to produce distinguishable blue fluorescence at 445 nm and green fluorescence at 542 nm. Accordingly, a single green emission was attributed to H_2S_n, whereas the simultaneous appearance of the two well-separated emissions confirmed the presence of H_2S. Moreover, probe **83** could even determine the relative amounts of H_2S_n and H_2S in mixed solutions with varying Na_2S_2/Na_2S concentrations, *via* detection of the ratio of the fluorescence intensities at 542 nm and 452 nm respectively. Dual-channel imaging experiments could be carried out to detect H_2S and H_2S_n selectively *in situ* in living HeLa cells. The 2-fluoro-5-nitrobenzoic group was used as the reaction unit for the detection of H_2S_n in another ratiometric fluorescent probe, NRT-HP (**84**).[145] With an adjustable ICT system, the fluorescence of **84** could be modulated by H_2S_n through the sequential formation of benzodithiolone and the release of an electron-donating amino group, which is able to image exogenous and endogenous H_2S_n in living cells.

Other principles for selectively detecting H_2S_n have also been reported and two typical examples are given here. The first is based on the unique reversible reaction of sulfane sulfur with other sulfur atoms, as illustrated in the probe SSip-1 (**85**),[148] a reversible fluorescent probe modulated by the FRET mechanism. Specifically, the target sulfane sulfur reversibly binds to the sulfur atom in **85** to generate intramolecular spirocyclization of the xanthene dye, turning on the fluorescence of the system, which could be employed to visualize reversibly the concentration fluctuation of sulfane sulfur *in situ* in living cells. Considering the potential cross-talk between different reactive species to maintain the redox balance, multicomponent detection is preferred as one of the prominent advantages for the future development of fluorescent probes, as illustrated in the following example with the probes Hcy-Mito and Hcy-Biot (**86**).[149] With a heptamethine cyanine dye as NIR fluorescent skeleton, **86** is first responsive to intracellular $O_2^{\bullet-}$ to produce a conjugated system with low fluorescence and then the nitro group is reduced to an amino group by intracellular H_2S_n to cancel the PET process of the structure, emitting much stronger fluorescence at longer wavelength. *Via* masking with a different substituent, the probe could be located in mitochondria or tumour tissue. This sensitive probe could selectively image $O_2^{\bullet-}$ and H_2S_n in cells and *in vivo*. Fluorescence imaging experiments also indicated that the release of $O_2^{\bullet-}$ in cells would induce the rapid generation of H_2S_n.

8.6 Fluorescent Probes for Cys-SSH

Cysteine hydropersulfide (Cys-SSH), considered to be the source of the whole profile of R-SSH derivatives in living organisms at the micromolar level,[150] has attracted increasing attention in recent years because of its leading role in the synthesis of sulfur-containing cofactors, regulating cellular signalling, modulating enzyme activities and adjusting cellular redox milieu.[150,151] However, compared with the thiol analogues, the fundamental chemistry and chemical biology of Cys-SSH in cells are yet to be elucidated. Highly selective detection methods other than modified biotin and tag-switch assays are urgently needed for the analysis of Cys-SSH efficiently and conveniently, especially *in situ* in a living cellular environment. The fluorescence sensing principles for Cys-SSH are far from systematic and the following two examples are introduced here, in an attempt to provide some advice for the design of more Cys-SSH probes. The first example, the probe Cy-Dise (**87**),[152] is employed for analysing Cys-SSH in living cells and *in vivo* with an NIR ratiometric fluorescence signal, based on the selenium–sulfur exchange reaction. Specifically, intracellular Cys-SSH reduces the diselenide group within minutes, triggering an ICT process in the heptamethine cyanine skeleton of the probe, and provides a spectral blue shift of 48 nm (from 797 to 749 nm). Ratiometric imaging ($R = F_{750\text{–}800\ \text{nm}}/F_{690\text{–}740\ \text{nm}}$) could be achieved in living cells for the quantitative detection of CBS enzyme-mediated Cys-SSH biosynthesis. The second example, probe **88**,[153] is based on the decrease in FRET efficiency after recognition of R-SSHs, including H_2S_2, Cys-SSH and GSSH. The conjugated structure of the xanthene ring is interrupted upon the reverse reaction between RSSHs and the pyronine fluorophore, inducing a dual-emission change at 479 and 584 nm. Notably, ratiometric imaging ($R = F_{430\text{–}480\ \text{nm}}/F_{550\text{–}630\ \text{nm}}$) with probe **88** could dynamically visualize the concentration of R-SSHs *in situ* in living cells.

8.7 Fluorescent Probes for SO_2

Sulfur dioxide (SO_2) is an air pollutant, and exposure to SO_2 has become increasingly widespread because of the combustion of fossil fuels. The inspiratory SO_2 will hydrate and generate sulfurous acid and then hydrolyse immediately to the corresponding bisulfite and sulfite derivatives, which mainly determine the toxicity of SO_2.[154] Epidemiological studies indicated that exposure to SO_2 not only could give rise to a variety of respiratory tract reactions,[155] but also was related to cancers, neurological disorders (migraine headaches, stroke, brain tumours)[156] and angiocardiopathy. It was further shown, based on toxicological studies, that SO_2 and its derivatives could change the voltage-gated features of the sodium and potassium channels in rat hippocampus neurons,[157] affect the intracellular thiol content[158] and cause neuron damage.[159] However, in contrast to its toxicological effects, some studies showed that the intake of SO_2 and its derivatives (Na_2SO_3 and $NaHSO_3$) or intraperitoneal injection of these compounds could lower the blood pressure of male Wistar mice effectively.[160] Furthermore, SO_2 can modulate numerous biological processes as a gaseous signal transmitter. For example, SO_2 is recognized to cooperate with NO as vasodilator substance to regulate the vascular smooth muscle.[161,162] Although it is known that SO_2 can be produced by the oxidation of hydrogen sulfide of sulfur-containing amino acid metabolism *in vivo*,[163] the associated physiological metabolic processes still deserve more research.

Recently, various types of fluorescent probes with high sensitivity and selectivity for SO_2 and its derivatives (bisulfite and sulfite) have been exploited. The molecular structures of probes for SO_2 derivatives are presented in Figure 8.11 and can be divided into three categories. In general, these probes are designed by taking advantage of the nucleophilic attack of SO_2 derivatives on unsaturated bonds. It is worth noting that colorimetric and ratiometric fluorescence signals can be obtained on account of the moderately nucleophilic addition of SO_2 derivatives to α,β-unsaturated bonds, which normally generate new blue-shifted fluorescence emission. The probes **89–98** (Figure 8.11) are designed according to this strategy.[164–173] FRET, as the second strategy for SO_2 detection, was applied to construct probes **99–103**.[174–177] Initially, these probes give the emission derived from the FRET acceptors. After the attack of SO_2 derivatives on the nucleophilic sites, the conjugation of the acceptors is interrupted, which results in the cancellation of the FRET process and the quenching of the fluorescence. As a result, the emission belonging to the FRET donors gradually increases, giving the ratiometric fluorescence signal.

The third strategy for the detection of SO_2 derivatives is based on nucleophilic addition to aldehydes which are exploited as the reaction site. The probe DMPCA (**103**) was used to detect HSO_3^- by undergoing two-photon excitation.[178] The mechanism involves the inhibition of the TICT of probe **103** from the *N,N*-dimethyl group serving as a TICT donor to the coumarin skeleton serving as a TICT acceptor after the nucleophilic addition to aldehyde. As a result, upon excitation at 700 nm, the two-photon fluorescence at 415 nm is increased dramatically by 65-fold. Further, the detection of HSO_3^- in HeLa cells was achieved. Recently, Tang's group developed a mitochondria-targeting probe, FMHI (**104**), for imaging SO_2.[179] The aldehyde group that served as the recognition unit and fluorescence quencher was removed upon treatment with HSO_3^- and thus the ICT process in probe **104** was activated. With a pyridinium mitochondria-targeting group, probe **104** was used to reveal the regulation of mitochondrial SO_2 in HeLa cells.

Two kinds of probes for SO_2 derivatives with different specific functionalities were introduced. The probe SP-2 (**105**) was designed as a phosphorescent probe derived from a dinuclear iridium(III) complex linked by an azo group that served as the responsive site for SO_2 derivatives.[180] It was reported that probe **105** exhibits a turn-on signal with a long lifetime towards SO_3^{2-} and HSO_3^-. Moreover, both extraneous and endogenous SO_3^{2-}/HSO_3^- could be identified in living cells. The other, the probe TAPQC (**106**), is characterized by a fast responsiveness (within 15 s) and excellent water solubility

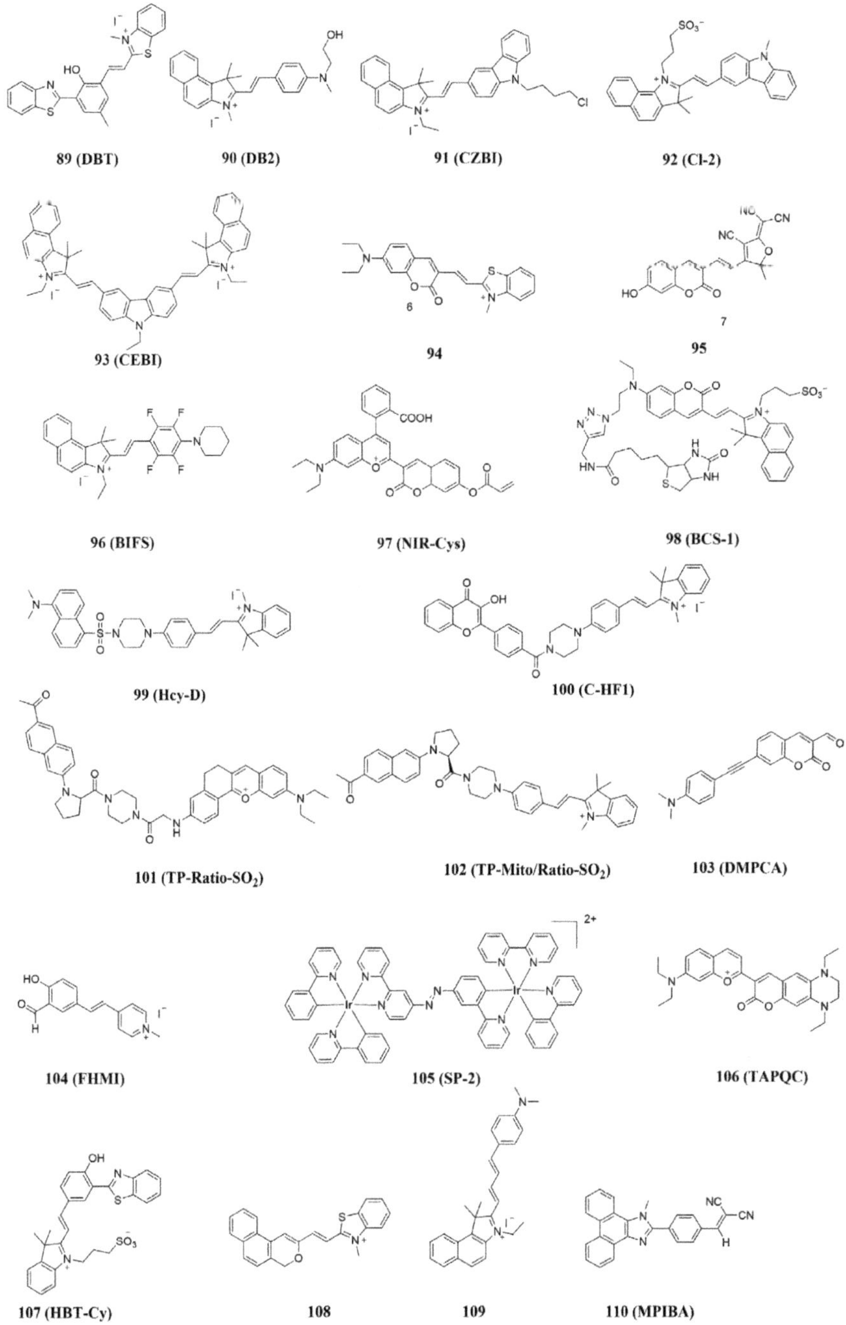

Figure 8.11 Structures of fluorescent probes for SO_2 derivatives.

ascribed to the benzopyrylium moiety.[181] It is important to note that probe **106** is capable of the real-time sensing and bioimaging of HSO_3^- in living A549 cells and HepG2 cells.

In addition, several reversible probes for HSO_3^- have been exploited. For example, the probe HBT-Cy (**107**), with conjugation of benzothiazole and a sulfonated cyanine dye, was synthesized with a ratiometric response mode.[182] On reaction with HSO_3^-, the high-contrast blue emission from the HBT moiety of probe **107** dominated. The resulting adduct of HSO_3^- and HBT-Cy dissociated to the original structure (HBT-Cy) again on treatment with H_2O_2, accompanied by native red emission with a high yield. Subsequent fluorescence imaging experiments indicated that this probe could sense HSO_3^- reversibly both in living cells and in aqueous solution. Additionally, probe **108**, based on the conjugation of naphthopyran and benzothiazolium, was applied to detect the bisulfate–H_2O_2 redox cycle in live cells and zebrafish.[183]

Finally, two multi-detection probes are presented. First, SO_3^{2-} and SO_4^{2-}/HSO_4^- were detected simultaneously at different emission wavelengths.[184] The non-fluorescent probe **109** was turned on at 442 nm because of the shortening of the conjugated structure, which led to the restriction of ICT on addition of SO_3^{2-}. For the detection of SO_4^{2-}/HSO_4^-, the AIE mechanism was realized to turn on the fluorescence of the aggregate induced by SO_4^{2-}/HSO_4^- at 511 nm. Importantly this probe is capable of differentiating intracellular SO_3^{2-} and SO_4^{2-} in live HeLa cells. The probe MPIBA (**110**), consisting of phenanthroimidazole and a benzylidene malononitrile moiety designed as the nucleophilic attack site, was developed to detect SO_2 and ClO^- simultaneously.[185] Specifically, the nucleophilic addition of SO_2 turns on the emission at 410 nm within 60 s and ClO^- oxidizes the SO_2-mediated adduct and then gives rise to emission at 500 nm within seconds. This probe was used to detect endogenous SO_2 and ClO^- in RAW264.7 cells and HeLa cells. According to the results from the visualization of SO_2/ClO^- homeostasis, the oxidative and antioxidative effects of intracellular SO_2 were confirmed.

Acknowledgement

Dr Xu Wang and Fanpeng Kong are thanked for expert assistance with the writing of this chapter.

References

1. D. Procházková, N. Wilhelmová and M. Pavlík, Reactive Nitrogen Species and Nitric Oxide, in *Nitric Oxide Action in Abiotic Stress Responses in Plants*, ed. M. Khan, M. Mobin, F. Mohammad and F. Corpas, Springer, Cham, Switzerland, 2015.
2. M. H. Zaki, T. Okamoto, T. Sawa, S. Fujii and T. Akaike, *Clin. Exp. Allergy Rev.*, 2007, 7, 19.

3. A. Weidinger and A. V. Kozlov, *Biomolecules*, 2015, **5**, 472.
4. X. Y. Jiao, Y. Li, J. Y. Niu, X. L. Xie, X. Wang and B. Tang, *Anal. Chem.*, 2018, **90**, 533.
5. N. Lau and M. D. Pluth, *Curr. Opin. Chem. Biol.*, 2019, **49**, 1.
6. H. M. Kim and B. R. Cho, *Chem. Rev.*, 2015, **115**, 5014.
7. Z. R. Lou, P. Li and K. L. Han, *Acc. Chem. Res.*, 2015, **48**, 1358.
8. A. C. Sedgwick, L. L. Wu, H. H. Han, S. D. Bull, X. P. He, T. D. James, J. L. Sessler, B. Z. Tang, H. Tian and J. Yoon, *Chem. Soc. Rev.*, 2018, **47**, 8842.
9. P. F. Bove and A. van der Vliet, *Free Radical Biol. Med.*, 2006, **41**, 515–527.
10. X. Chen, F. Wang, J. Y. Hyun, T. Wei, J. Qiang, X. Ren, I. Shin and J. Yoon, *Chem. Soc. Rev.*, 2016, **45**, 2976–3016.
11. N. Paolocci, T. Katori, H. C. Champion, M. E. St. John, K. M. Miranda, J. M. Fukuto, D. A. Wink and D. A. Kass, *Proc. Natl. Acad. Sci. U. S. A.*, 2003, **100**, 5537–5542.
12. R. H. Ritchie, N. Cao, Y. G. Wong, S. Rosli, H. Kiriazis, K. Huynh, C. Qin, X. J. Du and B. K. Kemp-Harper, *Circulation*, 2014, **130**, 16485.
13. L. E. McQuade and S. J. Lippard, *Curr. Opin. Chem. Biol.*, 2010, **14**, 43–49.
14. G. J. Mao, X. B. Zhang, X. L. Shi, H. W. Liu, Y. X. Wu, L. Y. Zhou, W. Tan and R. Q. Yu, *Chem. Commun.*, 2014, **50**, 5790–5792.
15. C. Liu, H. Wu, Z. Wang, C. Shao, B. Zhu and X. Zhang, *Chem. Commun.*, 2014, **50**, 6013–6016.
16. X. Jin, X. Sun, X. Di, X. Zhang, H. Huang, J. Liu, P. Ji and H. Zhu, *Sens. Actuators, B*, 2016, **224**, 209–216.
17. C. Liu, Z. Cao, Z. Wang, P. Jia, J. Liu, Z. Wang, B. Han, X. Huang, X. Li, B. Zhu and X. Zhang, *Sens. Actuators, B*, 2015, **220**, 27–733.
18. Z. Miao, J. A. Reisz, S. M. Mitroka, J. Pan, M. Xian and S. B. King, *Bioorg. Med. Chem. Lett.*, 2015, **25**, 16–19.
19. K. Zheng, W. Lin, D. Cheng, H. Chen, Y. Liu and K. Liu, *Chem. Commun.*, 2015, **51**, 5754–5757.
20. H. Li, Q. Yao, F. Xu, N. Xu, X. Ma, J. Fan, S. Long, J. Du, J. Wang and X. Peng, *Anal. Chem.*, 2018, **90**, 4641–4648.
21. Y. Tan, R. Liu, H. Zhang, R. Peltier, Y. W. Lam, Q. Zhu, Y. Hu and H. Sun, *Sci. Rep.*, 2015, **5**, 16979.
22. X. Gong, X. F. Yang, Y. Zhong, Y. Chen and Z. Li, *Dyes Pigm.*, 2016, **131**, 24–32.
23. B. Dong, X. Song, X. Kong, C. Wang, N. Zhang and W. Lin, *J. Mater. Chem. B*, 2017, **5**, 5218–5224.
24. X. Jing, F. Yu and L. Chen, *Chem. Commun.*, 2014, **50**, 14253–14256.
25. M. Ren, B. Deng, K. Zhou, J. Y. Wang, X. Kong and W. Lin, *J. Mater. Chem. B*, 2017, **5**, 1954–1961.
26. J. Rosenthal and S. J. Lippard, *J. Am. Chem. Soc.*, 2010, **132**, 5536–5537.

27. Y. Zhou, Y. W. Yao, J. Y. Li, C. Yao and B. P. Lin, *Sens. Actuators, B*, 2012, **174**, 414–420.
28. Y. Zhou, K. Liu, J. Y. Li, Y. Fang, T. C. Zhao and C. Yao, *Org. Lett.*, 2011, **13**, 1290–1293.
29. A. T. Wrobel, T. C. Johnstone, A. D. Liang, S. J. Lippard and P. Rivera-Fuentes, *J. Am. Chem. Soc.*, 2014, **136**, 4697–4705.
30. U.-P. Apfel, D. Buccella, J. J. Wilson and S. J. Lippard, *Inorg. Chem.*, 2013, **52**, 3285–3294.
31. P. N. Coneski and M. H. Schoenfisch, *Chem. Soc. Rev.*, 2012, **41**, 3753–3758.
32. Y. Yang, S. K. Seidlits, M. M. Adams, V. M. Lynch, C. E. Schmidt, E. V. Anslyn and J. B. Shear, *J. Am. Chem. Soc.*, 2010, **132**, 13114–13116.
33. C. G. Dai, J. L. Wang, Y. L. Fu, H. P. Zhou and Q. H. Song, *Anal. Chem.*, 2017, **89**, 10511–10519.
34. X. Lv, Y. Wang, S. Zhang, Y. Liu, J. Zhang and W. Guo, *Chem. Commun.*, 2014, **50**, 7499–7502.
35. T. Nagano and T. Yoshimura, *Chem. Rev.*, 2002, **102**, 1235–1270.
36. E. Sasaki, H. Kojima, H. Nishimatsu, Y. Urano, K. Kikuchi, Y. Hirata and T. Nagano, *J. Am. Chem. Soc.*, 2005, **127**, 3684–3685.
37. H. X. Zhang, J. B. Chen, X. F. Guo, H. Wang and H. S. Zhang, *Anal. Chem.*, 2014, **86**, 3115–3123.
38. L. Liu, F. Zhang, B. Xu and W. Tian, *J. Mater. Chem. B*, 2017, **5**, 9197–9203.
39. J. Tang, Z. Guo, Y. Zhang, B. Bai and W. H. Zhu, *Chem. Commun.*, 2017, **53**, 10520–10523.
40. Q. Wang, X. Jiao, C. Liu, S. He, L. Zhao and X. Zeng, *J. Mater. Chem. B*, 2018, **6**, 4096–4103.
41. Z. Mao, W. Feng, Z. Li, L. Zeng, W. Lv and Z. Liu, *Chem. Sci.*, 2016, **7**, 5230–5235.
42. S. J. Li, D. Y. Zhou, Y. Li, H. W. Liu, P. Wu, J. O. Yang, W. L. Jiang and C. Y. Li, *ACS Sens.*, 2018, **3**, 2311–2319.
43. F. Wang, S. Yu, Z. Xu, L. Li, Y. Dang, X. Xu, Y. Luo, Z. Cheng, H. Yu, W. Zhang, A. Zhang and C. Ding, *Anal. Chem.*, 2018, **90**, 7953–7962.
44. H. Takakura, R. Kojima, M. Kamiya, E. Kobayashi, T. Komatsu, T. Ueno, T. Terai, K. Hanaoka, T. Nagano and Y. Urano, *J. Am. Chem. Soc.*, 2015, **137**, 4010–4013.
45. Y. Huo, J. Miao, L. Han, Y. Li, Z. Li, Y. Shi and W. Guo, *Chem. Sci.*, 2017, **8**, 6857–6864.
46. J. Miao, Y. Huo, X. Lv, Z. Li, H. Cao, H. Shi, Y. Shi and W. Guo, *Biomaterials*, 2016, **78**, 11–19.
47. Z. Mao, H. Jiang, X. Song, W. Hu and Z. Liu, *Anal. Chem.*, 2017, **89**, 9620–9624.
48. C. J. Reinhardt, E. Y. Zhou, M. D. Jorgensen, G. Partipilo and J. Chan, *J. Am. Chem. Soc.*, 2018, **140**, 1011–1018.
49. Z. Mao, H. Jiang, Z. Li, C. Zhong, W. Zhang and Z. Liu, *Chem. Sci.*, 2017, **8**, 4533–4538.

50. P. Pacher, J. S. Beckman and L. Liaudet, *Physiol. Rev.*, 2007, **87**, 315–424.
51. X. Xie, F. Tang, G. Liu, Y. Li, X. Su, X. Jiao, X. Wang and B. Tang, *Anal. Chem.*, 2018, **90**, 11629.
52. Y. Li, X. Xie, X. Yang, M. Li, X. Jiao, Y. Sun, X. Wang and B. Tang, *Chem. Sci.*, 2017, **8**, 4006–4011.
53. J. Li, L. Chen, Q. Wang, H. W. Liu, X. X. Hu, L. Yuan and X. B. Zhang, *Anal. Chem.*, 2018, **90**, 4167–4173.
54. D. Cheng, W. Xu, L. Yuan and X. Zhang, *Anal. Chem.*, 2017, **89**, 7693–7700.
55. E. W. Miller, A. E. Albers, A. Pralle, E. Y. Isacoff and C. J. Chang, *J. Am. Chem. Soc.*, 2005, **127**, 16652–16659.
56. M. Weber, A. B. Mackenzie, S. D. Bull and T. D. James, *Anal. Chem.*, 2018, **90**, 10621–10627.
57. L. Wu, Y. Wang, M. Weber, L. Liu, A. C. Sedgwick, S. D. Bull, C. Huang and T. D. James, *Chem. Commun.*, 2018, **54**, 9953–9956.
58. A. C. Sedgwick, H. H. Han, J. E. Gardiner, S. D. Bull, X. P. He and T. D. James, *Chem. Commun.*, 2017, **53**, 12822–12825.
59. J. Zhang, Y. Li, J. Zhao and W. Guo, *Sens. Actuators, B*, 2016, **237**, 67–74.
60. P. Cheng, J. Zhang, J. Huang, Q. Miao, C. Xu and K. Pu, *Chem. Sci.*, 2018, **9**, 6340–6347.
61. D. Wu, J. C. Ryu, Y. W. Chung, D. Lee, J. H. Ryu, J. H. Yoon and J. Yoon, *Anal. Chem.*, 2017, **89**, 10924–10931.
62. H. Li, X. Li, X. Wu, W. Shi and H. Ma, *Anal. Chem.*, 2017, **89**, 5519–5525.
63. J. T. Hou, J. Yang, K. Li, Y. X. Liao, K. K. Yu, Y. M. Xie and X. Q. Yu, *Chem. Commun.*, 2014, **50**, 9947–9950.
64. Z. Li, S. H. Yan, C. Chen, Z. R. Geng, J. Y. Chang, C. X. Chen, B. H. Huang and Z. L. Wang, *Biosens. Bioelectron.*, 2017, **90**, 75–82.
65. D. Y. Zhou, Y. Li, W. L. Jiang, Y. Tian, J. Fei and C. Y. Li, *Chem. Commun.*, 2018, **54**, 11590–11593.
66. H. Zhang, J. Liu, Y. Q. Sun, Y. Huo, Y. Li, W. Liu, X. Wu, N. Zhu, Y. Shi and W. Guo, *Chem. Commun.*, 2015, **51**, 2721–2724.
67. A. C. Sedgwick, X. Sun, G. Kim, J. Yoon, S. D. Bull and T. D. James, *Chem. Commun.*, 2016, **52**, 12350–12352.
68. T. Peng, X. Chen, L. Gao, T. Zhang, W. Wang, J. Shen and D. Yang, *Chem. Sci.*, 2016, **7**, 5407–5413.
69. J. Miao, Y. Huo, Q. Liu, Z. Li, H. Shi, Y. Shi and W. Guo, *Biomaterials*, 2016, **107**, 33–43.
70. J. Mia, Y. Huo, H. Shi, J. Fang, J. Wang and W. Guo, *J. Mater. Chem. B*, 2018, **6**, 4466–4473.
71. K. Xu, H. Chen, J. Tian, B. Ding, Y. Xie, M. Qiang and B. Tang, *Chem. Commun.*, 2011, **47**, 9468–9470.
72. F. Yu, P. Li, G. Li, G. Zhao, T. Chu and K. Han, *J. Am. Chem. Soc.*, 2011, **133**, 11030–11033.
73. F. Yu, P. Li, B. Wang and K. Han, *J. Am. Chem. Soc.*, 2013, **135**, 7674–7680.

74. S. V. Mulay, Y. Kim, K. J. Lee, T. Yudhistira, H. S. Park and D. G. Churchill, *New J. Chem.*, 2017, **41**, 11934–11940.
75. J. Cao, W. An, A. G. Reeves and A. R. Lippert, *Chem. Sci.*, 2018, **9**, 2552–2558.
76. N. M. Giles, A. B. Watts, G. I. Giles, F. H. Fry, J. A. Littlechild and C. Jacob, *Chem. Biol.*, 2003, **10**, 677.
77. W. A. Kleinman and J. P. Richie, *Biochem. Pharmacol.*, 2000, **60**, 19.
78. K. Umezawa, M. Yoshida, M. Kamiya, T. Yamasoba and Y. Urano, *Nat. Chem.*, 2016, **9**, 279.
79. Z. Liu, X. Zhou, Y. Miao, Y. Hu, N. Kwon, X. Wu and J. Yoon, *Angew. Chem.*, 2017, **129**, 5906.
80. J.-Y. Xie, C.-Y. Li, Y.-F. Li, J. Fei, F. Xu, J. Ou-Yang and J. Liu, *Anal. Chem.*, 2016, **88**, 9746.
81. L.-Y. Niu, Y.-S. Guan, Y.-Z. Chen, L.-Z. Wu, C.-H. Tung and Q.-Z. Yang, *J. Am. Chem. Soc.*, 2012, **134**, 18928.
82. J. Yin, Y. Kwon, D. Kim, D. Lee, G. Kim, Y. Hu, J.-H. Ryu and J. Yoon, *J. Am. Chem. Soc.*, 2014, **136**, 5351.
83. S.-Y. Lim, K.-H. Hong, D. I. Kim, H. Kwon and H.-J. Kim, *J. Am. Chem. Soc.*, 2014, **136**, 7018.
84. M. Cao, H. Chen, D. Chen, Z. Xu, S. H. Liu, X. Chen and J. Yin, *Chem. Commun.*, 2016, **52**, 721.
85. L. Palego, L. Betti and G. Giannaccini, *Biochem. Pharmacol.*, 2015, **4**, 1.
86. S. Shahrokhian, *Anal. Chem.*, 2001, **73**, 5972.
87. H. Tong, J. Zhao, X. Li, Y. Zhang, S. Ma, K. Lou and W. Wang, *Chem. Commun.*, 2017, **53**, 3583.
88. D. Kand, A. M. Kalle, S. J. Varma and P. Talukdar, *Chem. Commun.*, 2012, **48**, 2722.
89. T. Liu, F. Huo, J. Li, J. Chao, Y. Zhang and C. Yin, *Sens. Actuators, B*, 2016, **237**, 127.
90. J. Zhang, J. Wang, J. Liu, L. Ning, X. Zhu, B. Yu, X. Liu, X. Yao and H. Zhang, *Anal. Chem.*, 2015, **87**, 4856.
91. Z. Guo, S. Nam, S. Park and J. Yoon, *Chem. Sci.*, 2012, **3**, 2760.
92. Y. Yue, F. Huo, P. Ning, Y. Zhang, J. Chao, X. Meng and C. Yin, *J. Am. Chem. Soc.*, 2017, **139**, 3181.
93. Z. Yang, N. Zhao, Y. Sun, F. Miao, Y. Liu, X. Liu, Y. Zhang, W. Ai, G. Song, X. Shen, X. Yu, J. Sun and W.-Y. Wong, *Chem. Commun.*, 2012, **48**, 3442.
94. X. Liu, N. Xi, S. Liu, Y. Ma, H. Yang, H. Li, J. He, Q. Zhao, F. Li and W. Huang, *J. Mater. Chem.*, 2012, **22**, 7894.
95. D. Gong, Y. Tian, C. Yang, A. Iqbal, Z. Wang, W. Liu, W. Qin, X. Zhu and H. Guo, *Biosens. Bioelectron.*, 2016, **85**, 178.
96. S. Seshadri, A. Beiser, J. Selhub, P. F. Jacques, I. H. Rosenberg, R. B. D'Agostino, P. W. F. Wilson and P. A. Wolf, *N. Engl. J. Med.*, 2002, **346**, 476.
97. Y.-W. Wang, S.-B. Liu, W.-J. Ling and Y. Peng, *Chem. Commun.*, 2016, **52**, 827.

98. H. Peng, K. Wang, C. Dai, S. Williamson and B. Wang, *Chem. Commun.*, 2014, **50**, 13668.
99. H. Y. Lee, Y. P. Choi, S. Kim, T. Yoon, Z. Guo, S. Lee, K. M. Swamy, G. Kim, J. Y. Lee, I. Shin and J. Yoon, *Chem. Commun.*, 2014, **50**, 6967.
100. J. Liu, Y.-Q. Sun, Y. Huo, H. Zhang, L. Wang, P. Zhang, D. Song, Y. Shi and W. Guo, *J. Am. Chem. Soc.*, 2014, **136**, 574.
101. Y. Li, W. Liu, P. Zhang, H. Zhang, J. Wu, J. Ge and P. Wang, *Biosens. Bioelectron.*, 2017, **90**, 117.
102. K. Wang, T. Leng, Y. Liu, C. Wang, P. Shi, Y. Shen and W.-H. Zhu, *Sens. Actuators, B*, 2017, **248**, 338.
103. H. Chen, Y. Tang, M. Ren and W. Lin, *Chem. Sci.*, 2016, **7**, 1896.
104. J. Zhang, A. Shibata, M. Ito, S. Shuto, Y. Ito, B. Mannervik, H. Abe and R. Morgenstern, *J. Am. Chem. Soc.*, 2011, **133**, 14109.
105. W. Chen, H. Luo, X. Liu, J. W. Foley and X. Song, *Anal. Chem.*, 2016, **88**, 3638.
106. X. Xie, M. Li, F. Tang, Y. Li, L. Zhang, X. Jiao, X. Wang and B. Tang, *Anal. Chem.*, 2017, **89**, 3015.
107. H. Zhang, R. Liu, J. Liu, L. Li, P. Wang, S. Q. Yao, Z. Xu and H. Sun, *Chem. Sci.*, 2016, **7**, 256.
108. M. H. Lee, J. H. Han, P.-S. Kwon, S. Bhuniya, J. Y. Kim, J. L. Sessler, C. Kang and J. S. Kim, *J. Am. Chem. Soc.*, 2012, **134**, 1316.
109. C. S. Lim, G. Masanta, H. J. Kim, J. H. Han, H. M. Kim and B. R. Cho, *J. Am. Chem. Soc.*, 2011, **133**, 11132.
110. J. H. Lee, C. S. Lim, Y. S. Tian, J. H. Han and B. R. Cho, *J. Am. Chem. Soc.*, 2010, **132**, 1216.
111. M. M. Pires and J. Chmielewski, *Org. Lett.*, 2008, **10**, 837.
112. L. Li, P. Rose and P. K. Moore, *Annu. Rev. Pharmacol. Toxicol.*, 2011, **51**, 169.
113. D. C. Dorman, F. J.-M. Moulin, B. E. McManus, K. C. Mahle, R. A. James and M. F. Struve, *Toxicol. Sci.*, 2002, **65**, 18.
114. J. L. Wallace, *Trends Pharmacol. Sci.*, 2007, **28**, 501.
115. Y.-H. Liu, M. Lu, L.-F. Hu, P. T.-H. Wong, G. D. Webb and J.-S. Bian, *Antioxid. Redox Signaling*, 2012, **17**, 141.
116. A. Papapetropoulos, A. Pyriochou, Z. Altaany, G. Yang, A. Marazioti, Z. Zhou, M. G. Jeschke, L. K. Branski, D. N. Herndon and R. Wang, *Proc. Natl. Acad. Sci. U. S. A.*, 2009, **106**, 21972.
117. K. Módis, Y. Ju, A. Ahmad, A. A. Untereiner, Z. Altaany, L. Wu, C. Szabo and R. Wang, *Pharmacol. Res.*, 2016, **113**, 116.
118. B. D. Paul, J. I. Sbodio, R. Xu, M. S. Vandiver, J. Y. Cha, A. M. Snowman and S. H. Snyder, *Nature*, 2014, **509**, 96.
119. L. F. Hu, M. Lu, C. X. Tiong, G. S. Dawe, G. Hu and J. S. Bian, *Aging Cell*, 2010, **9**, 135.
120. D. Giuliani, A. Ottani, D. Zaffe, M. Galantucci, F. Strinati, R. Lodi and S. Guarini, *Neurobiol. Learn. Mem.*, 2013, **104**, 82.
121. X. Feng, T. Zhang, J.-T. Liu, J.-Y. Miao and B.-X. Zhao, *Chem. Commun.*, 2016, **52**, 3131.

122. M. Ren, B. Deng, X. Kong, K. Zhou, K. Liu, G. Xu and W. Lin, *Chem. Commun.*, 2016, **52**, 6415.
123. Y. Liu, F. Meng, L. He, K. Liu and W. Lin, *Chem. Commun.*, 2016, **52**, 7016.
124. L. He, X. Yang, K. Xu, X. Kong and W. Lin, *Chem. Sci.*, 2017, **8**, 6257.
125. H. Li, W. Peng, W. Feng, Y. Wang, G. Chen, S. Wang, S. Li, H. Li, K. Wang and J. Wang, *Chem. Commun.*, 2016, **52**, 4628.
126. Y. L. Pak, J. Li, K. C. Ko, G. Kim, J. Y. Lee and J. Yoon, *Anal. Chem.*, 2016, **88**, 5476.
127. J. Kang, F. Huo, P. Ning, X. Meng, J. Chao and C. Yin, *Sens. Actuators, B*, 2017, **250**, 342.
128. K. Zhang, J. Zhang, Z. Xi, L.-Y. Li, X. Gu, Q.-Z. Zhang and L. Yi, *Chem. Sci.*, 2017, **8**, 2776.
129. X. Jin, S. Wu, M. She, Y. Jia, L. Hao, B. Yin, L. Wang, M. Obst, Y. Shen and Y. Zhang, *Anal. Chem.*, 2016, **88**, 11253.
130. E. Karakus, M. Üçüncü and M. Emrullahoğlu, *Anal. Chem.*, 2016, **88**, 1039.
131. S. I. Reja, N. Sharma, M. Gupta, P. Bajaj, V. Bhalla, R. D. Parihar, P. Ohri, G. Kaur and M. Kumar, *Chem. – Eur. J.*, 2017, **23**, 9872.
132. J. Lv, F. Wang, J. Qiang, X. Ren, Y. Chen, Z. Zhang, Y. Wang, W. Zhang and X. Chen, *Biosens. Bioelectron.*, 2017, **87**, 96.
133. Z. Wu, D. Liang and X. Tang, *Anal. Chem.*, 2016, **88**, 9213.
134. N. Velusamy, A. Binoy, K. N. Bobba, D. Nedungadi, N. Mishra and S. Bhuniya, *Chem. Commun.*, 2017, **53**, 8802.
135. B. Deng, M. Ren, J.-Y. Wang, K. Zhou and W. Lin, *Sens. Actuators, B*, 2017, **248**, 50.
136. A. K. Steiger, S. Pardue, C. G. Kevil and M. D. J. Pluth, *J. Am. Chem. Soc.*, 2016, **138**, 7256.
137. S. Wang, S. Xu, G. Hu, X. Bai, T. D. James and L. Wang, *Anal. Chem.*, 2016, **88**, 1434.
138. L. Chen, D. Wu, C. S. Lim, D. Kim, S.-J. Nam, W. Lee, G. Kim, H. M. Kim and J. Yoon, *Chem. Commun.*, 2017, **53**, 4791.
139. D. Gong, X. Zhu, Y. Tian, S.-C. Han, M. Deng, A. Iqbal, W. Liu, W. Qin and H. Guo, *Anal. Chem.*, 2017, **89**, 1801.
140. J. I. Toohey, *Anal. Biochem.*, 2011, **413**, 1.
141. V. Vitvitsky, O. Kabil and R. Banerjee, *Antioxid. Redox Signaling*, 2012, **17**, 22.
142. R. Greiner, Z. Pálinkás, K. Bäsell, D. Becher, H. Antelmann, P. Nagy and T. P. Dick, *Antioxid. Redox Signaling*, 2013, **19**, 1749.
143. W. Chen, A. Pacheco, Y. Takano, J. J. Day, K. Hanaoka and M. Xian, *Angew. Chem., Int. Ed.*, 2016, **55**, 9993.
144. Y. Fang, W. Chen, W. Shi, H. Li, M. Xian and H. Ma, *Chem. Commun.*, 2017, **53**, 8759.
145. Q. Han, Z. Mou, H. Wang, X. Tang, Z. Dong, L. Wang, X. Dong and W. Liu, *Anal. Chem.*, 2016, **88**, 7206.

146. J. Zhang, X.-Y. Zhu, X.-X. Hu, H.-W. Liu, J. Li, L.-L. Feng, X. Yin, X.-B. Zhang and W. Tan, *Anal. Chem.*, 2016, **88**, 11892.
147. K.-B. Li, F.-Z. Chen, Q.-H. Yin, S. Zhang, W. Shi and D.-M. Han, *Sens. Actuators, B*, 2018, **254**, 222.
148. Y. Takano, K. Hanaoka, K. Shimamoto, R. Miyamoto, T. Komatsu, T. Ueno, T. Terai, H. Kimura, T. Nagano and Y. Urano, *Chem. Commun.*, 2017, **53**, 1064.
149. Y. Huang, F. Yu, J. Wang and L. Chen, *Anal. Chem.*, 2016, **88**, 4122.
150. T. Ida, T. Sawa, H. Ihara, Y. Tsuchiya, Y. Watanabe, Y. Kumagai, M. Suematsu, H. Motohashi, S. Fujii and T. Matsunaga, *Proc. Natl. Acad. Sci. U. S. A.*, 2014, **111**, 7606.
151. J. L. Luebke, J. Shen, K. E. Bruce, T. E. Kehl-Fie, H. Peng, E. P. Skaar and D. P. Giedroc, *Mol. Microbiol.*, 2014, **94**, 1343.
152. X. Han, F. Yu, X. Song and L. Chen, *Chem. Sci.*, 2016, **7**, 5098.
153. R. Kawagoe, I. Takashima, S. Uchinomiya and A. Ojida, *Chem. Sci.*, 2017, **8**, 1134.
154. Z. Q. Meng, G. H. Qin, B. Zhang and J. L. Bai, *Mutagenesis*, 2004, **19**, 465.
155. S. Iwasawa, Y. Kikuchi, Y. Nishiwaki, M. Nakano, T. Michikawa and T. Tsuboi, *J. Occup. Health*, 2009, **51**, 38.
156. N. Sang, Y. Yun, H. Li, L. Hou, M. Han and G. K. Li, *Toxicol. Sci.*, 2010, **114**, 226.
157. G. K. Li and N. Sang, *Ecotoxicol. Environ. Saf.*, 2009, **72**, 236.
158. L. Migliore and F. Coppede, *Mutat. Res.*, 2009, **674**, 73.
159. N. Sang, Y. Yun, G. Yao, H. Li, L. Guo and G. Li, *Toxicol. Sci.*, 2011, **124**, 400.
160. Z. Meng, H. Geng, J. Bai and G. Yan, *Inhalation Toxicol.*, 2003, **15**, 951.
161. J. Li and Z. Meng, *Nitric oxide*, 2009, **20**, 166.
162. J. Li, R. Li and Z. Meng, *Eur. J. Pharmacol.*, 2010, **645**, 143.
163. M. H. Stipanuk, J. E. Dominy Jr, J. Lee and R. M. Coloso, *J. Nutr.*, 2006, **136**, 1652.
164. L. Tang, P. He, X. Yan, J. Sun, K. Zhong, S. Hou and Y. Bian, *Sens. Actuators, B*, 2017, **247**, 421.
165. D. Li, Y. Shi, X. Tian, M. Wang, B. Huang, F. Li, S. Li, H. Zhou, J. Wu and Y. Tian, *Sens. Actuators, B*, 2016, **233**, 1.
166. J. Xu, J. Pan, X. Jiang, C. Qin, L. Zeng, H. Zhang and J. F. Zhang, *Biosens. Bioelectron.*, 2016, **77**, 725.
167. Y. Liu, K. Li, K. X. Xie, L. L. Li, K. K. Yu, X. Wang and X. Q. Yu, *Chem. Commun.*, 2016, **52**, 3430.
168. W. Xu, P. Ma, Q. Diao, L. Xu, X. Liu, Y. Sun, X. Wang and D. Song, *Sens. Actuators, B*, 2017, **252**, 86.
169. Y. Q. Sun, J. Liu, J. Zhang, T. Yang and W. Guo, *Chem. Commun.*, 2013, **49**, 2637.
170. M. Y. Wu, K. Li, C. Y. Li, J. T. Hou and X. Q. Yu, *Chem. Commun.*, 2014, **50**, 183.

171. L. Zhu, J. Xu, Z. Sun, B. Fu, C. Qin, L. Zeng and X. Hu, *Chem. Commun.*, 2015, **51**, 1154.
172. X. Yang, W. Liu, J. Tang, P. Li, H. Weng, Y. Ye, M. Xian, B. Tang and Y. Zhao, *Chem. Commun.*, 2018, **54**, 11387.
173. J. Yang, K. Li, J. T. Hou, L. L. Li, C. Y. Lu, Y. M. Xie, X. Wang and X. Q. Yu, *ACS Sens.*, 2016, **1**, 166.
174. D. P. Li, Z. Y. Wang, X. J. Cao, J. Cui, X. Wang, H. Z. Cui, J. Y. Miao and B. X. Zhao, *Chem. Commun.*, 2016, **52**, 2760.
175. D. P. Li, Z. Y. Wang, H. Su, J. Y. Miao and B. X. Zhao, *Chem. Commun.*, 2017, **53**, 577.
176. S. Yang, X. Wen, X. Yang, Y. Li, C. Guo, Y. Zhou, H. Li and R. H. Yang, *Anal. Chem.*, 2018, **90**, 14514.
177. X. Yang, Y. Zhou, X. Zhang, S. Yang, Y. Chen, J. Guo, X. Li, Z. Qing and R. Yang, *Chem. Commun.*, 2016, **52**, 10289.
178. S. Yu, X. Yang, Z. Shao, Y. Feng, X. Xi, R. Shao, Q. Guoand and X. Meng, *Sens. Actuators, B*, 2016, **235**, 362.
179. G. Chen, W. Zhou, C. Zhao, Y. Liu, T. Chen, Y. Li and B. Tang, *Anal. Chem.*, 2018, **90**, 12442.
180. G. Y. Li, Y. Chen, J. Q. Wang, Q. Lin, J. Zhao, L. N. Ji and H. Chao, *Chem. Sci.*, 2013, **4**, 4426.
181. W. Chen, Q. Fang, D. Yang, H. Zhang, X. Song and J. Foley, *Anal. Chem.*, 2015, **87**, 609.
182. Y. Zhang, L. Guan, H. Yu, Y. Yan, L. Du, Y. Liu, M. Sun, D. Huang and S. Wang, *Anal. Chem.*, 2016, **88**, 4426.
183. W. Zhang, T. Liu, F. Huo, P. Ning, X. Meng and C. Yin, *Anal. Chem.*, 2017, **89**, 8079.
184. S. Samanta, P. Dey, A. Ramesh and G. Das, *Chem. Commun.*, 2016, **52**, 10381.
185. K. Dou, Q. Fu, G. Chen, F. Yu, Y. Liu, Z. Cao, G. Li, X. Zhao, L. Xia, L. Chen, H. Wang and J. You, *Biomaterials*, 2017, **133**, 82.

CHAPTER 9

Imaging of the Tumour Microenvironment

9.1 Introduction

Intracellular pH plays a pivotal role in many biological processes, including cell growth, apoptosis, autophagy, ion transport, homeostasis and enzymatic activities.[1–6] The intracellular microenvironment shows different pH values; for example, cytosol is slightly basic (pH 7.2) whereas the interiors of lysosomes and endosomes are slightly acidic (pH 4.0–6.0). Studies indicated that the range of pH variation may exceed 0.1–0.2 units, which not only causes cardiopulmonary and neurological diseases, but may even endanger life.[7,8] However, abnormal cellular pH values are known features in some common diseases such as cancer and Alzheimer's disease.[9,10] Hence monitoring of pH fluctuations would afford important information about cellular analysis or diagnosis. Various systems have been developed for the detection of pH values, such as microelectrodes, nuclear magnetic resonance and optical methods.[11–13] Among them, fluorescence analysis has been extensively used owing to the high spatiotemporal resolution, non-destructive operation and real-time monitoring in cancer cells.[14,15]

Hypoxia is an inevitable physiological feature because the oxygen from the blood vessels is insufficient to support the rapid growth of tumour cells without limitation. As a result, the oxygen content in a tumour cell (1–5%) is lower than that in a normal cell (5–20%), which causes many related diseases. In some places, the oxygen concentration even falls to 0%.[16] Hypoxia features of tumour cells are thought to block the effective therapy of tumours and in some cases may contribute to failure to control the disease. Hence both before and during therapy, measurement of the hypoxic cell fraction in tumours could be of considerable clinical significance. Therefore, diverse invasive methods have been proposed for sensing the hypoxic cell fraction in

Detection Science Series No. 16
In Situ Analysis of Cellular Functional Molecules
By Huangxian Ju, Bo Tang and Lin Ding

Published by the Royal Society of Chemistry, www.rsc.org

tumours, such as oxygen electrodes, histomorphometric analysis and DNA strand breaks.[17] A simpler and easier and especially non-invasive method for identifying hypoxic cells is suggested to be the use of turn-on fluorescence with far-red and near-infrared (NIR) excitation and emission wavelengths.

Acidity and hypoxia are concurrent physiological conditions in common diseases such as cardiovascular disease, inflammation, ischaemia, stroke, chronic lung disease[18,19] and other diseases. This chapter focuses on fluorescent probes for the detection of acidity and hypoxia *in vivo* and their application to therapy monitoring in tumour areas.

9.2 Fluorescent Probes for Intracellular pH and Hypoxia Imaging

In order to implement the simultaneous detection of intracellular pH and hypoxia, Zhao and co-workers[20] designed a rhodamine B-based fluorescent probe (RML) by inserting a lysosome-targetable group for measuring lysosomal pH. The H-bonding interaction and the electrostatic attraction between the protonated morpholine N atom and the rhodamine lactam O atom could induce a change from a spirocyclic (non-fluorescent) form to a ring-opened (fluorescent) form in response to a pH change (Figure 9.1). The fluorescence intensity of RML at 583 nm increased more than 80-fold when the pH decreased from 7.4 to 4.0. Moreover, RML could selectively detect the pH change in lysosomes induced by bafilomycin A1 within HeLa cells.

Of course, single detection of a pH value is possible. Ge's group[21] developed a lysosome-targetable off–on type of pH sensor for the selective detection of cancer cells. Three Darrow Red-based pH probes (**1a–1c**, Figure 9.2) were demonstrated to show pH-inducible emissions at 580–850 nm upon excitation at 570 nm by the comparison of cellular imaging assays using HeLa, KB and V79 cells. Of these, the pyridinium-2-yl Darrow Red analogue **1a** with a pK_a of 2.4 was found to be a lysosome tracker for the optical identification and distinction of cancer cells from normal cells.

For the monitoring of hypoxia alone in tumour cells, Huang and co-workers[22] developed the fluorescent probe AZO-DCM by conjugating the NIR dicyanomethylene-4*H*-pyran dye DCM for detection of cytochrome P450 reductase and hypoxia in tumour cells (Figure 9.3). The NIR probe is based on an azo bond as probe quencher and hypoxic trigger with the fluorophore DCM. In the presence of NADPH, the probe showed high selectivity towards

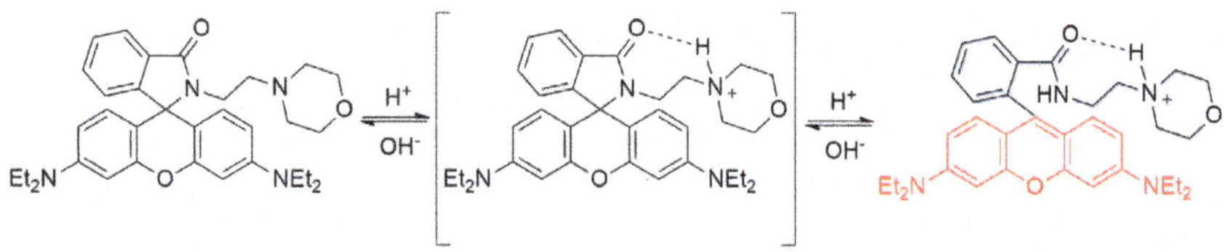

Figure 9.1 Spirolactam ring-opening process of RML upon addition of H^+.

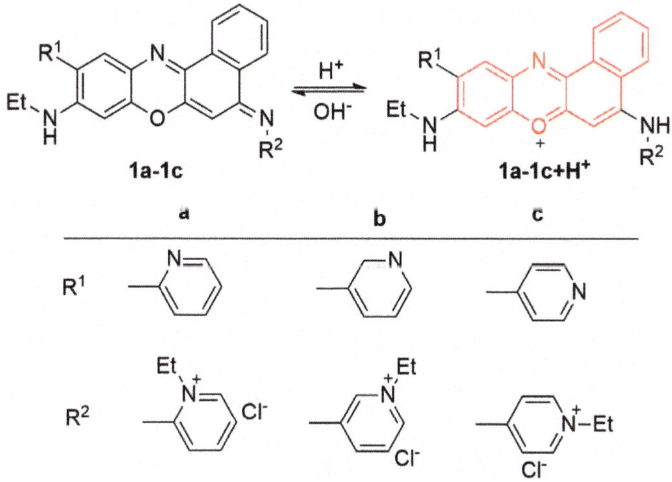

Figure 9.2 Detection mechanism of the Darrow Red-type pH probes (**1a–1c**).

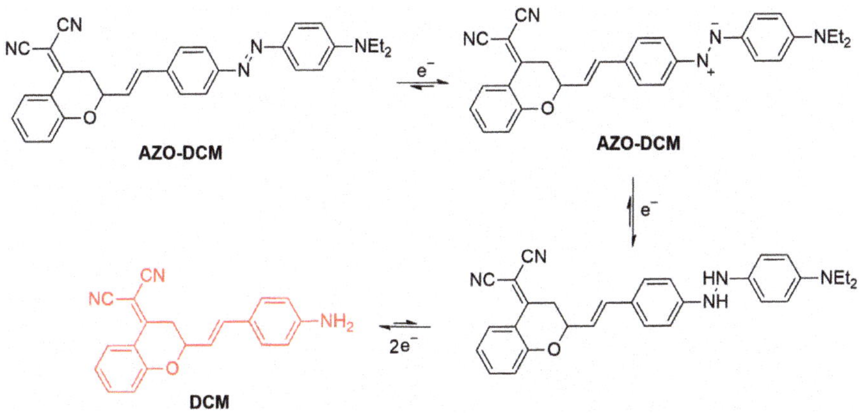

Figure 9.3 Mechanism of probe AZO-DCM for reductase and NADPH.

cytochrome P450 reductase under hypoxia with a rapid increase in fluorescence intensity of more than 150-fold, which meant that AZO-DCM could be a candidate for hypoxia imaging.

A common strategy for intracellular hypoxic detection in tumours is based on bioreduction of the nitro group. The nitro group quenches the fluorescence of the aromatic ring system, but the compound becomes more fluorescent after bioreduction of the nitro group by reduced nicotinamide adenine dinucleotide (NADH) in hypoxic cells.[23] Shao and co-workers[24] developed a turn-on BODIPY(a boron dipyrromethene) fluorescent probe by using a 5-nitrofuran moiety for the rapid detection of hypoxic status in tumour cells with the help of nitroreductase. The detection was based on reduction of the nitrofuran moiety catalysed by NADH as an electron donor, followed by 1,6-rearrangement–elimination and the release of free BODIPY dye **2** for fluorescent imaging (Figure 9.4). The probewas successfully applied to the real-time determination of the hypoxic status of A549 cells.

A naphthalimide fluorescent probe[25] with two heterocyclic side chains of 2-nitroimidazole for bioreductive binding as fluorescent markers was also designed for hypoxic cells *in vitro*. For further imaging the hypoxic status in tumour cells with an NIR fluorescent probe, Tang and co-workers[26] developed the NIR fluorescent probe Cy-NO_2 bearing a nitroimidazole group (Figure 9.5) for detecting hypoxic levels. The probe displays a highly selective and sensitive response of fluorescence enhancement towards hypoxia, due to the specific enzyme-catalysed reduction of the nitroimidazole moiety and the characteristics of the probe *per se*, such as a large Stokes shift (55 nm), an excellent fluorescence-quenched moiety (nitro group) and NIR excitation and emission wavelengths.

Based on the recognition site of 2-nitroimidazole with Indocyanine Green (ICG), several NIR probes[27] were developed for targeting hypoxia in tumour cells. Most pH-based probes rely on pH-dependent fluorescence intensity changes with single-emission windows.[28] The fluorescence signal output is possibly perturbed by the probe concentration, instrument sensitivity and environmental conditions.[29] Ratiometric probes, which respond to pH variations *via* the self-calibration of two emission bands, were developed to alleviate the above problems. They simultaneously record fluorescence

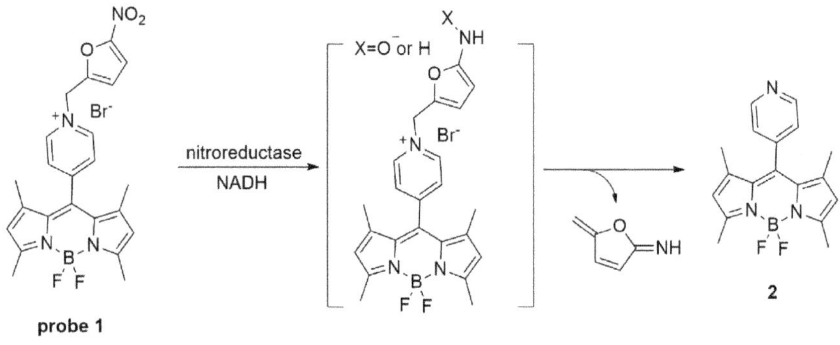

Figure 9.4 Reaction of probe **1** with nitroreductase for hypoxic imaging.

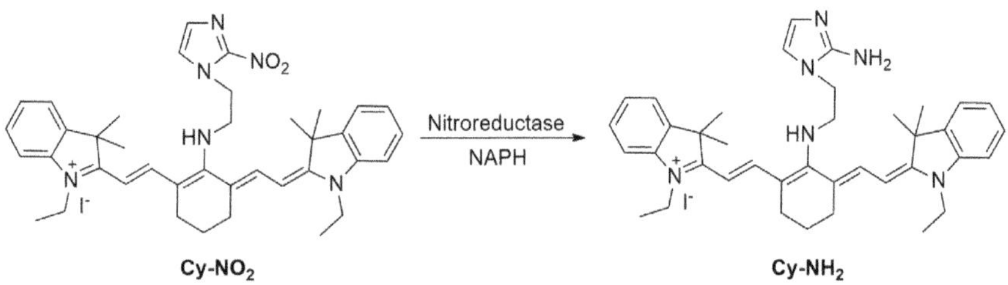

Figure 9.5 Reaction of Cy-NO_2 with nitroreductase.

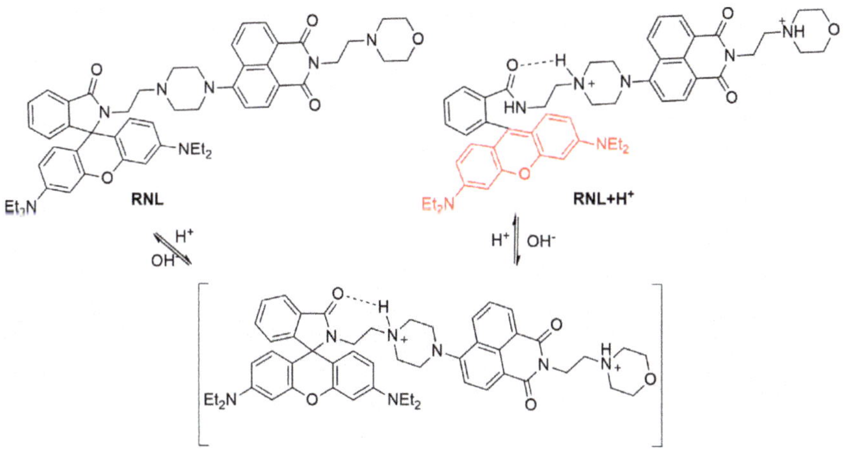

Figure 9.6 Mechanism of RNL for sensing H^+.

intensities at two wavelengths and calculate their ratios instead of a single wavelength,[30] which permits the quantitative measurement of pH even in complicated biological systems.

Mohr's group[28] developed the novel ratiometric pH probe RNL based on fluorescence resonance energy transfer (FRET) by integrating a naphthalimide moiety as a FRET donor with a rhodamine moiety as a FRET acceptor (Figure 9.6). With a pK_a of 4.82, the fluorescence intensity ratio (I_{529}/I_{580}) of the probe changed significantly within the pH range 4.50–5.50. This probe, with low cytotoxicity and satisfactory selectivity, could detect lysosomal pH changes in HeLa cells by a lysosome-locating group. In 2018, Zhao and co-workers[31] further used a naphthalimide moiety as a ratiometric fluorescence probe for the measurement of the acidic microenvironment and the concentration of nitroreductase at the same time and the same site. Wong and co-workers[32] also developed a carbazole–benzimidazole-based two-photon ratiometric fluorescence probe for the quantitative detection and imaging of lysosomal pH changes in the linearly proportional pH range 3.8–5.0 in live cells and living cancer tissues.

Qian and co-workers[33] reported the novel long-wavelength fluorescence probe NBP for the detection of hypoxia in the presence of nitroreductase (NTR) (Figure 9.7). NBP could be activated by NTR to release the NIR fluorophore NBF to detect tumour hypoxia in A549 cells.

9.3 pH and Hypoxia Activatable Tumour Imaging

Based on the hypoxia prodrug *p*-nitrobenzyl moiety, Qian and co-workers[34] designed a ratiometric fluorescent sensor (RHP) for the detection of the microenvironment of hypoxia in solid tumours (Figure 9.8). RHP can be selectively activated by bioreductive enzymes (NTR), resulting in a blue to green change in fluorescence emission wavelength both in solution phases and in cell lines.

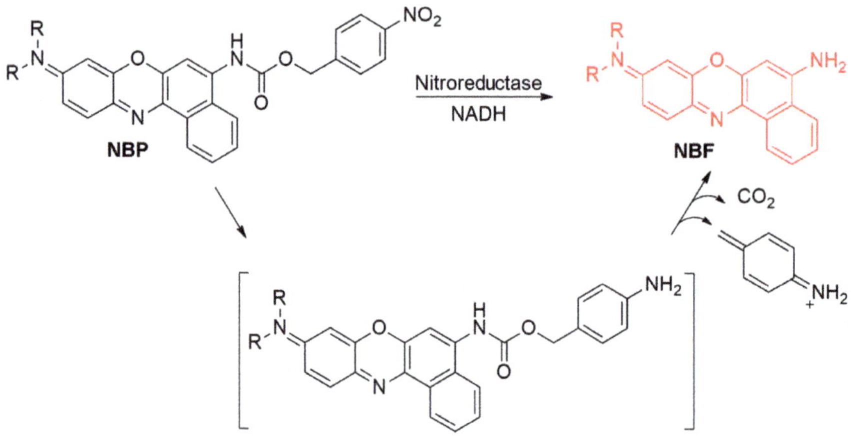

Figure 9.7 Mechanism of hypoxia probe NBP activated by NTR.

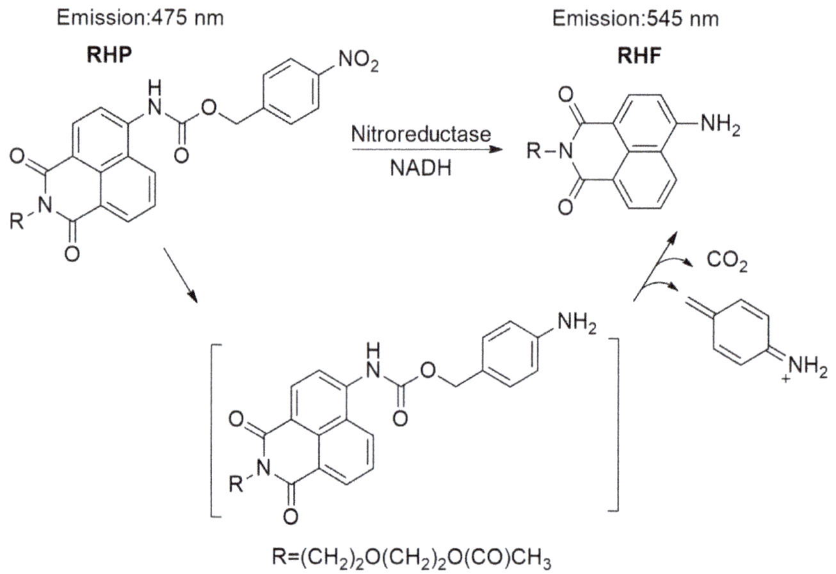

Figure 9.8 Detection mechanism of RHP under hypoxia.

Recently, Ning and co-workers[35] developed a diaminobenzene-based probe with the same recognition site for real-time monitoring of the hypoxia state in 2D cell monolayers, 3D tumour spheroids and even solid tumours in mice. Urano *et al.*[36] formulated pH-activatable reversible fluorescent probes to visualize cancer *in vivo* (Figure 9.9). The probes are almost non-fluorescent at neutral pH and highly fluorescent at acidic pH. Another advantage of these probes is that activation is reversible, which enables the probe to produce a signal only in viable cancer cells. These probe conjugates

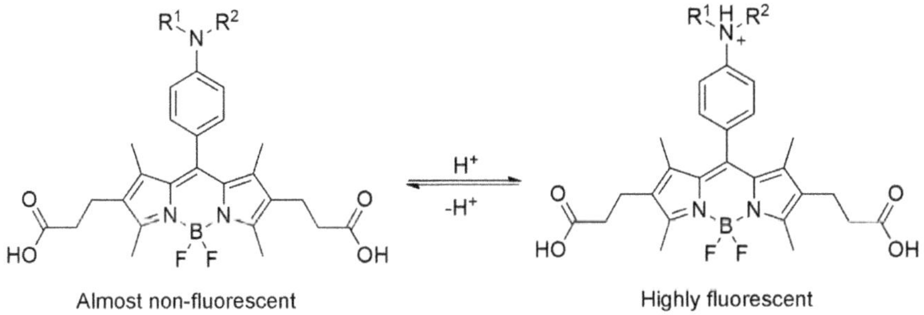

Figure 9.9 Reversible and acidic pH-induced fluorescence activation of a probe.

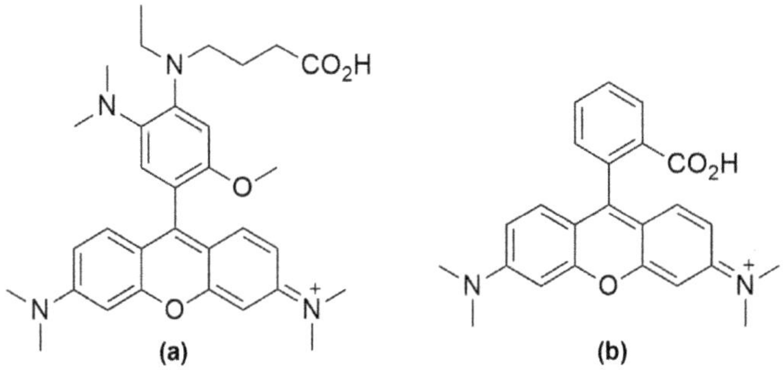

Figure 9.10 Proposed structure of pHrodo (a) and the structure of TAMRA (b).

can potentially be used as *in vitro* tools for evaluating intracellular receptor kinetics and cell viability and for real-time monitoring of cell death.

Kobayashi and co-workers[37] demonstrated dual activatable xanthene optical probes, pHrodo (Figure 9.10a) and TAMRA (Figure 9.10b), *via* the combination of H-dimer formation and pH activation, which could achieve a high target-to-background signal ratio at early time points during *in vivo* molecular imaging.

An NIR pH-activatable probe was formulated by Achilefu and co-workers[38] by conjugating a pH-sensitive cyanine dye to a cyclic arginine–glycine–aspartic acid (cRGD) peptide targeting $\alpha_v\beta_3$ integrin (ABIR) for *in vivo* imaging of tumours (Figure 9.11). This pH-sensitive dye exhibits a high spectral sensitivity to pH changes. It has negligible fluorescence above pH 6 but becomes highly fluorescent below pH 5, with a pK_a of 4.7. This binding of the cRGD probe to ABIR is useful for imaging primary and metastatic breast tumours. Subsequently, they continued to use cyanine dyes and described an NIR molecule, LS662, with a pH-sensitive NH group for non-invasive reporting of the cellular and extracellular pH of tumours[39] (Figure 9.12).

In 2017, a rhodamine-based pH probe was synthesized to detect the pH in cancer cells and lysosomal pH *via* "on–off" reversibility.[40] This probe also worked well in the imaging of ovarian tumours, excised mouse stomach and living mice under acidic stimulation. Li and co-workers[41] explored five NIR cyanine dyes, Cy7-1–Cy7-5 (Figure 9.13), with different nitro aromatic groups

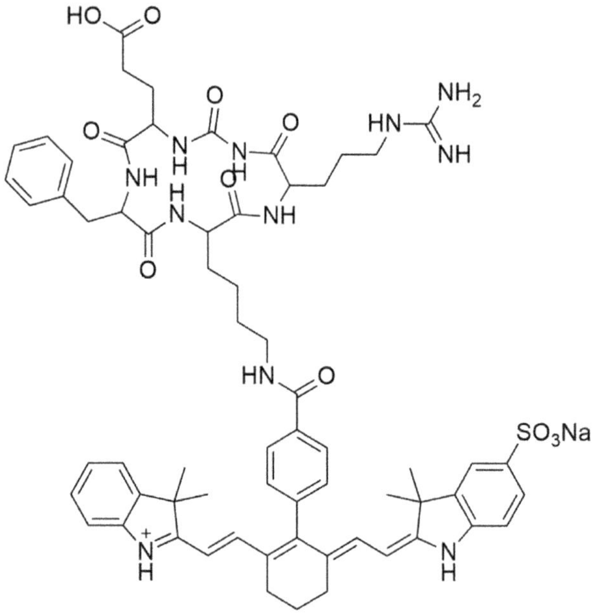

Figure 9.11 $\alpha_v\beta_3$-targeting pH-sensitive dye.

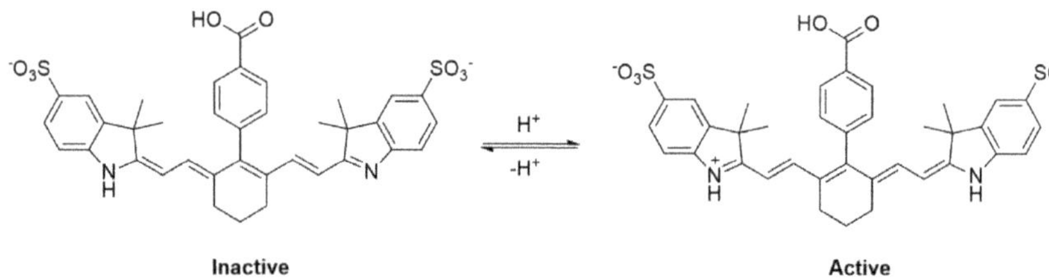

Figure 9.12 pH-activated mechanism of LS662 probe.

as the fluorescence reporting site for the rapid detection of NTR. They found that only cyanine probe Cy7-1 with a *p*-nitrobenzoate group could act as a fast NIR fluorescence-enhanced probe for the monitoring of NTR. Based on a theoretical study, they showed that the linker connecting the detecting and fluorescence reporting groups and the nitro group position is a key factor for the formation of hydrogen bonds and spatial structure match, inducing the NTR catalytic ability enhancement. Cy7-1 could detect tumour hypoxia in A549 cells and a murine hypoxic tumour model.

For the rapid imaging of all human hypoxic tumours, Liu and co-workers[42] designed two universal NIR fluorescent probes with a marker of carbonic anhydrase IX (CA IX). Regarding pH sensing, most fluorescent probes developed so far rely solely on the variation of fluorescence intensity, which is usually compromised by many non-pH factors such as the local concentration of chromophores.[43] In contrast, ratiometric fluorescent probes[44] can

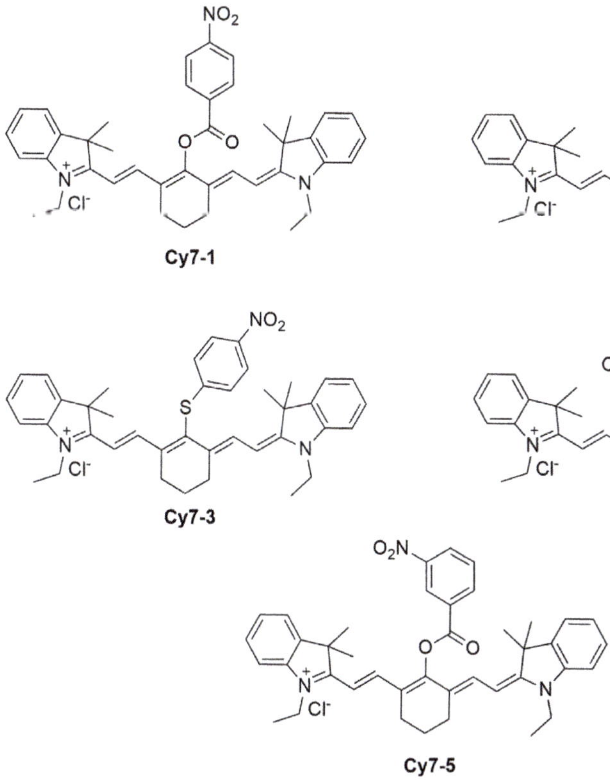

Figure 9.13 Structures of five fluorescent probes Cy7-1–Cy7-5.

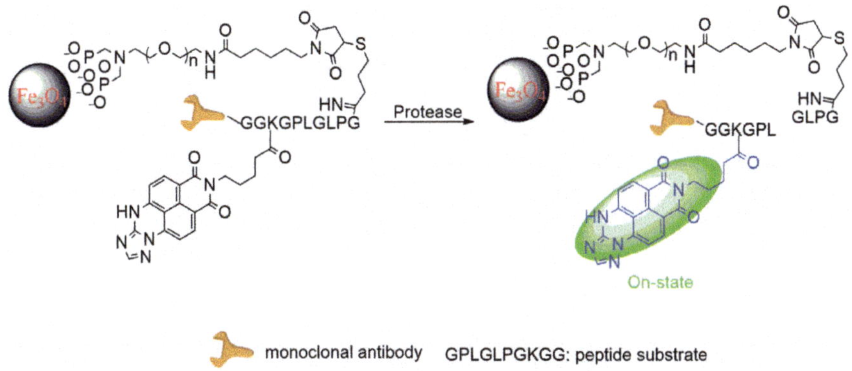

Figure 9.14 Off–on fluorescent probe triggered by MMP-9.

avoid such problems and thus allow the quantitative determination of pH. Hence ratiometric fluorescent probes for pH mapping are a better choice.

In 2015, Gao and co-workers[45] constructed a ratiometric fluorescent probe based on FRET between a pH-sensitive fluorescent dye and biocompatible Fe_3O_4 nanocrystals (Figure 9.14), which covalently coupled through an

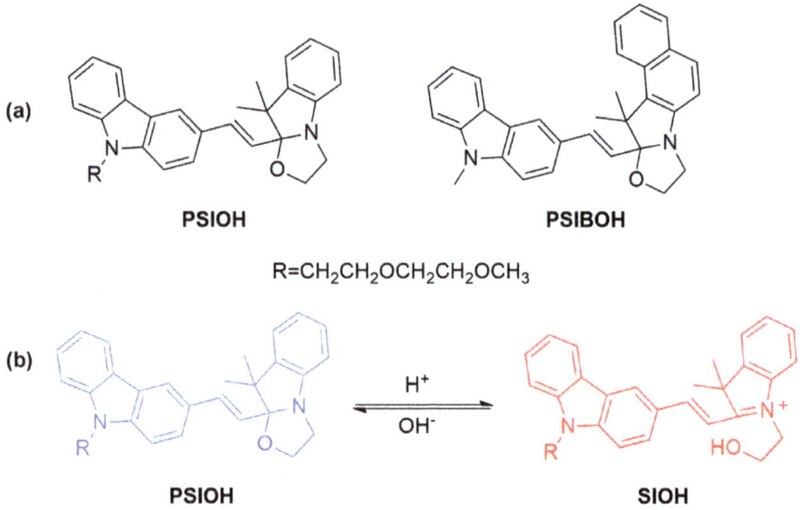

Figure 9.15 (a) Structures of PSIOH and PSIBOH; (b) pH-sensing response of PSIOH.

MMP-9-responsive peptide linker. The probe can be used for imaging the pH of subcutaneous tumour xenografts *in vivo*.

Two carbazole–oxazolidine fluorophores, PSIOH and PSIBOH, were developed as effective ratiometric two-photon probes (Figure 9.15a) for pH sensing and imaging.[46] Oxazolidine ring opening was induced by the protonation of the probe under acidic pH conditions in a biological system (Figure 9.15b). Because of the exceptionally large shift in the emission bands (169–189 nm) of the two emissive states upon protonation/deprotonation, these ratiometric probes can be used for the high-resolution imaging of pH variations in living cells and tissues.

Another intracellular pH (pH_i) ratiometric probe with the genetically encoded indicator SypHer2 was developed by Shirmanova *et al.*[47] for monitoring cancer cells *in vitro* and tumours *in vivo*. This probe could evaluate the tumour metabolic state and progression in xenograft models. In order to detect simultaneously both pH and hypoxia in tumours, Jiang and co-workers[48] designed a macromolecular NIR probe, Ir-Im-PEG, to respond to pH and oxygen levels in primary tumours and metastatic tumour nodules *in vivo* while amplifying the detection sensitivity *via* signal propagation.

9.4 Tumour pH and Hypoxia Targeting Nanotechnology

Nanomaterials with light stability, a high quantum yield and adjustable emission spectrum have been efficiently applied in tumour visualization. Liu and co-workers[49] successfully fabricated an albumin-based ratiometric photoacoustic nanoprobe, HSA-BPOx-IR825(C), for the non-invasive real-time pH imaging of tumours (Figure 9.16). In this system, human serum

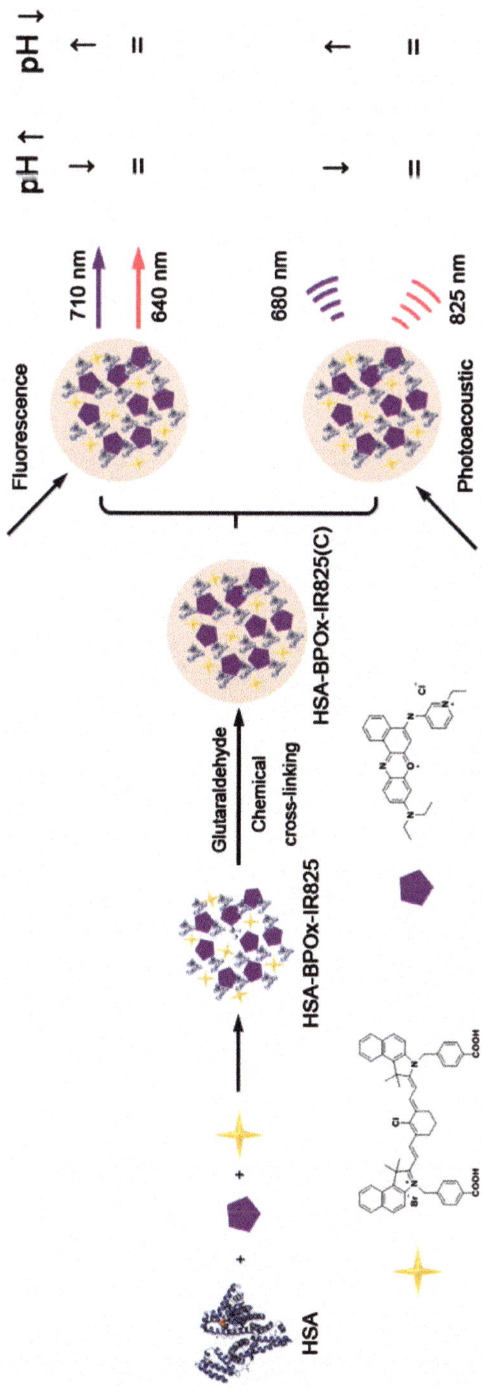

Figure 9.16 HSA-BPOx-IR825(C) for pH imaging *via* ratiometric fluorescence and photoacoustic imaging.

albumin (HSA) is a biocompatible nanocarrier that can simultaneously load two different organic dyes: BPOx as a pH-sensitive dye for real-time pH sensing and IR825 as a pH-inert NIR dye for internal reference. Ratiometric photoacoustic imaging for pH sensing appears to be fairly accurate even for targets covered by tissues with considerable thicknesses. Based on NIR fluorescence imaging, further nanoparticles for *in vivo* tumour detection have also been reported.[50,51] For multispectral optoacoustic tomographic identification, Gurka *et al.*[52] developed a theranostic nanoplatform for monitoring pancreatic cancer with targeting of both acidic pH and urokinase plasminogen activator receptor (UPAR). Nishiyama and co-workers[53] formulated HSA-coated liposomal Indocyanine Green for *in vivo* magnetic resonance imaging (MRI) *via* a selective response to pH in a precise contrast off–on mode.

9.5 pH and Hypoxia Imaging for Tumour Therapeutic Monitoring

Abnormal pH values and hypoxia are known to be an inevitable physiological feature in the growth process of malignant tumours, and the detection of pH and hypoxia paves the way for the treatment of tumours as the ultimate goal. During the exploration of fluorescence probes for tumours, several cancer diagnosis and therapy strategies have been developed using photothermal therapy (PTT), photodynamic therapy (PDT) and drug delivery with NIR fluorescent probes and nanocomposites.

In 2016, Li and co-workers[54] developed four cyanine-based theranostic probes, IR1–IR4, that demonstrated pH-switchable NIR fluorescence and photothermal efficiency in physiological environments (Figure 9.17). By fine tuning the pK_a values by substitutions at the terminal indole of the probes, it was found that IR2 with a narrow pH range specifically illuminated and eradicated multiple types of cancer cells by maximizing its NIR fluorescence intensity in the acidic lysosomal lumen and photothermal effects in the alkaline mitochondrial matrix.

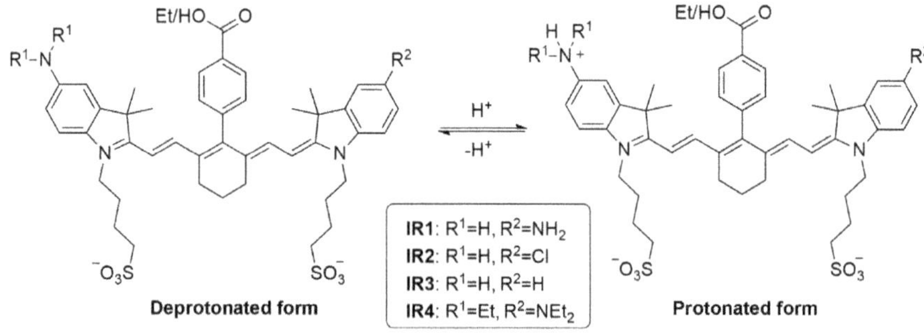

Figure 9.17 Structures of probes IR1–IR4 and mechanism of pH tuning.

A small-molecule probe, RhoSSCy, with "four-in-one" multifunctions for sensing, targeting, multi-modal imaging and therapy, was constructed by Cai and co-workers.[55] The probe conjugated a 5′-carboxyrhodamine (Rho) and heptamethine cyanine IR765 (Cy) *via* a disulfide linker and a pH-tuneable amino group, which resulted in dual ratiometric detection of free thiols and intracellular H^+ and thiol concentrations, as shown in Figure 9.18.

The same group subsequently developed another smart cyanine–pyridine-type probe for dual-modal pH-targeted imaging in tumours.[56] Yi and co-workers[57] reported the probe pH-PTT assembled with bovine serum albumin (BSA) and pH-responsive cyanine dyes for PTT *via* Golgi apparatus activation (Figure 9.19). The nanoparticles formed by BSA and pH-PTT preferentially accumulated in the Golgi apparatus of cancer cells both *ex vivo* and *in vivo*.

Wang and co-workers[58] designed an anticancer prodrug, FDU-DB-NO_2, for selective two-photon imaging of hypoxia status, real-time tracking of drug release and solid tumour therapy (Figure 9.20). This probe is "unlocked" specifically by the hypoxic environment in solid tumour cells and releases the fluorescent dye 7-(diethylamino)coumarin (CM) along with floxuridine (FDU) generation. Wang and co-workers[59] engineered functional nanoparticles, D_m-NP, for siRNA delivery using a functionalized maleic anhydride (Figure 9.21). The surface function of D_m-NP enabled a selective response to pH in tumours with the help of the precise introduction of a degradable bridged bond (D_m) group. D_m-NP achieved both prolonged circulation and effective accumulation in tumour cells and resulted in the safe and enhanced inhibition of non-small-cell lung cancer growth.

Liu and co-workers[60] assembled pH- and H_2O_2-responsive HSA-MnO_2-Ce6&Pt (HMCP) nanoparticles for effective combination therapy (Figure 9.22). Once inside the tumour, MnO_2 nanoclusters within HMCP nanoparticles would react with endogenous H_2O_2 to produce oxygen and relieve hypoxia in solid tumours. Further, the reaction between MnO_2 and H^+ and H_2O_2 would lead to therapeutic albumin complexes with sub-10 nm particle size for optimized tumour treatment. Nie and co-workers[61] constructed a tumour microenvironment-responsive nanosystem based on gold nanoparticles coated with PEGylated polyethylenimine to co-deliver all-*trans*-retinoic acid (ATRA, a quiescence inducer) and siRNA-targeting heat-shock protein 47 (HSP47, a molecular chaperone) to silence pancreatic stellate cells (PSCs) (Figure 9.23). The nanosystem also simultaneously effectively inhibits extracellular matrix (ECM) hyperplasia both *in vitro* and *in vivo*. This combination strategy restored homoeostatic stromal function by targeting activated PSCs and led to the effective suppression of tumour progression in two desmoplastic pancreatic tumour models.

Given that a low concentration of oxygen is vital to cancer growth, killing tumours through the consumption of intratumoural oxygen is a potentially useful strategy in fighting cancer. Injectable polyvinylpyrrolidone (PVP)-modified Mg_2Si nanoparticles as a deoxygenation agent (DOA) to starve tumours were developed by Shi and co-workers[62] (Figure 9.24). In the acidic

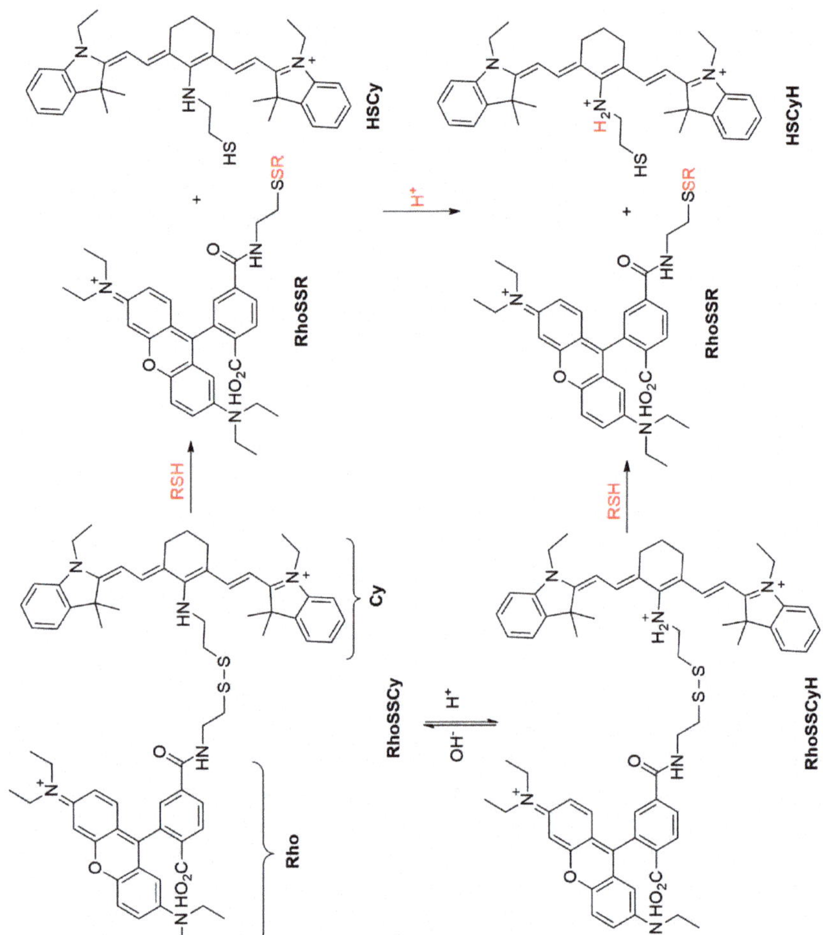

Figure 9.18 Response mechanisms of RhoSSCy to thiols and pH.

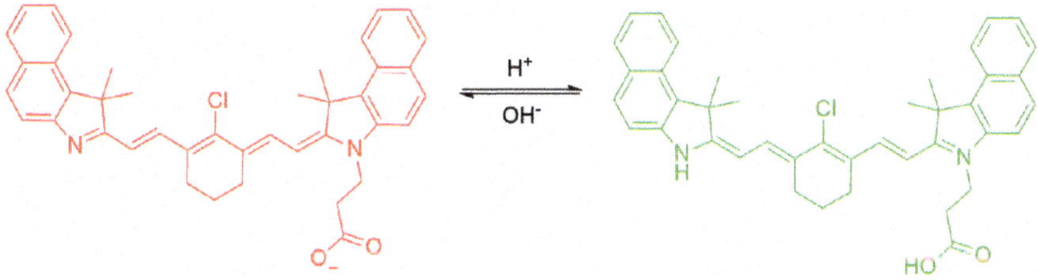

Figure 9.19 Illustration of pH-PTT for specific Golgi apparatus-activated PTT.

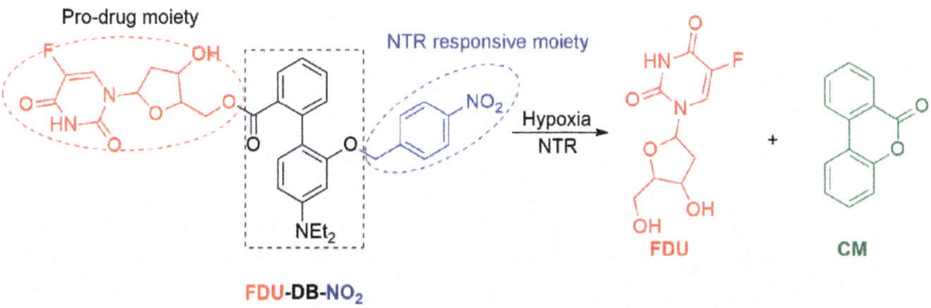

Figure 9.20 Structure of FDU-DB-NO_2 and hypoxia-activated mechanism.

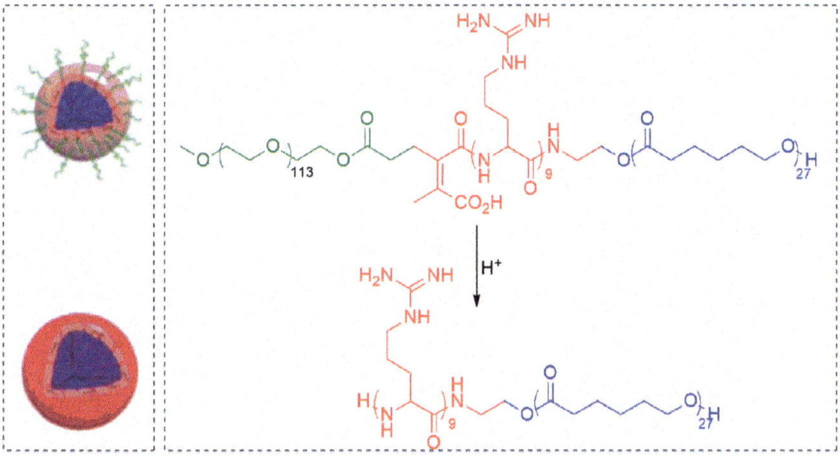

Figure 9.21 Acidity recognition of D_m-NPsiRNA in a tumour cell.

tumour microenvironment, the Mg_2Si releases silane, which efficiently consumes intratumoural oxygen to form silicon dioxide (SiO_2) aggregates. This *in situ* formation of SiO_2 blocks the tumour blood capillaries and prevents tumours from being reoxygenated *via* SiO_2 microaggregate by-products.

Liu and co-workers[63] formulated a cancer treatment strategy by combining glucose oxidase (GOx)-based starvation therapy and banoxantrone

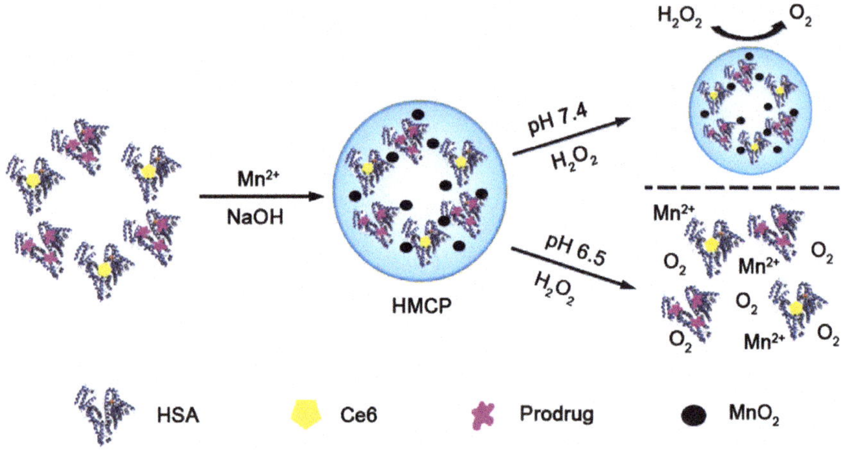

Figure 9.22 pH-/H_2O_2-responsive nanocarrier HMCP.

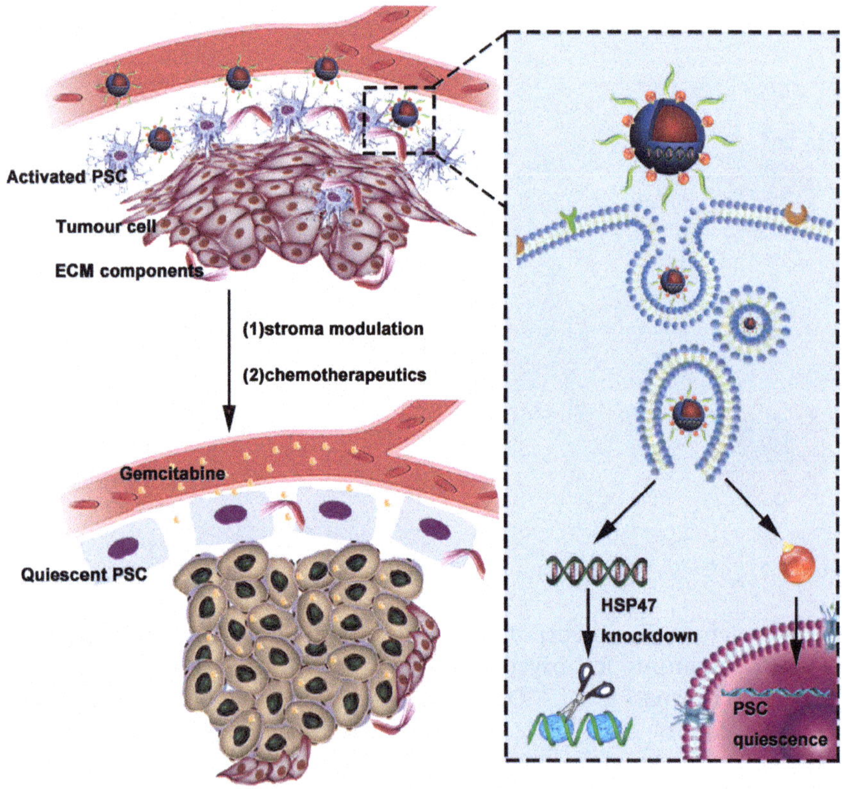

Figure 9.23 Re-education of PSCs and stroma modulation by the nanosystem.

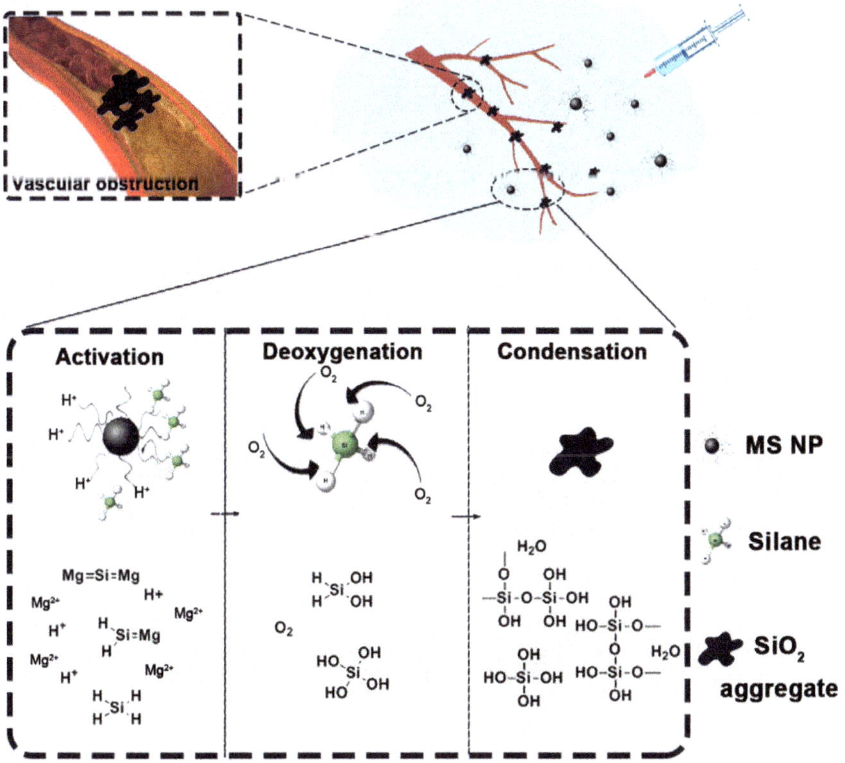

Figure 9.24 Acidic-activated mechanism of Mg_2Si (MS) nanoparticles as a DOA.

dihydrochloride (AQ4N, a hypoxia-activated pro-drug)-based hypoxia-activated therapy using liposomal nanocarriers (Figure 9.25). The liposome–GOx could specifically block the tumorous glucose supply, exhaust tumorous oxygen for hypoxia enhancement and produce toxic H_2O_2 inside the tumours. Further, liposome–AQ4N, once inside the tumour with enhanced hypoxia, would be further activated to achieve a strong synergistic anti-tumour effect. Tang and co-workers[64] developed NIR light/H_2O_2-triggered nanocomposites based on the solid LRET (luminescence resonance energy transfer) from an Eu^{3+} complex for highly selective synergistic PDT processes against hypoxic tumour cells.

9.6 Perspective

Acidic pH and hypoxia are the most common characteristics of tumours. The development of molecular and nanoprobes with low cytotoxicity and satisfactory photostability offers new fluorescent tools for the *in situ* analysis of pH and hypoxia-related biological tumour events by tumour imaging and therapeutic monitoring. However, probes with high resolution in time and

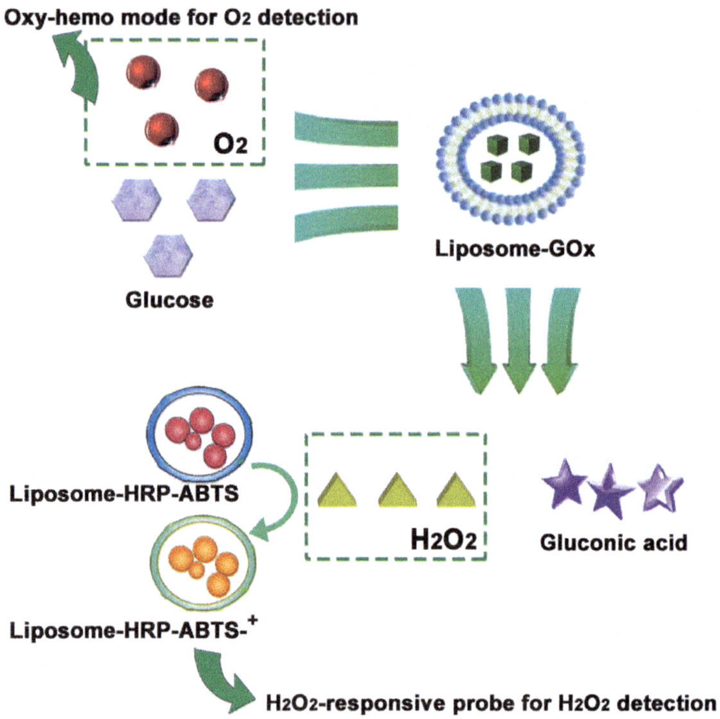

Figure 9.25 Liposomes for cancer starvation and hypoxia-activated therapy.

space need to be further developed for complicated native tumour conditions. The simultaneous or continuous detection of pH and hypoxia by a single probe with precise chemoselectivity also holds great promise. Additionally, if the probes are clinically compatible with actual cases, then theoretical research will effectively promote the progress of tumour imaging and treatment.

Acknowledgement

Dr Wen Gao and Zhenhua Liu are thanked for expert assistance with the writing of this chapter.

References

1. D. Perez-Sala, D. Collado-Escobar and F. J. Mollinedo, *Biol. Chem.*, 1995, **270**, 6235.
2. A. P. Lieberman, R. Puertollano, N. Raben, S. Slaugenhaupt, S. U. Walkley and A. Ballabio, *Autophagy*, 2012, **8**, 719.
3. M. H. Lee, J. H. Han, J. H. Lee, N. Park, R. Kumar, C. Kang and J. S. Kim, *Angew. Chem., Int. Ed.*, 2013, **52**, 6206.

4. R. Huang, S. Y. Yan, X. L. Zheng, F. L. Luo, M. G. Deng, B. S. Fu, Y. J. Xiao, X. Zhao and X. Zhou, *Analyst*, 2012, **137**, 4418.
5. X. D. Liu, Y. Xu, R. Sun, Y. J. Xu, J. M. Lu and J. F. Ge, *Analyst*, 2013, **138**, 6542.
6. L. Wang, X. Zhu, C. Xie, N. Ding, X. F. Weng, W. Y. Lu, X. B. Wei and C. Li, *Chem. Commun.*, 2012, **48**, 11677.
7. H. J. Kim, C. H. Heo and H. M. Kim, *J. Am. Chem. Soc.*, 2013, **135**, 17969.
8. X. Q. Chen, T. H. Pradhan, F. Wang, J. S. Kim and J. Yoon, *Chem. Rev.*, 2012, **112**, 1910.
9. H. Izumi, T. Torigoe, H. Ishiguchi, H. Uramoto, Y. Yoshida, M. Tanabe, T. Ise, T. Murakami, T. Yoshida, M. Nomoto and K. Kohno, *Cancer Treat. Rev.*, 2003, **29**, 541.
10. T. A. Davies, R. E. Fine, R. J. Johnson, C. A. Levesque, W. H. Rathbun, K. F. Seetoo, S. J. Smith, G. Strohmeier, L. Volicer and L. Delva, *Biochem. Biophys. Res. Commun.*, 1993, **194**, 537.
11. D. Ellis and R. C. Thomas, *Nature*, 1976, **262**, 224.
12. R. G. Zhang, S. G. Kelen and J. C. LaManna, *J. Appl. Phys.*, 1990, **68**, 1101.
13. S. J. A. Hesse, G. J. G. Ruijter, C. Dijkema and J. J. Vissser, *J. Biotechnol.*, 2000, **77**, 5.
14. X. H. Li, X. H. Gao, W. Shi and H. M. Ma, *Chem. Rev.*, 2014, **114**, 590.
15. Q. Q. Wan, S. M. Chen, W. Shi, L. H. Li and H. M. Ma, *Angew. Chem., Int. Ed.*, 2014, **53**, 10916.
16. W. R. Wilson and M. P. Hay, *Nat. Rev. Cancer*, 2011, **11**, 393.
17. Y. Liu, Y. Xu, X. Qian, J. Liu, L. Shen, J. Li and Y. Zhang, *Bioorg. Med. Chem.*, 2006, **14**, 2935.
18. F. A. Gallagher, M. I. Kettunen, S. E. Day, D.-E. Hu, J. H. Ardenkjær-Larsen, R. Zandt, P. R. Jensen, M. Karlsson, K. Golman, M. H. Lerche and K. M. Brindle, *Nature*, 2008, **453**, 940.
19. G. L. N. Semenza, *N. Engl. J. Med.*, 2011, **365**, 537.
20. S.-L. Shen, X.-P. Chen, X.-F. Zhang, J.-Y. Miao and B.-X. Zhao, *J. Mater. Chem. B*, 2015, **3**, 919.
21. D.-D. He, W. Liu, R. Sun, C. Fan, Y.-J. Xu and J.-F. Ge, *Anal. Chem.*, 2015, **87**, 1499.
22. L. Cui, Y. Shi, S. Zhang, L. Yan, H. Zhang, Z. Tian, Y. Gu, T. Guo and J. Huang, *Dyes Pigm.*, 2017, **139**, 587.
23. M. H. Lee, A. Sharma, M. J. Chang, J. Lee, S. Son, J. L. Sessler, C. Kang and J. S. Kim, *Chem. Soc. Rev.*, 2018, **47**, 28.
24. J. Xu, S. Sun, Q. Li, Y. Yue, Y. Li and S. Shao, *Analyst*, 2015, **140**, 574.
25. Y. Liu, Y. Xu, X. Qian, J. Liu, L. Shen, J. Li and Y. Zhang, *Bioorg. Med. Chem.*, 2006, **14**, 2935.
26. K. Xu, F. Wang, X. Pan, R. Liu, J. Ma, F. Kong and B. Tang, *Chem. Commun.*, 2013, **49**, 2554.
27. (a) C. Pavlik, N. C. Biswal, F. C. Gaenzler, M. D. Morton, L. T. Kuhn, K. P. Claffey, Q. Zhu and M. B. Smith, *Dyes Pigm.*, 2011, **89**, 9; (b) Y. Xu, S. Zanganeh, I. Mohammad, A. Aguirre, T. Wang, Y. Yang, L. Kuhn, M. B. Smith and Q. Zhu, *J. Biomed. Opt.*, 2013, **18**(6), 066009;

(c) A. Abuteen, F. Zhou, C. Dietz, I. Mohammad, M. B. Smith and Q. Zhu, *Dyes Pigm.*, 2016, **126**, 251.
28. T. Doussineau, A. Schultz, A. L. Fernandez, A. Moro, S. Korsten, S. Trupp and G. J. Mohr, *Chem. - Eur. J.*, 2010, **16**, 10290.
29. J. Ge, L. Fan, K. Zhang, T. Ou, Y. Li, C. Zhang, C. Dong, S. Shuang and M. S. Wong, *Sens. Actuators, B*, 2018, **262**, 913.
30. (a) M. H. Lee, J. H. Han, J. H. Lee, N. Park, R. Kumar, C. Kang and J. S. Kim, *Angew. Chem., Int. Ed.*, 2013, **52**, 6206; (b) J. Y. Han, A. Loudet, R. Barhoumi, R. C. Burghardt and K. Burgess, *J. Am. Chem. Soc.*, 2009, **131**, 1642.
31. X.-F. Zhang, T. Zhang, S.-L. Shen, J.-Y. Miao and B.-X. Zhao, *J. Mater. Chem. B*, 2015, **3**, 3260.
32. J. Ge, L. Fan, K. Zhang, T. Ou, Y. Li, C. Zhang, C. Dong, S. Shuang and M. S. Wong, *Sens. Actuators, B*, 2018, **262**, 913.
33. T. Guo, L. Cui, J. Shen, W. Zhu, Y. Xu and X. Qian, *Chem. Commun.*, 2013, **49**, 10820.
34. L. Cui, Y. Zhong, W. Zhu, Y. Xu, Q. Du, X. Wang, X. Qian and Y. Xiao, *Org. Lett.*, 2011, **13**, 928.
35. L. Zhang, X. Shan, L. Guo, J. Zhang, J. Ge, Q. Jiang and X. Ning, *Analyst*, 2019, **144**, 284.
36. Y. Urano, D. Asanuma, Y. Hama, Y. Koyama, T. Barrett, M. Kamiya, T. Nagano, T. Watanabe, A. Hasegawa, P. L. Choyke and H. Kobayashi, *Nat. Med.*, 2009, **15**, 104.
37. M. Ogawa, N. Kosaka, C. A. S. Regino, M. Mitsunaga, P. L. Choyke and H. Kobayashi, *Mol. BioSyst.*, 2010, **6**, 888.
38. H. Lee, W. Akers, K. Bhushan, S. Bloch, G. Sudlow, R. Tang and S. Achilefu, *Bioconjugate Chem.*, 2011, **22**, 777.
39. R. C. Gilson, R. Tang, A. Som, C. Klajer, P. Sarder, G. P. Sudlow, W. J. Akers and S. Achilefu, *Mol. Pharmaceutics*, 2015, **12**, 4237.
40. B. Li, G. Ge, L. Wen, Y. Yuan, R. Zhang, X. Peng, F. Liu and S. Sun, *Dyes Pigm.*, 2017, **139**, 318.
41. Y. Li, Y. Sun, J. Li, Q. Su, W. Yuan, Y. Dai, C. Han, Q. Wang, W. Feng and F. Li, *J. Am. Chem. Soc.*, 2015, **137**, 6407.
42. S. M. Mahalingam, H. Chu, X. Liu, C. P. Leamon and P. S. Low, *Bioconjugate Chem.*, 2018, **29**, 3320.
43. J. Zhou, C. L. Fang, T. J. Chang and D. H. Shangguan, *J. Mater. Chem. B*, 2013, **1**, 661.
44. L. Albertazzi, B. Storti, L. Marchetti and F. Beltram, *J. Am. Chem. Soc.*, 2010, **132**, 18158.
45. Y. Hou, J. Zhou, Z. Gao, X. Sun, C. Liu, D. Shangguan, W. Yang and M. Gao, *ACS Nano*, 2015, **9**, 3199.
46. D. Xu, Y. Li, C.-Y. Poon, H.-N. Chan, H.-W. Li and M. S. Wong, *Anal. Chem.*, 2018, **90**, 8800.
47. M. V. Shirmanova, I. N. Druzhkova, M. M. Lukina, M. E. Matlashov, V. V. Belousov, L. B. Snopova, N. N. Prodanetz, V. V. Dudenkova,

S. A. Lukyanov and E. V. Zagaynova, *Biochim. Biophys. Acta*, 2015, **1850**, 1905.

48. X. Zheng, H. Mao, D. Huo, W. Wu, B. Liu and X. Jiang, *Nat. Biomed. Eng.*, 2017, **1**, 0057.
49. Q. Chen, X. Liu, J. Chen, J. Zeng, Z. Cheng and Z. Liu, *Adv. Mater.*, 2015, **27**, 6820.
50. W. Hou, X. Zhao, X. Qian, F. Pan, C. Zhang, Y. Yang, J. M. de la Fuentea and D. Cui, *Nanoscale*, 2016, **8**, 104.
51. S. Chen, G. Yu, B. Zhang, Y. Wang, N. Zhang and Y. Chen, *RSC Adv.*, 2016, **6**, 15220.
52. M. K. Gurka, D. Pender, P. Chuong, B. L. Fouts, A. Sobelov, M. W. McNally, M. Mezera, S. Y. Woo and L. R. McNally, *J. Controlled Release*, 2016, **231**, 60.
53. P. Mi, D. Kokuryo, H. Cabral, H. Wu, Y. Terada, T. Saga, I. Aoki, N. Nishiyama and K. Kataoka, *Nat. Nanotechnol.*, 2016, **11**, 724.
54. J. Zhang, Z. Liu, P. Lian, J. Qian, X. Li, L. Wang, W. Fu, L. Chen, X. Wei and C. Li, *Chem. Sci.*, 2016, **7**, 5995.
55. X. Meng, Y. Yang, L. Zhou, L. Zhang, Y. Lv, S. Li, Y. Wu, M. Zheng, W. Li, G. Gao, G. Deng, T. Jiang, D. Ni, P. Gong and L. Cai, *Theranostics*, 2017, **7**, 1781.
56. X. Meng, W. Li, Z. Sun, J. Zhang, L. Zhou, G. Deng, P. Gong and L. Cai, *J. Mater. Chem. B*, 2017, **5**, 9405.
57. F. Xue, Y. Wen, P. Wei, Y. Gao, Z. Zhou, S. Xiao and T. Yi, *Chem. Commun.*, 2017, **53**, 6424.
58. W. Liu, H. Liu, X. Peng, G. Zhou, D. Liu, S. Li, J. Zhang and S. Wang, *Bioconjugate Chem.*, 2018, **29**, 3332.
59. C. Y. Sun, S. Shen, C. F. Xu, H. J. Li, Y. Liu, Z. T. Cao, Y. X. Zhu, J. X. Xia and J. Wang, *J. Am. Chem. Soc.*, 2015, **137**, 15217.
60. Q. Chen, L. Feng, J. Liu, W. Zhu, Z. Dong, Y. Wu and Z. Liu, *Adv. Mater.*, 2016, **28**, 7129.
61. X. Han, Y. Li, Y. Xu, X. Zhao, Y. Zhang, X. Yang, Y. Wang, R. Zhao, G. J. Anderson, Y. Zhao and G. Nie, *Nat. Commun.*, 2018, **9**, 3390.
62. C. Zhang, D. Ni, Y. Liu, H. Yao, W. Bu and J. Shi, *Nat. Nanotechnol.*, 2017, **12**, 378.
63. R. Zhang, L. Feng, Z. Dong, L. Wang, C. Liang, J. Chen, Q. Ma, R. Zhang, Q. Chen, Y. Wang and Z. Liu, *Biomaterials*, 2018, **162**, 123.
64. Y. Zhang, T. Shen, A. M. Kirillov, W. S. Liu and Y. Tang, *Chem. Commun.*, 2016, **52**, 7939.

CHAPTER 10

Visualization of Dynamic Intermolecular Interactions in Living Cells

10.1 Introduction

Intracellular bioactive molecules are extensively associated with cell functions and their intermolecular interactions regulate multiple signal transduction pathways and pathophysiological events, including cell proliferation, differentiation and apoptosis and occurrence and the development and migration of major diseases.[1–3] These bioactive molecules include not only biological macromolecules such as proteins and nucleic acids, but also a variety of small molecules such as hormones and signal transduction molecules capable of regulating cell functions. The expression, distribution, transmission, interaction and cross-talking of these molecules are of great significance to maintaining physiological homeostasis and also facilitate pathological progression. Therefore, studying the pathophysiological processes of cells requires not only an understanding of the specific location of each type of component, but also an understanding of the complex processes of their interactions and cross-talk in live cells. Therefore, the deep elucidation of the dynamic intermolecular interactions within living cells opens up a new field of research and also great challenges in chemical biology.

Owing to their real-time, high spatiotemporal resolution and non-invasive characteristics, fluorescent probes have been considered efficient imaging tools for the *in situ* visualization of the molecular events within live cells.[4–9] However, it is still a major challenge to design a single fluorescent probe capable of the simultaneous detection of multiple biomolecules and

Detection Science Series No. 16
In Situ Analysis of Cellular Functional Molecules
By Huangxian Ju, Bo Tang and Lin Ding

Published by the Royal Society of Chemistry, www.rsc.org

elucidation of their interactions. This chapter focuses on recent advances in the fluorescence imaging of dynamic intermolecular interactions within living cells and is organized into three main sections. The first section discusses the dynamic interactions among active small molecules, including reactive oxygen species (ROS), reactive sulfur species (RSS), reactive nitrogen species (RNS), microRNA (miRNA) and H^+. Next, the dynamic interactions among biomacromolecules, such as proteins, DNAs and RNAs, are illustrated. Finally, the interactions between active small molecules and biomacromolecules are discussed, including phosphatase and H_2S, miRNA and mRNA, *etc.* One of the challenges in the subject matter of this chapter is to see how various molecules work with others in a complicated *in vivo* system. The crucial aspects of this chapter are based on the rational design of fluorescent probes capable of the dual detection of bioactive molecules, and the chapter starts with relationships between active small molecules and how the interactions of the molecules are described and determined.

10.2 Dynamic Interactions among Active Small Molecules

Bioactive small molecules, such as ROS, RNS and RSS, are extensively involved in the normal and abnormal conditions of biological systems. They experience interactions and interconversions to maintain physiological homeostasis and facilitate pathological progression. Elucidation of their cross-talking will provide critical insights into the molecular regulatory mechanisms in live cells and organisms.

10.2.1 Dual Detection of ROS

ROS, including superoxide radicals ($O_2^{\bullet -}$), hydrogen peroxide (H_2O_2), hydroxyl radicals ($^{\bullet}OH$), hypochlorous acid (HClO) and singlet oxygen ($^1O_2$), commonly coexist in live cells.[10] Transformations among ROS occur widely within cells, hence there is an urgent need to develop a powerful tool to discriminate ROS and reveal their correlations. Probe **1** was constructed by chemically integrating a hydrazide group and a triethylene glycol chain into the parent fluorescein, which turned off the green fluorescence of fluorescein because the newly formed five-membered ring made the probe form a non-fluorescent spirocyclic structure (Figure 10.1).[11] $^{\bullet}OH$ could simultaneously cleave the five-membered ring and the bottom-left aromatic ring of probe **1** to form a final product with two conjugated aromatic rings, giving out cyan fluorescence. On the other hand, HClO could only break the five-membered ring to form a final product with three conjugated aromatic rings as in fluorescein, emitting green fluorescence. Cell imaging experiments demonstrated that this probe provided the real-time discrimination and quantitative analysis of $^{\bullet}OH$ and HClO and that it was a powerful tool for investigating the dynamic interaction between $^{\bullet}OH$ and HClO.

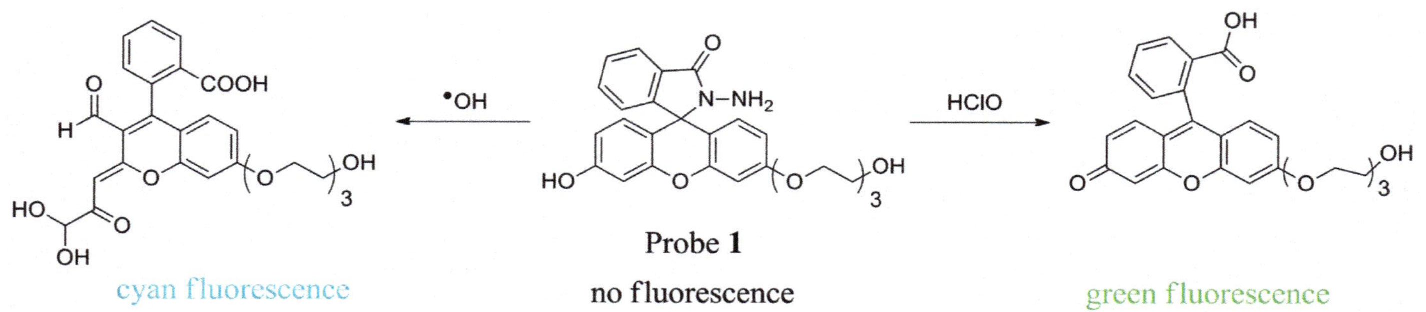

Figure 10.1 Reaction mechanisms of probe **1** with •OH and HClO.

10.2.2 Dual Detection of ROS and RSS

Cellular redox status is precisely modulated by ROS and reducing species [*e.g.* glutathione (GSH), thioredoxins, NADPH, flavins, ascorbate]. The imbalance of cellular redox status contributes to a wide range of pathological processes, so this new field has attracted increasing attention. GSH is the most abundant non-protein biothiol within live cells (1–10 mM) and thus has been considered to be the best antioxidant defence against the cytotoxicity of ROS.[12] H_2O_2 is the main member of the ROS family and its intracellular concentration is much higher than that of its ROS congeners. Therefore, the simultaneous detection of GSH and H_2O_2 can reveal their dynamic intermolecular interactions and further monitor the redox homeostasis.[13] Accordingly, probe **2** was developed by combination of a heptamethine cyanine platform and an ebselen moiety, which were chosen as the near-infrared (NIR) fluorophore and the specific reaction trigger response to GSH/H_2O_2, respectively (Figure 10.2).[14] The five-membered ring of ebselen can be switched between open and closed forms by GSH/H_2O_2, which induces off–on cyanine fluorescence. This probe can respond reversibly to GSH and H_2O_2, which enables the changes in redox status during cell apoptosis to be monitored. The probe provides an excellent imaging tool for monitoring redox homeostasis and studying dynamic intermolecular interactions between GSH and H_2O_2.

HClO, a highly reactive oxygen species, plays a significant role in innate host defence. However, aberrant HClO fluctuations contribute to organ injury or dysfunction.[15] H_2S exhibits protective or curative effects against organ injuries, including kidney dysfunction, cardiac damage and liver injury.[16] To investigate the HClO fluctuations and the H_2S-mediated antidotal effect during drug-induced liver injury, probe **3** was designed for the discrimination and detection of HClO and H_2S simultaneously (Figure 10.3).[17] Rhodamine B and 7-aminocoumarin were linked by a rigid piperazine to construct the skeleton of the probe. Alternatively, the authors selected a

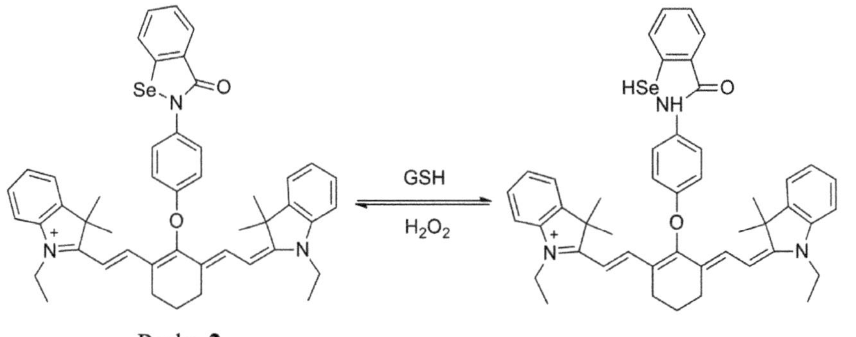

Probe **2**
fluorescence on

fluorescence off

Figure 10.2 Reversible reaction of probe **2** with GSH/H_2O_2.

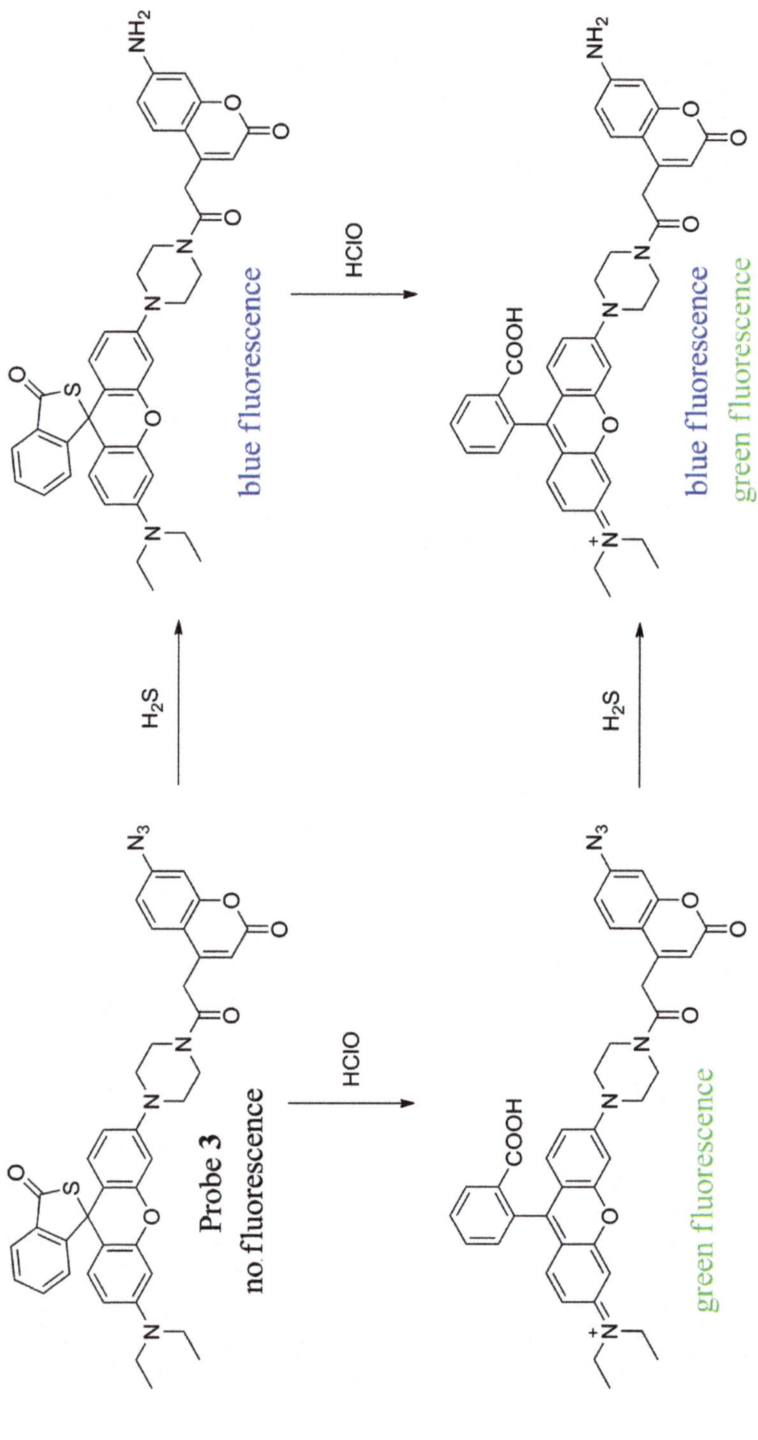

Figure 10.3 Reaction mechanisms of probe **3** with HClO and H_2S.

thiolactone and an azide specific for HClO and H_2S to mask rhodamine B and 7-aminocoumarin, respectively, to generate probe **3**. The spirocyclization of rhodamine B and azidation of 7-aminocoumarin resulted in a low fluorescence of the probe, which would emit green or blue fluorescence upon incubation with HClO or H_2S, guaranteeing dual-channel detection of HClO and H_2S. Confocal imaging revealed HClO upregulation after combined antidepressant administration. Meanwhile, an increased H_2S level was also observed. Additionally, *N*-acetylcysteine (NAC) pretreatment further increased the endogenous H_2S concentration and the HClO level was effectively suppressed. With the aid of probe **3**, the cross-talking relationship between HClO and H_2S in keeping the injury-antidotal balance during antidepressant-induced hepatotoxicity and NAC remission was revealed.

10.2.3 Dual Detection of ROS and pH

Mitochondrial ROS and pH variations are closely linked with mitochondrial dysfunctions, which are involved in multiple diseases including cancers and neurodegenerative disorders.[18] To monitor simultaneously the fluctuations of H_2O_2 and pH in mitochondria, a novel fluorescent nanosensor was developed based on mesoporous silica nanoparticles (MSNs).[19] The molecular fluorescent probes for detecting H_2O_2 and pH were loaded and polyethylenimine was attached to cap the pores of MSNs, then the triphenylphosphonium moiety was further modified to target mitochondria in live cells. It was demonstrated that the nanosensor could visualize mitochondrial H_2O_2 and pH variations in live cells in real time, providing a promising strategy for investigating the interplaying roles of multiple bioactive molecules in subcellular compartments.

ROS and pH variations are of great significance for promoting and regulating autophagy and apoptosis, which are extensively involved in a range of pathophysiological processes.[20] To differentiate quantitatively $O_2^{\bullet-}$ and pH changes in autophagy and apoptosis in live cells, a dual-ratiometric fluorescent nanoprobe was constructed using rhodamine B as the reference, doping two molecular fluorescent probes for $O_2^{\bullet-}$ and pH measurement in the outer layer shell of SiO_2 and modifying chitosan and triphenylphosphonium on the surface of SiO_2.[21] This nanoprobe was capable of simultaneously and ratiometrically visualizing the changes in $O_2^{\bullet-}$ and pH in mitochondria. The results indicated that the kinetic changes in $O_2^{\bullet-}$ and pH during cell autophagy and apoptosis were remarkably distinct. This study provides a powerful tool for disentangling the interplay between mitochondrial $O_2^{\bullet-}$ and pH to understand better the related pathology and physiology.

10.2.4 Dual Detection of ROS and MicroRNA

miRNAs and ROS are concurrently involved in cardiac ischaemia-reperfusion (IR) injury.[22,23] To elucidate the mutual cross-talk between miRNAs and ROS, a novel crown-like silica@polydopamine-DNA-CeO_2 nanocomposite was

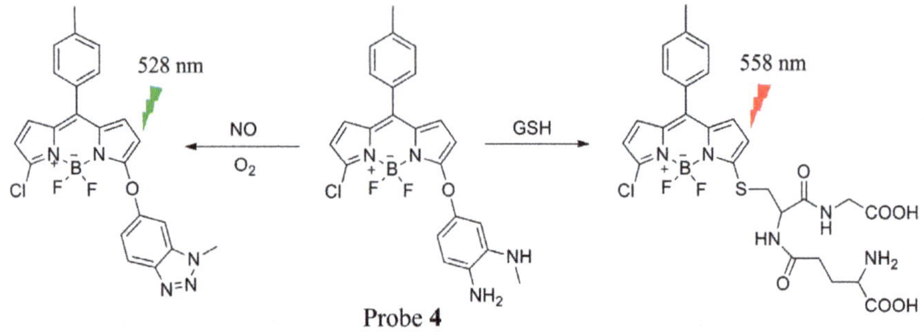

Figure 10.4 Reaction mechanisms of probe **4** with NO and GSH.

developed for simultaneously imaging miRNA-21 (miR-21) and H_2O_2 in live cells.[24] With the aid of the nanoprobe, it was observed that miR-21 and H_2O_2 were upregulated and H_2O_2 was found to regulate the miR-21 level *via* the PI3K/AKT signalling pathway in simulated IR-injured H9C2 cells. The study revealed that there is definite interplay between miR-21 and H_2O_2 and provides a promising imaging strategy for exploring the cross-talk between miRNAs and ROS in pathological events.

10.2.5 Dual Detection of RNS and RSS

Nitric oxide (NO) is a representative RNS and plays a significant role in multiple signalling events such as vasodilation, immune response and inflammation.[25] NO and GSH have interplaying roles in the oxidant–antioxidant balance. To disentangle the interplay between NO and GSH, 4-amino-3-(methylamino)phenol was introduced into a BODIPY (boron dipyrromethene) scaffold to generate probe **4**, which simultaneously displayed turn-on fluorescence responses toward NO and GSH through dual emission channels (Figure 10.4).[26] Specifically, the NO-induced chemical conversion of the diamine into the corresponding triazole led to green fluorescence enhancement and the GSH-mediated S_NAr substitution triggered red-shifted emission in the red channel. Probe **4** revealed that NO upregulation elicited by interferon-γ, lipopolysaccharide and L-arginine could further induce intracellular GSH augmentation, providing a deep insight into the cross-talk relationship between NO and GSH in the inflammation cell model.

10.3 Dynamic Interactions among Biomacromolecules

Biological macromolecules such as proteins and nucleic acids (RNA and DNA) are the basic substances for living organisms and are also the main bearers of life activities.[27] Proteins are involved in almost all body tissues and cellular activities, whereas nucleic acids provide the basis for storing and transmitting genetic information.[28] The interaction between biological macromolecules forms the basis of life activities.

10.3.1 Proteins

Proteins are an important part of all organisms, playing a very important role in living organisms, and they participate in almost all life activities. Catalysis, metabolic processes, immunity, substance transport, information transmission, movement and biological regulation occurring in living organisms all have direct participation of proteins.[29] Without the replication of proteins, the transfer of information, the translation of genetic code, *etc.*, would be impossible. In short, proteins are the basic functional unit on which an organism's morphological structure and life activities depend.

There are many kinds of proteins and their chemical structures are extremely complicated. The results of long-term studies of protein structure have confirmed that it has different levels and is generally divided into primary, secondary, tertiary and quaternary structure. The primary structure is the most stable and basic structure of a protein and the amino acids are bound by an amino bond. The long chain of the primary structure contains different groups such as –H, –C=O, $-NH_2$ and –COOH,[30] and the interactions between them form the three-dimensional structure of the molecule. The non-bonding interactions between atomic groups in this molecule are much weaker than covalent bonds. This is called a secondary bond, mainly a hydrogen bond, hydrophobic action, salt bond, van der Waals force, *etc.*[31]

The secondary structure of a protein refers to various specific compositions of the backbone structural units such as helices, folds, turns and random coils. The tertiary structure of a protein involves the spatial arrangement of all atoms in the protein molecule or subunit and is the final state of folding and curling of a multi-skin chain.[32] Regardless of the complexity of the protein's spatial structure, it is always composed of various secondary structural units combined in a variety of ways to form larger folding units or even larger domains. The quaternary structure is involved in the spatial arrangement of the various subunits in the oligomeric protein and the interaction between the various subunits, excluding the spatial structure inside the subunit.[33]

Hence the interactions between biological protein macromolecules are complicated. Early research was mainly focused on biological methods, such as western blotting and polymerase chain reaction (PCR), to observe the interactions between proteins. However, these methods are often too complicated to operate and involve destruction of cells.[34] In more recent research, fluorescent nanoprobes have allowed real-time dynamic *in situ* monitoring of a protein's evolution, providing a powerful tool for the investigation of dynamic intermolecular interactions among proteins.[35]

Since Mirkin and co-workers designed an Au–S-bonded nanoprobe for the investigation of protein relationships in 1996, the development of fluorescent nanoprobes has been extremely rapid.[36] However, when such nanoprobes are applied in *in vitro* and *in vivo* tests, the Au–S bond can be interrupted by *in vivo* thiols such as GSH, leading to a false-positive result and difficulties in monitoring the evolution of proteins. In order to improve the fidelity of probes *in vivo*, Se-functionalized peptides were used to build Au–Se-bonded nanoprobes by Tang's group.[37–39] This type of probe successfully visualized the evolution of proteins *in vitro* for caspases (cas-3, cas-8 and cas-9), metallomatrix proteases (MMP-2, MMP-7 and MMP-9), caspases with cathepsin B, *etc.* Unlike western blotting, the nanoprobes can sequentially monitor the activation of proteins in cells and also exhibit significant stability against thiol interference.

Although various methods have been introduced for studying the intermolecular interactions of proteins, this field is still in a preliminary stage and such complicated systems still need further studies exploring the future development of accurate, fast and simple techniques to gain a deeper insight into the interactions of proteins.

10.3.2 DNAs

DNA is a multimer of four kinds of deoxyribonucleotide, connected by 3′,5′-monophosphate. There are four kinds of bases, namely adenine (A), thymine (T), guanine (G), cytosine (C). These four bases, A and T, and G and C, are paired by hydrogen bonding. The structure of DNA has primary, secondary and tertiary structures. The primary structure of DNA refers to the single-stranded sequence of the four deoxynucleotides in the molecule. The primary structure of DNA determines the information it carries and the functions it performs. The secondary structure of DNA refers to the double-helix structure.[40] Factors that maintain the stability of DNA secondary structure include (1) hydrogen bonding between specific base pairings between AT and GC in DNA and (2) base stacking forces in the vertical direction between base pairs, which is the main force from the electrostatic interaction between the negatively charged base on the phosphate group and the cation in the medium. The tertiary structure of DNA refers to conformational patterns formed by further twisting of the DNA double helix. These include twisted and supercoils, multiple helices in a linear double strand, loops formed by intramolecular single strands and various topological states such as knots, supercoils and loops in circular DNA. Since various twist angles and various base conformations can be formed in the arrangement of phosphoric acid and bases, different A–D and Z configurations are produced. The shapes of the different configurations of DNA differ in the spiral diameters of the large groove and the small groove. Therefore, the way in which the same compound is bonded to it will be different.[41]

In living organisms, most of the active DNA is generally thought to exist in the form of B-DNA, which can interact with bioactive small molecules to induce many biological effects. The interactions with targeted drug molecules mainly include non-covalent bond action, covalent bond action and shear action. The sites where the targeted bioactive small molecule binds to

DNA are the base, the phosphate backbone and the pentose ring of the DNA. DNA consists of parallel bases, polyanionic phosphate backbones and major and minor grooves that constitute sites for targeted drug molecule recognition.[42] Most targeted drug molecules interact with DNA *via* non-covalent bonds. Non-covalent bonds are generally referred to as secondary bonds and non-covalent binding between DNA and bioactive small molecules is generally associated with the following secondary bonds: (1) mutual matching of three-dimensional spatial arrangements; (2) van der Waals attraction; (3) dipole–dipole interactions; (4) ion bond electrostatic interactions; (5) hydrogen bonds; and (6) hydrophobic interactions. The most important are hydrogen bonds, ionic bonds, van der Waals forces and hydrophobic bonds.[43] The main modes of non-covalent bond interactions are as follows:

1. *Electrostatic binding.* DNA is a highly charged polyelectrolyte whose anionic phosphate backbone strongly influences the conformation of DNA and its response. That is, electrostatic binding has no selectivity and acts on the phosphate skeleton.
2. *Groove binding.* The bioactive small molecule acts directly on the base pair edge of the major groove or minor groove of the DNA. The large and minor groove areas have large differences in potential energy, hydrogen bonding characteristics, stereoscopic effects and hydration. Many macromolecular proteins bind to DNA specifically through the large groove. Bioactive small molecules with lower molecular weight act in the minor groove region. The typical small groove region-bound bioactive small molecule contains several simple aromatic heterocyclic structures such as furan, pyrrole or benzene rings, and the molecules bound to the minor groove region can act selectively on the DNA double-helix structure by hydrogen bonding, van der Waals attraction, *etc.* Such interaction prevents the template replication of DNA and plays an anti-viral and anti-tumour role.
3. *Intercalative binding.* The π–π stacking, dipole–dipole interaction and hydrophobic interaction between the aromatic fused ring and the base allow the planar or nearly planar aromatic fused ring structure of the molecules to be interposed between the base pairs. The molecules embedded in the DNA form an ordered stack of base pairs and the surface of the intercalated compound closely follows the aromatic heterocycle of the DNA base. In the double-helix π–π conjugate, the dipole–dipole interaction is electrically stable, which often makes the combining of DNA and other molecules result from a combination of multiple factors.

The interaction of other molecules with DNA will directly affect the expression of genes, and the regulation and control of gene expression have become one of the most important research topics. At present, almost all anti-tumour drugs use DNA as a target to achieve the goal by directly damaging DNA or inhibiting DNA synthesis. Studying the interaction between drug molecules and DNA is the molecular basis for drug design.[44] Therefore, studying the interaction between DNA and other molecules will help in understanding the nature of life phenomena at the molecular level and in understanding the pathogenesis of certain diseases such as cancer and AIDS at the genetic level, so that molecular design can be achieved.

10.3.3 RNAs

Biomacromolecules such as RNA are important components of living organisms and play an irreplaceable role in life activities. The function of these biomacromolecules is mainly achieved in an interactive manner. Various important physiological processes in the cell, such as signal transduction, immune response, protein synthesis and RNA transcription and translation, are all related to the interactions between biomacromolecules. Among these RNAs, messenger RNA (mRNA), transfer RNA (tRNA) and ribosomal RNA (rRNA) are the three major types involved in protein synthesis in the cytoplasm.[45] Therefore, understanding and studying the interaction mechanism of RNA will help us to understand the life processes more deeply and then influence and control biological processes by regulating the interaction process between biological macromolecules.

The *in vivo* stability of RNAs is usually limited, thereby making the investigation of their intracellular interactions difficult to achieve. Studies so far on the RNA pathway have revealed the mutual regulation mechanism among different types of RNAs, including mRNA and miRNA, Alzheimer's disease, RNA sequencing (RNA-Seq) in liver cancer, non-coding RNAs, *etc.*[46] Moreover, proteins also interact with RNAs and a typical example is the interaction of HIV-1 nuclecapsid protein with Tat-RNA.[47] These studies are still at a preliminary stage and more research interest needs to be focused on this field in the near future.

10.4 Dynamic Intermolecular Interactions Between Active Small Molecules and Biomacromolecules

10.4.1 Dual Detection of Phosphatase and H_2S

Phosphatase is a representative hydrolase that dephosphorylates other enzymes or proteins to regulate their biological activities. The phosphatase activity in live cells is dynamically controlled by bioactive small molecules. Hydrogen sulfide (H_2S) is the most important modulator that can affect the phosphatase activity through different pathways.[48] To visualize directly the cross-talk between phosphatase and H_2S, a dual-responsive fluorescent probe was developed by conjugating azido-modified coumarin and phosphate-modified rhodol with a piperazine bridge (Figure 10.5).[49] Probe **5** provides blue and green fluorescence responses together with their resonance energy transfer signal, allowing the simultaneous detection of H_2S level, phosphatase activity and their correlation through three-channel fluorescence imaging. Fluorescence analysis based on probe **5** revealed

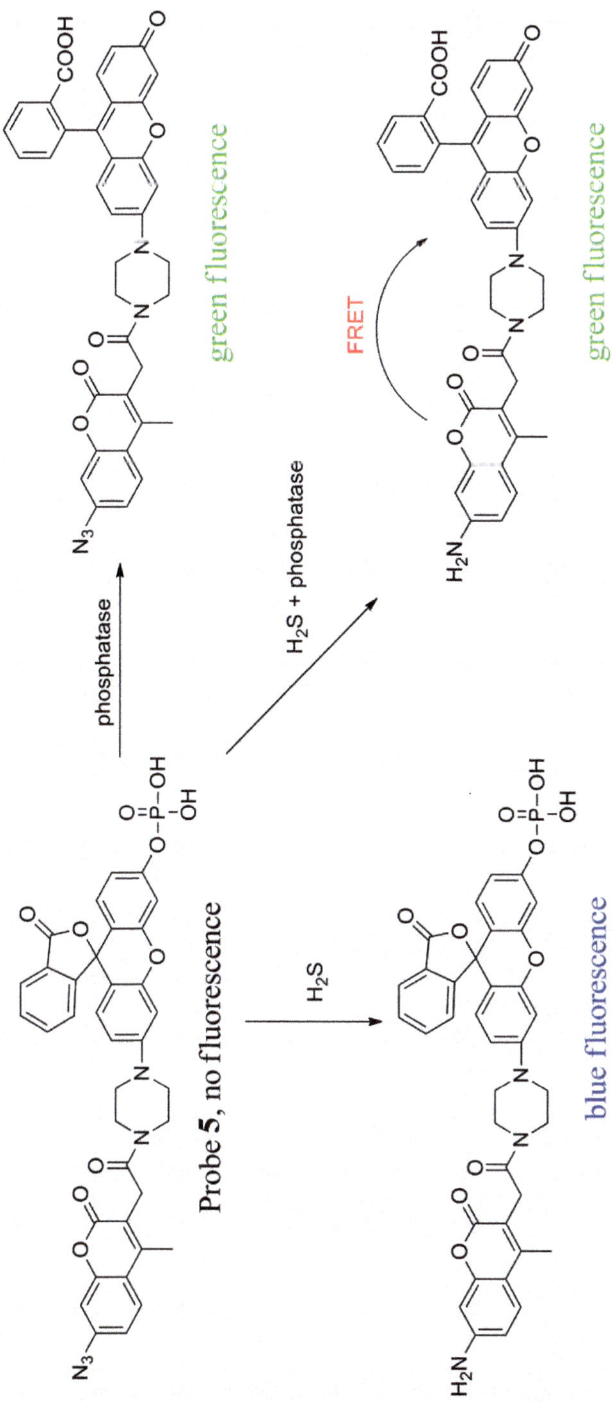

Figure 10.5 Reaction mechanisms of probe 5 with phosphatase and H_2S.

that an ideal H_2S level maintains the maximum phosphatase activity and a slight deviation of the H_2S level induces a significant decrease in phosphatase activity. This study provides a new strategy for exploring the cross-talk between bioactive small molecules and enzymes in live cells.

10.4.2 Dual Detection of miRNA and mRNA

As a cancer suppressor, programmed cell death protein 4 (PDCD4) inhibits cancer transformation and invasion and is downregulated in cancer cells.[50] The PDCD4 mRNA level is of great significance in evaluating the corresponding protein expression in cancer diagnosis and therapy. MicroRNA-21 (miR-21) is considered a valuable biomarker over-expressed in multiple malignant tumours and is tightly correlated with PDCD4 expression.[51] Therefore, simultaneous visualization of miR-21 and PDCD4 mRNA will provide crucial insights into research on their dynamic interactions within live cells. Two different single-stranded DNAs (ssDNAs) specifically targeting miR-21 and PDCD4 mRNA were labelled with different fluorescent dyes and then adsorbed on the surface of graphene oxide (GO) *via* π–π stacking to generate a dual-sensing nanoprobe, the fluorescence of which was quenched by GO.[52] When ssDNA is hybridized with its corresponding target, the duplex formed deviates from the surface of GO and induces fluorescence recovery for the detection of miR-21 and PDCD4 mRNA. The nanoprobe permitted the simultaneous fluorescence imaging of miR-21 and PDCD4 mRNA and verified an inverse correlation between miR-21 and PDCD4 mRNA in MCF-7 breast cancer cells.

10.5 Perspectives

Studying the interaction mechanism and function between biological small molecules and biological macromolecules such as proteins is of great significance for understanding the distribution and metabolism of drugs in the body. Over the years, significant progress has been made and practical applications have been developed for the interaction mechanism and roles of biological small molecules with nucleic acids and proteins. However, there are still many problems to be resolved. Future research will be focused on the following aspects: (1) further study of the interaction mechanism between biological small molecules and DNA, RNA and proteins, thereby providing theoretical and experimental models for new drug synthesis and *in vitro* screening; (2) searching for a variety of sensitive nucleic acid and protein probes to meet the needs of the rapid developments in the life sciences; and (3) development of research methods that are simple, fast, sensitive and cost-effective.

Acknowledgement

Dr Xilei Xie and Xiaonan Gao are thanked for expert assistance with the writing of this chapter.

References

1. B. D'Autréaux and M. B. Toledano, *Nat. Rev. Mol. Cell Biol.*, 2007, **8**, 813–824.
2. D. M. Townsend, K. D. Tew and H. Tapiero, *Biomed. Pharmacother.*, 2003, **57**, 145–155.
3. A. Razgulin, N. Ma and J. Rao, *Chem. Soc. Rev.*, 2011, **40**, 4186–4216.
4. X. Jiao, Y. Li, J. Niu, X. Xie, X. Wang and B. Tang, *Anal. Chem.*, 2017, **90**, 533–555.
5. X. Xie, X. Yang, T. Wu, Y. Li, M. Li, Q. Tan, X. Wang and B. Tang, *Anal. Chem.*, 2016, **88**, 8019–8025.
6. X. Xie, M. Li, F. Tang, Y. Li, L. Zhang, X. Jiao, X. Wang and B. Tang, *Anal. Chem.*, 2017, **89**, 3015–3020.
7. Y. Li, X. Xie, X. Yang, M. Li, X. Jiao, Y. Sun, X. Wang and B. Tang, *Chem. Sci.*, 2017, **8**, 4006–4011.
8. X. Xie, F. Tang, X. Shangguan, S. Che, J. Niu, Y. Xiao, X. Wang and B. Tang, *Chem. Commun.*, 2017, **53**, 6520–6523.
9. L. Qian, L. Li and S. Q. Yao, *Acc. Chem. Res.*, 2016, **49**, 626–634.
10. L. A. Sena and N. S. Chandel, *Mol. Cell*, 2012, **48**, 158–167.
11. R. Zhang, J. Zhao, G. Han, Z. Liu, C. Liu, C. Zhang, B. Liu, C. Jiang, R. Liu, T. Zhao, M. Han and Z. Zhang, *J. Am. Chem. Soc.*, 2016, **138**, 3769–3778.
12. A. Meister and M. E. Anderson, *Annu. Rev. Biochem.*, 1983, **52**, 711–760.
13. B. Kalyanaraman, *Redox Biol.*, 2013, **1**, 244–257.
14. K. Xu, M. Qiang, W. Gao, R. Su, N. Li, Y. Gao, Y. Xie, F. Kong and B. Tang, *Chem. Sci.*, 2013, **4**, 1079–1086.
15. X. Xie, T. Wu, X. Wang, Y. Li, K. Wang, Z. Zhao, X. Jiao and B. Tang, *Chem. Commun.*, 2018, **54**, 11965–11968.
16. J. L. Wallace and R. Wang, *Nat. Rev. Drug Discovery*, 2015, **14**, 329–345.
17. X. Jiao, Y. Xiao, Y. Li, M. Liang, X. Xie, X. Wang and B. Tang, *Anal. Chem.*, 2018, **90**, 7510–7516.
18. S. Matsuyama, J. Llopis, Q. L. Deveraux, R. Y. Tsien and J. C. Reed, *Nat. Cell Biol.*, 2000, **2**, 318–325.
19. L. Yang, N. Li, W. Pan, Z. Yu and B. Tang, *Anal. Chem.*, 2015, **87**, 3678–3684.
20. M. C. Maiuri, E. Zalckvar, A. Kimchi and G. Kroemer, *Nat. Rev. Mol. Cell Biol.*, 2007, **8**, 741–752.
21. L. Yang, Y. Chen, Z. Yu, W. Pan, H. Wang, N. Li and B. Tang, *ACS Appl. Mater. Interfaces*, 2017, **9**, 27512–27521.
22. S. Roy, S. Khanna, S. R. A. Hussain, S. Biswas, A. Azad, C. Rink, S. Gnyawali, S. Shilo, G. J. Nuovo and C. K. Sen, *Cardiovasc. Res.*, 2009, **82**, 21–29.
23. M. G. Perrelli, P. Pagliaro and C. Penna, *World J. Cardiol.*, 2011, **3**, 186–200.
24. L. Yang, Y. Ren, W. Pan, Z. Yu, L. Tong, N. Li and B. Tang, *Anal. Chem.*, 2016, **88**, 11886–11891.

25. F. C. Fang, *Nat. Rev. Microbiol.*, 2004, **2**, 820–832.
26. X. X. Chen, L. Y. Niu, N. Shao and Q. Z. Yang, *Anal. Chem.*, 2019, **91**, 4301–4306.
27. W. Zhang and H. T. Liu, *Cell Res.*, 2002, **12**, 9–18.
28. J. Arias, A. S. Alberts, P. Brindle, F. X. Claret, T. Smeal, M. Karin, J. Feramisco and M. Montminy, *Nature*, 1994, **370**, 226–229.
29. A. Mantovani, F. Marchesi, A. Malesci, L. Laghi and P. Allevena, *Nat. Rev. Clin. Oncol.*, 2016, **7**, 399–416.
30. D. L. Nelson and M. M. Cox, *Lehninger's Principles of Biochemistry*, W. H. Freeman and Company, New York, 4th edn, 2005.
31. C. Branden and J. Tooze, *Introduction to Protein Structure*, Garland Pub, New York, 1999.
32. S. Boccalett, V. Latora, Y. Mereno, M. Chavez and D. U. Hwang, *Phys. Rep.*, 2006, **424**, 175–308.
33. A. D. MacKerell, Jr., D. Bashford, M. Bellott, R. L. Dunbrack, J. D. Evanseck, M. J. Field, S. Fischer, J. Gao, H. Guo, S. Ha, D. Joseph-McCarthy, L. Kuchnir, K. Kuczera, F. T. K. Lau, C. Mattos, S. Michnick, T. Ngo, D. T. Nguyen, B. Prodhom, W. E. Reiher, B. Roux, M. Schlenkrich, J. C. Smith, R. Stote, J. Straub, M. Watanabe, J. Wiórkiewicz-Kuczera, D. Yi and M. Karplus, *J. Phys. Chem. B*, 1998, **102**, 3586–3616.
34. R. Bakalova, Z. Zhelev, H. Ohba and Y. Baba, *J. Am. Chem. Soc.*, 2005, **127**, 9328–9329.
35. J. Liu, L. Zhang, J. Lei and H. Ju, *ACS Appl. Mater. Interfaces*, 2015, **7**, 19016–19023.
36. C. A. Mirkin, R. L. Letsinger, R. C. Mucic and J. J. Storhoff, *Nature*, 1996, **382**, 607–609.
37. B. Hu, F. P. Kong, X. Gao, L. L. Jiang, X. F. Li, W. Gao, K. H. Xu and B. Tang, *Angew. Chem., Int. Ed.*, 2018, **57**, 5306–5309.
38. X. Liu, X. Song, D. Luan, B. Hu, K. Xu and B. Tang, *Anal. Chem.*, 2019, **91**, 5994–6002.
39. M. Luan, M. Shi, W. Pan, N. Li and B. Tang, *Chem. Commun.*, 2019, **55**, 5817–5820.
40. M. R. Arkin, Y. Y. Tang and J. A. Wells, *Chem. Biol.*, 2014, **9**, 1102–1114.
41. F. T. Yu, V. M. Cangelosi, M. L. Zastrow, M. Tegoni, J. S. Plegaria, A. G. Tebo, C. S. Mocny, L. Ruckthong, H. Qayyum and V. L. Pecoraro, *Chem. Rev.*, 2014, **114**, 3495–3578.
42. G. Manning, D. B. Whyte, R. Martinez, T. Hunter and S. Sudarsanam, *Science*, 2002, **298**, 1912–1934.
43. T. Lindahl, *Nature*, 1993, **362**, 709–715.
44. M. Ferrari, *Nat. Rev. Cancer*, 2005, **5**, 161–171.
45. D. Wang and S. J. Lippard, *Nat. Rev. Drug Discovery*, 2005, **4**, 307–320.
46. A. Fire, S. Xu, M. K. Montgomery, S. A. Kostas, S. Driver and C. C. Mello, *Nature*, 1998, **391**, 806–811.
47. J. M. Lanchy, Q. N. Szafran and J. S. Lodmell, *Nucleic Acids Res.*, 2004, **32**, 4585–4595.

48. N. Krishnan, C. Fu, D. J. Pappin and N. K. Tonks, *Sci. Signalling*, 2011, **4**, ra86.
49. P. Ou, R. Zhang, Z. Liu, X. Tian, G. Han, B. Liu, Z. Hu and Z. Zhang, *Angew. Chem., Int. Ed.*, 2019, **58**, 2261–2265.
50. Y. Chen, T. Knösel, G. Kristiansen, A. Pietas, M. E. Garber, S. Matsuhashi, I. Ozaki and I. Petersen, *J. Pathol.*, 2003, **200**, 640–646.
51. A. Esquela-Kerscher and F. J. Slack, *Nat. Rev. Cancer*, 2006, **6**, 259–269.
52. W. Pan, B. Liu, X. Gao, Z. Yu, X. Liu, N. Li and B. Tang, *Nanoscale*, 2018, **10**, 14264–14271.

Subject Index

α-tocopherol, 159
$\alpha_v\beta_3$ integrin (ABIR), 217

A549 cells, 214, 218
acidity, 212, 225
adeno-associated virus (AAV), 51–53
aggregation-caused quenching (ACQ) effect, 78
aggregation-induced emission (AIE), 78–79, 84, 90–92, 179
AIEgen, 78–79, 84
3-aminophenylboronic acid (APBA), 12, 43, 135
Angeli's salt (AS), 172
anodic stripping signal, 12
apoptosis, 101, 104
arachidonic acid, 157
artemisinin (ART), 102, 104
AZO-DCM, 212, 213

benzothiazoline, hydrogen abstraction reaction, 147–148
bioactive small molecules
- dual detection, 233–238
- dynamic interactions, 233–238, 242–244
- perspectives, 244

bioluminescence imaging, 99–101
biomacromolecules
- DNA, 240–242
- dual detection, 242–244
- dynamic interactions, 238–244
- of miRNA and mRNA, 244
- of phosphatase and H_2S, 242–244
- proteins, 239–240
- RNA, 242

black phosphorus–manganese dioxide nanoplatform, 14
boron dipyrromethene (BODIPY) fluorophore, 159, 176, 213, 214

CAI probe, 154
caspase-3, 97, 104
caspase-7, 97
caspase-9, 104
caspase activity, living cells, 83–101
caspase cascade activation, real-time monitoring, 104–108
caspases, simultaneous imaging, 101–104
caspase-sensitive nanoaggregation fluorescent probe (C-SNAF), 97, 98
cathepsin B (CaB), 5, 6
cell-surface glycans, 21
- chemoenzymatic glycan labelling (CeGL), 27
- chemoselective labelling, 24–25
- functions of, 22
- global analysis of, 27–32
- labelling and recognition of, 22–27
- lectin recognition, 22, 24
- metabolic oligosaccharide engineering, 25–26
- perspective, 43
- protein-specific imaging of, 35–43
- quantification, competition systems, 33–35
- structures and classification of, 21–23

chemoenzymatic glycan labelling (CeGL), 27
chitosan, 53
concanavalin A (Con A), 2, 34
copper-catalysed azide–alkyne cycloaddition (CuAAC), 26
curcumin (CCM), 102
Cy-PFS probe, 154
cysteine (Cys), fluorescent probes, 189–190
  acrylate and acrylate derivatives addition, 190
  ketone or aldehyde addition, 190
  maleimide addition, 190
cysteine hydropersulfide (Cys-SSH), 199

Darrow Red-type pH probes, 213
diazotization, 177
Dicer, enzyme, 49
6-(dimethylamino)quinoline-2-benzothiazoline, 148
DNA cascade reaction (DCR), 10
DNA "nano string light" (DNSL) responsive, 10, 11
Doxil, 100
dual detection, active small molecules, 233–238
  HClO and H_2S, 235–237
  of miRNA and mRNA, 244
  of phosphatase and H_2S, 242–244
  of RNS and RSS, 238
  of ROS, 233–234
  of ROS and microRNA, 237–238
  of ROS and pH, 237
  of ROS and RSS, 235–237
"dual lock-and-key" mode, 2
duplexed luminescence resonance energy transfer system (D-LRET), 16, 41

edaravone, 154
electrostatic binding, 241
enhanced green fluorescence protein (EGFP), 114
entropy-driven, three-dimensional DNA amplifier (EDTD), 57
ethanoylcoumarin, 190
excited-state intramolecular proton-transfer (ESIPT) mechanism, 172

ferrocene–concanavalin A (Fc–Con A), 29
ferrocenylmethanol (FMA), 29
4-ferrocenyl(vinyl)pyridine, 153
fluorescein amidite (FAM), 85, 102
fluorescein isothiocyanate (FITC), 7, 15, 87
fluorescence lifetime (FLT), 87
fluorescence resonance energy transfer (FRET) effect, 72–76, 114, 130, 133, 200, 215
  based on AuNPs, 75–76
  based on organic quenchers, 73–75
  Jablonski diagram, 73
fluorescent probes
  cysteine (Cys), 189–190
  cysteine hydropersulfide (Cys-SSH), 199
  glutathione (GSH), 187–189
  H_2S_n, 197–198
  HOBr, 163–164
  homocysteine (Hcy), 190–191
  hydrogen peroxide, 149–154
  hydrogen sulfide (H_2S), 192–197
  hydroxyl radical, 154–157
  hypochlorous acid, 160–163
  hypoxia, 212–215
  intracellular pH, 212–215
  lipid peroxy radical, 157–160
  O_3, 165–166
  singlet oxygen ($^1O_2$), 165
  S_NAr substitution, 191–192
  sulfur dioxide (SO_2), 199–202
  superoxide anion, 144–149
  α,β-unsaturated addition, 191
fluorescent proteins (FPs), 86
fragile X mental retardation protein (FMRP), 48

gene delivery
  adeno-associated virus vectors for, 51–53
  adenovirus-mediated vectors for, 51
  lenti-/retrovirus-mediated vectors for, 51
glutathione (GSH), fluorescent probes, 187–189
  Michael reaction, 189
  nucleophilic substitution, 189
glycans. *see* cell-surface glycans
glycoproteins, 22
glycosylation, 21, 22
gold nanoparticles (AuNPs), 6, 29, 55, 86
green fluorescent protein (GFP), 37, 52
groove binding, 241

hairpin DNA cascade amplifier (HDCA), 58
HeLa cells, 131, 133, 174, 184, 197
heptamethine cyanine, 223
hierarchical coding (HieCo) strategy, 16, 17, 40
homocysteine (Hcy), fluorescent probes, 190–191
horseradish peroxidase (HRP), 12, 29
human colon tumour (HCT-15) cells, 135
hybridization chain reaction (HCR), 31, 57
hydrogen peroxide, fluorescent probes, 149–154
hydrogen sulfide (H_2S), 192–197
  azide group reduction, 196–197
  H_2S-induced thiolysis, 195–196
  nucleophilic addition of, 193–195
  other assay principles, 197
hypochlorous acid (HClO), fluorescent probes, 160–163
hypoxia, 211–215

Indocyanine Green (ICG), 214
inorganic gene vectors, 55–57
  gold-based nanomaterials, 55
  transition metal dichalcogenide (TMD) nanosheets, 56–57
  two-dimensional graphene oxide, 56–57
in situ cytosensing strategies
  detection specificity enhancement, 2–4
  general principles of, 1
  multi-channel analysis design, 13–17
  off–on signal switch design, 4–7
  perspective, 17–18
  signal amplification strategy design, 7–13
intercalative binding, 241
intracellular bioactive molecules, 232
intracellular caspase activity
  AIE probe, 90–92
  bioluminescence imaging, 99–101
  bioorthogonal cyclization, 97–99
  fluorescence lifetime imaging microscopy for, 87–90
  FRET-based sensors, 83–87
  nanomaterial-based probe for, 94–97
  small-molecule probes, *in situ* self-assembly, 97–99
  two-photon (TP) imaging of, 92–94
intracellular caspase family, 82–83
  perspective, 108
intracellular glycosylation, 112
  O-GlcNAcylation, 112–113
  sialylation, 113–114
intracellular miRNA detection
  "one-to-one" signal detection, 59
  perspective and challenges, 61
  sensitive detection, signal amplification, 59–61

intracellular mRNA detection, 57–58
intracellular Neus, 133–139
intracellular *O*-GlcNAcylation
  general *O*-GlcNAcylation, 119–125
  protein-specific *O*-GlcNAcylation, 114–119
  visualization of, 114–125
intracellular pH, 212–215
intracellular reactive nitrogen species, 171
  cysteine (Cys), fluorescent probes, 189–190
  glutathione (GSH), fluorescent probes, 187–189
  homocysteine (Hcy), fluorescent probes, 190–191
  nitric oxide (NO), probes, 175–181
  nitroxyl (HNO), probes, 171–175
  peroxynitrite ($ONOO^-$), probes, 182–187
intracellular reactive oxygen species
  benzothiazoline, privileged hydrogen abstraction reaction, 147–148
  chemiluminescence (CL) methods, 148
  HOBr, fluorescent probes, 163–164
  hydrogen peroxide, fluorescent probes, 149–154
  hydroxyl radical, fluorescent probes, 154–157
  hypochlorous acid, fluorescent probes, 160–163
  lipid peroxy radical, fluorescent probes, 157–160
  nucleophilic substitution reaction, 149
  O_3, fluorescent probes, 165–166
  phenol–quinone interconversion, 144–146
  singlet oxygen ($^1O_2$), fluorescent probes, 165
  superoxide anion, fluorescent probes, 144–149
intracellular sialylation-related enzymes
  intracellular Neus, 133–139
  intracellular STs, 125–133
  perspective, 139
  visualization of, 125–139
intracellular STs, 125–133
intracellular telomerase activity
  AIEgen-based bioprobes, *in situ* monitoring, 78–79
  FRET, *in situ* imaging, 72–76
  functional nanoprobes, *in situ* imaging, 70–72
  perspective, 79
  surface-enhanced Raman scattering (SERS), 76–78
intramolecular charge-transfer (ICT), 153, 172

Knoevenagel condensation reaction, 186

lipid-based hybrids, 54
lipofectamine, 54
lipofectin, 54
localized chemical remodelling (LCM) strategy, 40

mercaptopropionic acid-terminated QDs (MPA-QDs), 89
mesoporous silica nanoparticles (MSNs), 71, 237
messenger RNA (mRNA) biofunction, 48–49
metabolic oligosaccharide engineering (MOE), 25–26, 38
metal–organic framework (MOF), 4–6
3-methylpyrazolone, 154
4-methylumbelliferyl-*N*-acetylneuraminic acid (4MUNA), 134, 135

microRNA (miRNA) biofunction, 49–50
MitoPerOx, 158–159
mNeonGreen (mNG), 99
molecular beacons (MBs), 59, 75
multi-channel analysis design, 13–17
multifunctional nanocapsule (MNC), 102
multiplex fluorescence bioimaging method, 108
multi-walled carbon nanotubes (MWNTs), 32
murine hypoxic tumour model, 218
myeloperoxidase (MPO), 160

N-acetylneuraminic acid (NA), 135
nanoamplicon comparator (NAC), 9
naphthalimide fluorescent probe, 214
NBP, 215
nitric oxide (NO), 238
2-nitroimidazole, 214
nitroreductase (NTR), 215, 216
nitrosation reactions, 180, 181

off–on signal switch design, 4–7
organic gene vectors
cationic lipids, 54–55
cationic polymers, 53–54

persistent luminescence nanoprobe (PLNP), 87
pH and hypoxia
perspective, tumour imaging, 227–228
tumour, targeting nano-technology, 220–222
tumour therapeutic monitoring, 222–227
phenol–quinone interconversion, 144–146
phenylboronic acid (PBA), 25
photoacoustic (PA) imaging, 180
photodynamic therapy (PDT), 5, 15
pHrodo, 217
POI, 37, 38
polyamidoamine (PAMAM), 53
polydopamine (PDA), 101, 102
programmed cell death protein 4 (PDCD4), 244
protein glycosylation, 43
protein-specific glycans, 37
protein-specific *O*-GlcNAcylation, 114–119
PSIBOH, 220
PSIOH, 220

quantum dots (QDs), 12

ratiometric fluorescent sensor (RHP), 215
reactive nitrogen species (RNS). *see* intracellular reactive nitrogen species
reactive oxygen species (ROS). *see* intracellular reactive oxygen species
reactive sulfur species (RSS). *see* hydrogen sulfide (H_2S)
recognition module, 1
recognition motifs, 2
recombinant adeno-associated viral (rAAV) vectors, 51, 52
rhodamine B-based fluorescent probe (RML), 212

signal amplification strategy design, 7–13
signal output module, 1
signal transduction module, 1
single-walled carbon nanotubes (SWNTs), 7, 8
siRNA-loaded oligonucleotide nano-vehicle (siRNA-ONV), 2, 3
sodium dodecyl sulfate poly-acrylamide gel electrophoresis (SDS-PAGE), 128
spirolactam ring-opening process, 212
Staudinger ligation reaction, 26
strain-promoted azide–alkyne cycloaddition (SPAAC), 26
sulfur dioxide (SO_2), 199–202

surface-enhanced Raman scattering (SERS), 76–78
SypHer2, 220

TAMRA, 217
telomerase
  activity, 9
  clinical significance of, 67–69
  traditional analytical methods for, 69–70
  working principle of, 68
telomerase reverse transcriptase, 67
telomere repeat amplification protocol (TRAP), 69, 70
tetraphenylethene, 151
tobacco mosaic virus (TMV), 52
toehold-mediated strand displacement (TMSD) reaction, 59
transcytosis, 96
tumour imaging, pH and hypoxia, 215–220
two-photon absorption (TPA), 92
two-photon (TP) imaging, 92–94

upconversion nanoparticles (UCNPs), 4, 9, 16

viral gene vectors, 50–53